INVISIBLE
NATURE

KENNETH WORTHY

INVISIBLE
NATURE

Healing the
Destructive Divide
between People
and the Environment

Prometheus Books

59 John Glenn Drive
Amherst, New York 14228–2119

Published 2013 by Prometheus Books

Cover image © 2013 Media Bakery
Cover design by Grace M. Conti-Zilsberger

Inquiries should be addressed to
Prometheus Books
59 John Glenn Drive
Amherst, New York 14228–2119
VOICE: 716–691–0133
FAX: 716–691–0137
WWW.PROMETHEUSBOOKS.COM

17 16 15 14 13   5 4 3 2 1

Library of Congress Cataloging-in-Publication Data

Worthy, Kenneth, 1961-
   Invisible nature : healing the destructive divide between people and the environment / Kenneth Worthy.
       p. cm.
   Includes bibliographical references and index.
   ISBN 978-1-61614-763-1 (pbk.)
   ISBN 978-1-61614-764-8 (ebook)
   1. Nature—Effect of human beings on. 2. Human ecology—Philosophy.
3. Philosophy of nature. I. Title.

GF75.W67 2013
304.2—dc23
                                                                    2013012123

Printed in the United States of America

For

Earth

&

Frank

CONTENTS

10   **CONTENTS**

For good or ill, I am an ignorant man, almost a poet, and I can only spread a feast of what everybody knows. Fortunately exact science and the books of the learned are not necessary to establish my essential doctrine . . . for it rests on public experience. It needs, to prove it, only the stars, the seasons, the swarm of animals, the spectacle of birth and death, of cities and wars. My philosophy is justified . . . by the facts before every man's eyes; and no great wit is requisite to discover it, only (what is rarer than wit) candour and courage.

—George Santayana,
Scepticism and Animal Faith, 1923*

PREFACE

As I hurtled across North America last week on my way home to California from Christmas in New England, I peered out the window onto a magical scene. The nearly full moon cast a bright, silvery glow on the blanket of clouds far below. Above, the sky was a deep, cold midnight blue, and it seemed more three-dimensional than usual— the stars wrapped around us. The view was beautifully surreal. Flying through the atmosphere at five hundred miles per hour, thirty-five thousand feet above ground is a bit unreal, after all, for a mammal with all his senses. I looked back at the jet engine and saw that it, too, was illuminated by the glistening moonlight, which cast a shadow on its devouring mouth, making it look slightly menacing.

Thinking back on the scene, I wondered about all of the ingredients going into it. The light had been emitted by the sun about eight minutes earlier. The moon reflected it back toward Earth, but not all of it—just the tones of its own rocky, dusty surface. Being so high up eliminated much of the optical filtering of Earth's atmosphere, rendering the brilliance of the moon and stars all the more vivid.

The sun supplied more than the light—it also provided the energy keeping us aloft. Millions of years ago, it powered the growth of trees and various plants that, along with dinosaurs and other animals who ate them, eventually decayed and became covered with layers of sediment for geologic time, only to be drilled and pumped up as petroleum in human time so that we could burn it in engines, such as the ones on this jet. Even further into the past, billions of years ago, the nuclear fusion reactions in ancient stars formed the metals that we'd later smelt from rocky ores to build modern equipment such as this airplane. And for that matter, the same star-foundries forged all the basic chemical elements such as carbon and nitrogen making up everything inside the plane, including me. So when I looked out through my reflection in the window at the glowing clouds lighting us from below, the moonlit engine, and the stars hanging in space, I was seeing different versions of the same stuff, really. I like thinking about such connections.

But this picture so far leaves out the immense contribution of people to the moment. It was awe-inspiring human ingenuity that harnessed the energy and materials minted by the ancient suns to produce the fantastically complex machine I was flying in and that learned how to dig up the oil and refine it to fuel such a device. Human inventiveness contrived the sophisticated navigation, radar, and telecommunications electronics to keep the airplane safe and headed in the right direction, as well as all the education and training that enable people to run and fix all this equipment. Untold complexities of science, engineering, and management allowed me to speed through the atmosphere and arrive at my destination in one piece.

I think often of the incredible, almost dreamlike technologies that civilization has devised over the last century or so—a tiny fraction of the human drama on Earth—and the power, comfort, and protection they give me and many of my fellow citizens of this modern world. So in the airplane, I could sit in relative warmth and repose as I traveled thousands of miles in a few hours, with air so frigid outside, only a few inches away, it could kill me in minutes. Likewise, just that morning I had sat in my parents' snug home in the cold Northeast feeling fully sheltered against the harsh weather outside and well fed with food from the reliable American agricultural system. My father was there, courtesy of cancer therapies and a heart-valve replacement procedure (my uncle, too, who'd had a valve replaced twice), and my mother, sustained by pharmaceuticals that many elders take. Food, shelter, transportation, healthcare—modern life certainly makes a lot of people comfortable, nourished, and secure while giving us the ability to travel fast and far, speak in real time to people on other continents, and enjoy products from all over. I relish these gifts of our era.

If you, like me, are thankful for these tremendous modern powers, shouldn't our gratitude lead us to want to know more about how they come about, beyond the simple sketch above? An enormous amount of energy and resources animate each of our modern industrial lives (though some more than others). A lot happens behind the scenes to make our lifestyles possible, shaping them into what we recognize as modern living. Decisions are made in massive bureaucracies, products are fabricated in factories large and small, food is grown on industrial-scale farms, and financial networks transfer privilege around the globe at the speed of light.

But pulling back the veil is risky. We may discover unpleasant facts that could taint the veritable thrill of modern living. As the warning goes, if you like laws or sausages, don't try to find out how they're made. But what if, like the inhabitants of the dystopian world of the 1973 film *Soylent Green*, we're participating in a degenerate, cannibalizing system, one that destroys the very foundation on which our lives are built? Shouldn't we *want* to know?

———◦•◦———

This book makes the case that knowing is better. It empowers us to participate in creating a world in which our enjoyment and comfort don't come at the expense of other people and the environment. Since you're reading this, you're probably more or less aware that there are some major environmental and social costs arising from the systems that we depend on that make all the exciting jet engines, computers, cars, and television shows. The benefits are not equitably shared by any means, and the full price paid—by nature and less wealthy people—is strategically hidden away from us by technological and economic systems that obscure our view. Revealing the costs allows us to incorporate them into our decisions. But that's just a start. As I explain, we're going to have to rearrange things a bit so that we can actually bear witness to nature and to our impacts on it.

To be truly thankful for all of the advantages that we modern people have to celebrate means connecting to their sources and admitting their full expenses. I offer this book as an initial effort in that direction with the hope that it will stimulate more thought in this area. I approach the subject from many angles—philosophy, psychology, geography, classics— each of which adds to an overall understanding of the origins and effects of our disconnections from nature and the consequences of our everyday actions. Feel free to jump directly to the chapters that most interest you or to read them in any order you wish. If you like psychology, for instance, you might skip to chapter 3. Think of the book as a sculpture that you can walk around, gazing at particularly compelling parts. Explore it in your own way, and I believe you'll gradually take in the whole.

INTRODUCTION

Returning to the United States after being gone for two and a half years in the early 1990s, I wondered whether I would be like a fish out of water. Having lived in villages in Bali for the latter two of those years, studying culture, playing Balinese music, and socializing, I knew full well that my life was about to change dramatically, though I didn't realize how immediate and visceral the difference would be. A friend met me at the airport (half-expecting to see me wearing a sarong) and whisked me away on Highway 101. As we sped onto the highway, the pavement and landscape gliding by in a blur, I noticed that drivers didn't seem quite aware of each other. They were in their own worlds, disconnected from each other and the land. I had been in cars in Bali, but drivers there, as in much of the world, engage in a constant negotiation for right of way, and speeds are usually much slower, so people see and interact more with people on the roadside and in other vehicles.

Over the following days and months, the texture of life in America stood out to me as it never had before. I noticed a kind of cool and detached easiness in how people moved through life, like pucks floating on an air hockey table. Urban anonymity lets you slip by with minimal interaction if you like. By contrast, walking down a village road in Bali or elsewhere in Indonesia involved answering a lot of questions: "Where are you from?" "Where are you going?" "Are you married yet?" Back in the States, economic transactions seemed smooth and super-efficient (even before Internet shopping); the stores were filled with a huge variety of goods, prices were scanned, and a credit card was swiped. In Bali, I might haggle with a market vendor over the price of some mangosteens. America felt like friction-free living. Even the air felt different—thinner and more ethereal, without the intense humidity and smells of frangipani, burning trash, and frying shrimp paste that locate you immediately in the sociocultural milieu of Bali. I felt like I was gliding *over* life rather than being enmeshed *in* it.

Although life was smoother stateside, social interactions seemed more circumscribed. On the sidewalks, I noticed people with their guards up in ways I rarely saw in villages throughout rural Asia. There was a palpable sense of a protective psychological barrier surrounding them: darting glances, avoidance of eye contact, and a feeling of cordoned-off personal spaces. That sense was reinforced by the life-threatening, hard shells of fast-moving cars, where people spend so much of their time, and by nuclear-family architecture that keeps extended families and neighbors at a distance. In Bali friends and others could walk into my home at any time, and I into theirs. Although my American home provided a sanctuary that felt like a relief after the intense social immersion of village life in Bali, it was disappointing to have to start making appointments to see friends rather than just running into them all the time or dropping in. The division between work and leisure time further divided relationships. In Bali, as in many older cultures, people often socialize while they work.

Perhaps most important, this sense of social fragmentation was underlined and reinforced by a disconnection from nature. The rich pulsations of wild, productive nature, on which our bodies and minds depend for clean water, oxygen, food, and spiritual nourishment, seemed a distant echo.

As in many less industrialized places, life for most residents of Bali means living closer to soil, plants, and animals and being surrounded by less manufactured stuff. In one family compound where I lived for a year, chickens strolled freely and pigs snorted in sties behind our dwellings. Expansive rice fields and river gorges were a few minutes' walk away. The road in front was dirt (mud in the rainy season). People, mostly women, cooked over open fires in the kitchens, creating smoke and aromas that wafted through the air of the neighborhood. They made oil from whole coconuts by shredding the flesh, soaking it in water, wringing the fluid out through a cloth, and boiling the liquid to separate the oil. Nearly every day, the women also made offerings to the ever-present spirits from palm leaves, flower petals, and other materials from the bounty of Bali's nature. Living a bit more easily, men worked in the rice fields a couple of hours per day, made wooden statues for global craft markets (where Balinese handicrafts are well represented), tended their fighting cocks, and played music. Joining in some of these activities, particularly music making, I learned to make Balinese flutes from local bamboo and play them. For ceremonies in the home, we made structures from bamboo and woven palm fronds.

I sometimes ate rice grown by friends and *paku*, a wild fern we gathered together. I helped others in the community as needed and in turn received help. With friends, I served the community by playing gamelan music for temple ceremonies. Far more than in America, living in these villages meant feeling engulfed in natural and social networks that sustain your life—an engulfment that can at times feel suffocating for visitors from more modern Western societies.

See, I don't mean to idealize life in Bali. On the streets, I often rode into billowing black clouds of diesel smoke from trucks; with the rise of tourism and economic development, major traffic jams clog some village roads and city streets with minivans, SUVs, and buses. Countless travelers in Bali have mentioned the happy, shiny, smiling people of the island, only to have me point out significant problems such as the mistreatment of women and the burning of plastics mixed in with trash, which releases poisonous dioxins into the air. Despite its famously sophisticated, elaborate, and well-studied culture, Bali is hardly a non-modern society from the distant past. Today, many, perhaps most, Balinese youths post to Facebook® daily, for example. Moreover, the more fully industrialized places such as the United States to which I compare Bali have their advantages as well, and you could argue that they're worth all the downsides that come with them, including these social and natural disconnections. Inhabitants of the most modernized societies generally live long, comfortable lives. All the food we want is easily available in local markets. Internet shopping delivers goods to our front doors in days. Our homes keep us warm and dry year round. Our lives are filled with luxuries that just two centuries ago would have been unimaginable to kings and queens: flat-panel televisions, the Internet, smartphones, rapid intercontinental travel.

Nevertheless, by being so amazingly different, the places around the world that I've explored have given me a renewed perspective on my own modern life. Indeed, on my return to more industrial life, I *was* a bit like a fish out of water, and that's a good thing because a fish can't see the water in which it swims. Years after returning, when I began formal study of the growing global environmental dilemmas that hang over modern life like a dark cloud, my shifted perspective on modern living proved helpful. It showed me that the very structure of the modern world, the way that its elements are divided up and separated—*dissociated*—drives our ecological crisis. This book tells how these dissociations influence us to be so destructive that we threaten the future of humanity against our own wishes and in spite of our efforts to slow the damage.

A MODERN STORY

The story I tell in this book lies at the heart of modern civilization. It's a story in which we—you, me, and the vast majority of people living in modern conditions—have become so disconnected from nature and the consequences of our own actions that we no longer really control how we affect the world beyond ourselves. Most of the outcomes of our actions are aggregated with many others and happen far away, affecting distant, unknown places and people. When we buy (and later dispose of) a computer, the toxic contaminations of the air, water, and workers' bodies happen elsewhere or are so dispersed as to be imperceptible. More broadly, as individuals we've become astonishingly cut off from the very ground and substrate of our existence. Few people are directly involved with nature in their work lives, for instance. These *dissociations*, as I call them, obscure our dependence on nature and other people. They carry with them the seeds of global environmental crisis and all the human and other suffering that goes with it.

I use the term *dissociation* broadly to capture all sorts of modern disconnections, different from the specialized meaning in psychology that some readers may be familiar with. The dissociations running through modern life set it apart from the way humankind has always lived. We're dissociated from nature and from the consequences of our actions but also from the production processes of products we depend on, such as food, which are mostly carried out in distant, anonymous factories and farms by people we don't know. We create dissociations in landscapes by dividing them with fences and other barriers, such as the giant fences between the United States and Mexico. We create ecological dissociations in deeply connected ecosystems by killing off—removing—an entire species, for example. Many natural areas decline rapidly in diversity when a key species such as the wolf is lost, in a process known as "trophic downgrading."[1] Our connections to place and community are becoming increasingly tenuous because we move so much. Of course, we also create new connections all the time, to new people and places. But as I show in this book, all the new connections we make can't undo the harms created when we sever important ones.

The dissociations that structure modern life are not accidental. They parallel a peculiarly Western way of perceiving and thinking about the world that is, as I explore in chapters 5 and 6, decisively fragmented. The West's intellectual patrimony teaches that humans are separate from nature and the mind is separate from the body. Westerners see themselves more as independent and autonomous self-defining beings and less as interconnected points in a web of life that includes nature, a view more typical in non-modern cultures. In the modern world we've come to think of nature as a collection of parts that can be taken apart and reassembled at will. So dissociation not only organizes our experiences; it infuses Western, modern thought. It may be fair to say that it's a key organizing principle of modern life. By showing the problems of dissociation, this book attests to the importance of connection.

In the next chapter I begin with a broad look at this global environmental crisis. Surveying the major problems all at once, as a whole, is daunting and sobering. In subsequent chapters I illustrate the dissociations that pervade modern lives, show how they lead to destructive choices, and identify the long trends in the West's intellectual past that support them. I explain why knowledge of environmental and social problems is necessary but not sufficient. It's not enough to know how many square miles of rainforest we've lost or the percentage of coral reef that's been destroyed. Instead, we need more direct and visceral experience of those places and the damages we inflict. Ultimately, that means reorganizing our economy and society on many levels, but in the meantime there are many ways to better connect us with nature and the harms we cause.

So take heart as you read about some of our devastations and our seemingly helpless situation, for the final chapter provides some direction for creating a more connected, healthier planet. As I describe there, the key is to reestablish contact with nature in our work and leisure lives and, over the long term, to re-organize our economic and technological systems to put us more directly in touch with the consequences of our actions. I call these *our* consequences, for even though they happen beyond our perceptual horizon, they're inexorably linked to our personal choices and actions.

This is a modern story because it's enmeshed in the confluence of economic growth, resource limits, empirical science, expansionist politics, and maritime technology that centuries ago launched the European Age of Discovery, propelling modern science and the colonization of much of the planet. Europeans massively mobilized resources and people around the planet for their own ends. Already by the sixteenth century CE, just decades after Columbus landed in the Americas, Europeans were trading silver mined in South America by American Indian and African slaves for Chinese silk, which they brought back to Europe. The redeployment of labor and natural materials accelerated tremendously with nineteenth-century industrialization, when fossil fuels and the newly invented steam engine powered transportation and manufacturing, ushering in our era of intensive remote influence and control. Although some older economies exercised far-flung influence, including those of cities in ancient Mesopotamia, Greece, and Rome, which redistributed natural resources and products over great distances, often causing widespread deforestation and soil erosion, modern society stands apart in the degree to which its technologies and institutions have extended the reach of its inhabitants, as well as the extent and severity of its environmental degradations. Modern people wield unprecedented power over the material world.

This is also our story—not exactly everyone's on the planet, but most of us who live fairly modern lives. You'll probably recognize it. It rings more true in the most industrialized societies. Yet it also describes the lives of people the world over to the extent that they participate in the global industrial economy. Tobacco farmers in Tanzania purchase cell phones made in a distant, anonymous place and use them to access data about the regional or world markets where they sell their crops.[2] Their technologies, markets, and products connect them to distant people and places much as those of a Wall Street day trader. In our complex and diverse world there are, of course, many exceptions to the observations I make. Some people grow their own food, some work intimately with nature, and some are warm and engaging with strangers they meet on urban streets. But here I'm more interested in the rule than the exception.

COMPREHENDING ENVIRONMENTAL DEGRADATION

Nearly every day, I see around me a strange and perplexing contradiction of modern living: people concerned about the environment making environmentally harmful choices. Polls routinely show that the large majority of Americans care quite a lot about environmental quality, yet they often behave as if they don't. In semi-arid California, droughts are common and major rivers are diverted for consumption in homes and on farms; therefore the habitats of various aquatic species are increasingly in peril. Nevertheless, I once met a California woman who flushes each individual facial tissue down the toilet, one every few minutes or so, when she has a cold—and with the tissue, gallons of fresh, clean mountain-river water. Recently, I saw a common grocery store sight: a shopper requesting that his two one-gallon plastic milk containers each be double-bagged in plastic shopping bags—even though the jugs themselves have handles. The bags are restricted or banned in many countries and some US cities. One problem is that plastic winds up in the oceans, where it's eaten by animals such as albatrosses and turtles who mistake it for food; they die of poisoning, suffocation, or a clogged digestive tract.

I've seen empty houses in the dead of winter in New England heated to tropical warmth while their owners are out; environmental scholars throwing printer paper into a trash can that sits beside a paper-recycling container; and huge car dealerships lit up like the midday sun at midnight. All around us are gas-guzzling vehicles and oversized American homes that require large amounts of energy to build and furnish (there's embodied energy in all building materials and every product) and to heat and cool. Meanwhile, the global climate is becoming unstable from greenhouse gases released by burning fossil fuels, and according to most climate experts, we're far behind where we should be to avoid various global disasters.

These strange contradictions in a society that generally cares about the environment aren't limited to private life, either: corporations and governments, made up of people well aware of and often concerned about environmental problems, make environmentally dubious choices all the time. I frequently see all the high-power lights over the vast Port of Oakland, California, glowing in the middle of the day, requiring Pacific Gas & Electricity's generating plants to burn more fossil fuels, thereby increasing global warming. On a recent cold, rainy winter day, I saw a sprinkler system spraying copious amounts of water over a baseball field at a university—water from over-taxed reservoirs, water that would benefit the ecology of the Mokelumne River, from which it was taken. Why do people make choices harmful to the environment even when they truly care about it and when other choices are available?

The problem runs even deeper than harmful individual choices freely made. Basically all of us moderners participate in destructive practices, intentionally or not. It's hard to live a modern life that fully aligns with your values if you care about nature and about the people—often the poor—who bear the brunt of environmental degradation. Many people attempt to lessen the damage by using public transit, buying less meat, flying less, line-drying clothing, turning the heat down in the winter—but these choices are no panacea. I often ride my bike for transportation and try to live modestly. But ultimately I can't avoid propagating harms around the globe because just leading a normal life in our society means, for instance, buying computers, which are made of various toxic materials; heating my home and thereby releasing global warming gases; and releasing even more such gases by flying on jets. Some people live "off grid" by disconnecting their homes from the electrical network. But just try to disconnect from all the material and energy production networks—for food, clothing, gasoline, building materials—that sustain our modern lives and that at the same time divorce us from Earth and our consequences.

Many writers have worked to get to the bottom of environmental crisis, and some have made great strides. A well-worn explanation for environmental crisis is overpopulation, made famous in 1968 by Paul and Anne Ehrlich in their book *The Population Bomb*. Following on the population theories of eighteenth- to nineteenth-century British counter-revolutionary economist Thomas Malthus, the Ehrlichs warned that the world's burgeoning population would quickly outstrip food supplies and mass starvation would occur (Malthus cynically argued that the poor should not be helped because doing so only perpetuates the problem). Paul Ehrlich and others later developed the famous "IPAT" equation: environmental impact is the product of population, affluence, and technology. Because wealth and technological development are paramount modern values, things we don't want to give up, many people focus on the remaining parameter, population.

But looking for a single cause of environmental destruction has proven fruitless. It's better to view population, wealth, and technology as a mutually reinforcing web of causes of environmental deterioration that includes various others. For instance, poverty, combined with lack of access to education, often leads to larger family sizes in economically disadvantaged areas such as parts of sub-Saharan Africa. Insecure land rights for developing-world farmers can lead to overexploitation of soils because under such conditions they have an incentive to grow intensively for short-term gain without replenishing the soils. And larger populations have a complex relationship with environmental degradation because sometimes they generate innovations that diminish environmental degradation. So it's simplistic to blame population size as *the* driver of environmental degradation.

Various other writers blame laissez-faire capitalism for environmental crisis. They point to industrial corporations exploiting natural resources all over the planet and leaving denuded landscapes, overfished oceans, and contaminated soils in their wake. In *The Bridge at the Edge of the World: Capitalism, the Environment, and Crossing from Crisis to Sustainability*, the environmental lawyer James Gustave Speth blames capitalism's dependence on growth and argues that Americans must rein in their consumption practices.[3] True enough. But the socialist regime in the former Soviet Union was infamous for massive environmental devastation resulting from nationalist fervor, not so much from ardent capitalist-style consumption. One sad result was that in 1990, standing beside Lake Karachay in western Russia, a radioactive waste disposal site, for one hour, you would receive a lethal dose of radiation. And communist China under Chairman Mao produced the failed "Great Leap Forward" that resulted in environmental ruin across that large country, including a loss of forest cover from which China is still recovering.[4] Attempting to give the socialist state a massive kick-start, Mao's program urgently converted forests to farmland throughout the country and shifted farmers to regions where their farming practices weren't well adapted and were thus ineffective and often damaging to local ecology. Tens of millions of people died of famine. Mao and his engineers and administrators were sufficiently disconnected from the ensuing tragedies to keep driving them forward. Dissociations can detrimentally influence people's choices under both capitalism and communism, in both rich countries and poor ones.

In his 2005 book *Collapse*, the scientist-geographer Jared Diamond describes how various societies throughout human history created their own demise through deforestation and soil erosion, which causes farming to fail, and the resulting starvation.[5] He uses a rational decision-making model to explain why societies take the directions they do: disastrous decisions based on failures to anticipate, perceive, and account for problems have led societies to overgraze, hunt animals to extinction, and overwork soils through intensive farming. But Diamond's conclusions don't seem to apply to modern situations in which environmental problems are well known and solutions often readily available. The environmental sociologist Kari Marie Norgaard studied those exact conditions. She found that denial plays a role in Norwegians' lack of response to climate change, as she explains in her 2011 book, *Living in Denial: Climate Change, Emotions, and Everyday Life*.[6] Social norms, emotions, and public discourse together create denial of the problem and thus inaction. Dissociations are relevant because they provide optimal conditions for denial: the more remote the problems, the easier it is to deny their reality.

The approach in this book has more in common with that of the environmental philosopher David Abram's groundbreaking 1996 book *The Spell of the Sensuous: Perception and Language in a More-Than-Human World*.[7] Abram explains the importance of phenomenal (of and related to perception and the senses) engagement with more-than-human nature. Only by being in sensuous, embodied contact with the rich, vibrant, complex realm of nature in landscapes and seascapes, with the air, soil, and water around us, can we begin to fully understand and experience nature's needs and thus be in a reciprocal and caring relationship with it. I expand on Abram's philosophy by tying in the history of ideas behind our separation from nature and the psychology of why our separations make us more likely to be destructive. In her landmark opus *The Death of Nature: Women, Ecology, and the Scientific Revolution*, the environmental historian Carolyn Merchant also provided an excellent foundation, which I build on, for understanding how new ideas in seventeenth-century Europe excised humans from nature and reconceived the natural world as a machine ready for manipulation by humanity.[8]

In the face of all of our efforts to understand the destructiveness of modern life, why does environmental crisis march on?

RECOGNIZING DISSOCIATIONS

We can look to overpopulation, denial, and lack of foresight for some answers, but existing explanations don't quite get to the bottom of why corporations, governments, and you and I persist in making the environmentally destructive choices we do, even when we care about nature. For that, we must more closely consider the context in which we make everyday choices. We live with divisions in various types of relationships that matter: between us and the consequences of our actions; between us and the people and nature we affect with our actions; and between us and the processes that create the products we consume. These dissociated conditions inform, shape, and constrain our choices. So, for instance, we can buy a new car blithely unaware of all of the pollution released in its manufacture. The fragmented relationships among the important elements of life in the modern world set us in stark contrast to societies throughout virtually all of human history. Being able to recognize them is a crucial first step to understanding how they lead to problems.

Although we depend on nature to sustain our lives—to put it another way, not to die—most people in modern societies know little about the natural environment. We see it and experience it remotely and in processed forms, such as food on the table or lumber. Few of us can name more than a handful of the local flora and fauna or the closest stream, for example, much less describe their subtle characteristics, including how they change over time, what they require for sustenance, or how they can be used. I've hiked with dozens of different companions over the past decades through the parks of the San Francisco East Bay Area, and all together we might be able to identify a few dozen local species there and probably know only a few uses for them. These days, few people besides professional naturalists really know local plants, animals, and geology; for most of us, they amount to objects of mere enjoyment.

Less than two centuries ago, people of the Ohlone Tribes living where I now live had intimate knowledge of and well-developed uses for abalone, melic grass, larkspur, geese, elk, salmon, and hundreds of other species they lived among. Today, when we learn about wildlife or fish, it's often out of a desire to hunt or fish for recreation rather than any need (though some people, particularly poorer ones, do subsist partly from foraging, hunting, and fishing, even in urban areas and their peripheries). Because few of us are compelled to work with surrounding nature, our knowledge of it remains superficial. What little we learn today usually revolves around facts and names rather than deeper ecological insights and appreciation. We can point to a coast redwood tree, *Sequoia sempervirens*, but what about its relationships to the many species living on it (does it need any of them?) and the ecological conditions it needs to thrive (such as fog, which could diminish with climate change)? Perhaps we know a salmon when we see one, but do we know its ecology? In the end we entrust anonymous scientists, engineers, and others with the deep, vital knowledge necessary for our subsistence.

Nature has become practically invisible in our daily experiences, replaced almost entirely by artifice, the things we manufacture from nature: cars, computers, houses, desks, televisions, pens, cell phones, and microwave ovens. Today in America, babies are born into a plastic environment: plastic car seats, plastic nipples, plastic mobiles dangling above their heads, plastic toys. Our adult minds are occupied with abstractions that have only distant connections with the natural world, such as income, spreadsheets, management problems, television programs, and new apps for our smartphone. Meanwhile, wild nature is segregated to parks and wilderness areas, and to a lesser degree to backyards and median strips. In some senses it's incarcerated, constrained to those assigned places, rather than inhabiting our regular lives and environs.[9] The idea that people must be separate from nature is seen in our history, in the forced, often violent, removal of Native Americans from the landscape in the creation of the US national parks.[10] In nineteenth-century America modern problems such as industrial pollution already gave people the idea that humans are incompatible with nature, even though Native Americans had been living for millennia where the parks arose. Now wilderness has become a museum display commemorating a past intimacy between people and nature.

The ties that bind our personal labor with the food, clothing, housing, and other things on which we depend are elongated and ethereal, reduced to simplified representations—wages and prices—and mediated by vast systems of finance, politics, and transportation. For many of us, social networks of family and friends are dispersed geographically and depend on cars, airplanes, phones, and email. Our connections to places are often contingent on job availability and other external factors and are less anchored in personal history, ancestry, family, and community. Much of the American rural landscape has been transformed into a vast Brunelleschi piazza of rectilinear, discontinuous monoculture farmlands easily seen from passenger jets over the American central plains.[11] Our urban landscapes are articulated by a rhythm of abstract names such as Tenth Street that signify no connection to local landscape or local history, punctuated with exceptions such as River Street and Market Street. Technological and economic systems propagate our consequences far and wide: you might buy plywood that comes from a faraway, irreplaceable tropical rainforest denuded partly for your benefit. Unmanned drones and video-game-like controls allow US Air Force personnel in Nevada to kill people seven thousand miles away in Afghanistan and Pakistan. Our lives are truly global. We've obtained godlike powers, but we still depend on nature and need our human senses to know nature's needs.

The groceries we buy are mostly produced in distant places, by people we don't know, using processes that we know little about and have little or no control over. In the United States many people would be surprised to learn that the hamburger they're eating comes from dozens or hundreds of different animals. Many food ingredients are made in chemical factories you've never seen. Various crops are genetically engineered, but we don't really know whether and how their engineering could affect our health. We depend on anonymous experts and bureaucrats to know and decide these things for us. Our clothing labels indicate country of origin, but we don't know much about who made the particular clothing we wear, under what conditions, and using what processes. Again, people we don't even know exist handle such matters, and they may or may not make choices in line with our values. Some people have actually built and do all the maintenance on the homes that they live in, but they're rarities. Most of us perform only the simplest maintenance tasks on the important things in our lives such as cars and homes. How could any one person be skilled and knowledgeable in all the production processes that he or she depends on when those processes are as complex as they are and work is compartmentalized into such specialized occupations as civil engineer, accountant, doctor, or machinist? Our ignorance of things that we depend on reflects our profound disconnection from the production processes of our life essentials and from nature. This dissociation might give us certain free-doms and material wealth, but it also allows us to destroy the very basis of those things without fully realizing we're doing so.

Of course, such disconnections are not entirely new. There have long been nomadic cultures wandering landscapes, sometimes having only fleeting connection to particular places (though they may travel to the same ones in cycles). Powerful rulers have long made decisions affecting large areas of nature and humanity in their kingdoms while having little direct interaction with the nature and people whose lives they affect. Trade net-works such as the Silk Road have extended across continents for thousands of years, communicating people's choices over long distances by delivering goods produced far away (along with diseases such as the plague).

Moreover, modern life has never been fully dissociated, and a counter-force is pushing against some modern dissociations. People are reclaiming and reconstituting connections with nature and community in myriad ways. Farmers' markets allow for a more direct connection between farmer and eater. Urban gardening movements including community gardens let people experience production of their own food. Religious and civic organizations build community. Slow-growth and urban renewal move-ments foster walking-oriented downtowns, cultivating casual interactions and connecting urbanites to local shop owners. Popular uprisings chal-lenge multilateral institutions such as the World Trade Organization, the International Monetary Fund, and the World Bank that systematically transgress local connections in the form of environmental and labor regu-lations in favor of a global order.

Yet the totality of the increasing, multiple, and pervasive ways that disconnection has become commonplace—inhabiting the everyday lives of most people—*is* unique to modernity. In daily life, many of us hardly need to cope with the climate other than to know what to wear and whether to take an umbrella. Automated heating and cooling systems produce detached weather zones inside our homes and workplaces, protecting us from the vagaries of nature. Production and transportation systems make the things we need or want available within an easy walk or drive, or deliver them to our door. Information travels to and from us wherever we are, through unimaginably intricate and far-reaching telecommunications net-works encircling the globe. Power arrives through long gas and electrical networks. Our garbage automatically disappears beyond our physical and conceptual horizon when we place it at the curb.[12] Clean, drinkable water flows into our homes from unknown sources with the turn of a valve and also carries away our bodily waste, which vanishes in an instant. Indeed, the modern life-world locates each of us at the nexus of a series of radically elongated material and informational networks that alienate us to an unprecedented degree from the origins of our sustenance, the destinations of our wastes, the sources of our knowledge, and the consequences of our decisions and actions. This is the world of dissociation.

CHAPTER 1

THE BANALITY OF EVERYDAY DESTRUCTION

How much have people living dissociated modern lives degraded the planet in recent centuries? Environmental problems such as global climate change, toxic waste contaminations, biodiversity loss, oil leaks, and water crises are regularly reported in the news media, but from individual reports it's difficult to gain a full picture of the human impacts on Earth's ecosystems. In 2000, United Nations secretary-general Kofi Annan called for a thorough assessment of the condition of Earth's ecosystems. From 2001 to 2005, more than two thousand scientists, social scientists, and domain experts from around the globe worked on the Millennium Ecosystem Assessment (MA),[1] the most comprehensive appraisal ever undertaken of the health of the planet and its implications for human well-being.

The MA tells a story of dramatic global-scale human-generated changes to nature and their major consequences for human societies.[2] It finds that the "structure and functioning of the world's ecosystems changed more rapidly in the second half of the twentieth century than at any time in human history," resulting in a "substantial and largely irreversible loss in the diversity of life on Earth."[3] A majority—60 percent—of the "eco-system services" (things nature produces that humans use, such as the wood produced in forests or the clean water provided by evaporation and pre-cipitation) examined in the MA are being degraded or used unsustainably, including fresh water, wild game, fisheries, and air and water purification. Such degradations are substantial and in many cases growing. They're also increasing the possibility of spontaneous, nonlinear ecological changes—abrupt changes with unpredictable outcomes—including accelerating or irreversible changes that may be catastrophic and may have major conse-quences for human well-being: "disease emergence, abrupt alterations in water quality, the creation of 'dead zones' in coastal waters, the collapse of fisheries, and shifts in regional climate."[4] The effects of ecosystem damages are often shifted onto classes of people who have little say, including future generations. They're borne disproportionately by the poor and thereby exacerbate poverty, social inequities, and conflict.[5]

Global climate change, perhaps the most devastating problem, could cause great turmoil in the coming century as cities are flooded, farmlands dry up and blow away, and new epidemic diseases emerge. Burning fossil fuels; cutting down forests; raising cattle, which emit methane, another "greenhouse gas"; and developing lands for human habitation have increased carbon dioxide in the atmosphere, a major driver of climate change, by about 32 percent. Most of the increase has taken place since 1960.[6] The US National Climate Assessment, the federal government's official report on the problem, confirms that human activity is the main driver of global climate change.[7]

Climate change isn't an abstract problem. Its consequences are already upon us and are worsening—and the wealthy United States isn't immune. Among other problems, it results in disappearing glaciers; rising sea levels; acidification of the oceans; and more frequent and intense weather events such as heat waves, floods, and droughts. Effects on American society are considerable already and are expected to get significantly worse. They "will be disruptive to society because our institutions and infrastructure have been designed for the relatively stable climate of the past, not the changing one of the present and future."[8] Incidences of heat stress, respiratory stress, and waterborne diseases, as well as "unfamiliar health threats" are increasing. Civil infrastructure is being damaged. The reliability of sources of drinking and irrigation water is declining. Because of the way that climate change affects ecosystems, their capacity to moderate further climate-change-induced disturbances is diminishing. In the United States, the report finds, the current level of effort to adapt to and mitigate the effects of climate change is insufficient to prevent these problems from becoming more serious in the future.[9]

Worldwide, the climate change picture is gloomier still. The organization DARA, based in Madrid, works with the United Nations and international donor agencies to assess the impacts of international policy and government aid on developing nations. It estimates that global climate change already kills as many as four hundred thousand people every year. Starvation and communicable diseases such as cholera, both of which hit children in developing nations particularly hard, are the most prevalent causes of death. Droughts reduce agricultural output and diminish water supplies, leading people to drink and cook from polluted water sources. Increased storm activity causes flooding, which can contaminate water supplies with human waste. The use of fossil fuels—coal, oil, gas—more directly causes another 4.5 million deaths per year through air pollution, indoor smoke, and other hazards that cause lung disease and cancer.[10]

The fifty scientists, economists, and policy experts who produced DARA's 2012 "Climate Vulnerability Monitor" report estimate that climate change and the carbon-based economy together reduce total world economic output by about 1.6 percent annually; climate change is responsible for about half of that. The researchers expect losses to increase rapidly, possibly reaching six million deaths and 3.2 percent of total world economic output each year by 2030.[11] Remember, these are only the human costs. Large-scale ecological change is under way, and habitat changes driven by climate change will jeopardize the viability of many species. The polar bear is just one highly visible example. It depends on arctic ice as a platform for hunting seal at certain times of the year.[12] Major economic, ecological, and human health improvements could be made by arresting global climate change, but little progress can be seen.

Climate change of course isn't our only significant environmental problem. Two of Earth's fourteen major terrestrial biomes (ecosystem or habitat types covering a large geographic area such as mangrove or temperate coniferous forests) have had two-thirds of their area converted—to farms and suburbs, mainly. Four other biomes have been more than half converted. The distribution of species on Earth is becoming more homogenous because we regularly move species between ecosystems, intentionally or not. For instance, the zebra mussel has been spreading since at least the nineteenth century, traveling on and in ships from its origins in lakes in Eastern Europe to waterways worldwide, including the Great Lakes and San Francisco Bay in the United States. Zebra mussels grow prolifically, destroying plumbing and displacing communities of native species. While we homogenize ecosystems by spreading species around, we're also killing off others, for example, by developing their habitat. We've increased the global species extinction rate to about a thousand times the "background" rate (the one typical throughout the planet's history and expected in the absence of human activity), so the total number of species on the planet is rapidly declining. Besides those already eradicated, 10 to 30 percent of mammal, bird, and amphibian species are currently threatened with extinction because of human economic activity.[13]

The scale of environmental change due to human action in just the last half of the twentieth century is breathtaking. Cultivated lands grew dramatically and now cover a quarter of Earth's terrestrial surface. Twenty percent of the world's coral reefs have been lost, and another 20 percent have been significantly damaged in just the last few decades. Thirty-five percent of mangrove areas (which can protect coastal areas from natural disasters such as hurricanes) have been lost. Fresh water impounded behind dams has quadrupled: reservoirs now contain three to six times as much fresh water as natural areas. Creating reservoirs wipes out some ecologies by flooding them and others by depriving them of water downstream. The MA authors state that although it's possible to partially reverse these major ecosystem degradations while meeting the needs of people throughout the world, doing so would involve significant changes in policies, institutions, and practices "that are not currently under way."[14]

The report's take-home message is that human activity has significantly degraded Earth's lands, waters, and atmosphere, and that these changes directly and indirectly create many problems for humankind, ranging from aesthetic and recreational losses to major illness outbreaks and considerable economic hardships. Although the problems impact poor people the most, they ultimately touch all of humanity. If you care about the rest of nature (besides humans) in its own right, the MA's depiction of the state of Earth looks even bleaker. The study and its conclusions are wholly anthropocentric: written from a human perspective and concerned almost entirely with human welfare, not the well-being of any of the other parts of the natural world, not even other sentient beings such as chimpanzees and wolves. Millions of years of rich ecological productivity and abundance are being wiped out, and many creatures are suffering. The scale, scope, number, and extent of problems and the difficulty of responding to them— these features taken together constitute the global environmental crisis of the late-modern period in which we live.

Although we may know about these problems, environmental degradations exist for us almost like a dream. They usually happen outside of our
direct experience, though we may see hints of them here and there: clear-
cut forests, vast agricultural monocultures, or mounds of plastic washed
up on beaches. The problems may seem abstract, but the unending accumulation of the consequences of our everyday actions nevertheless makes
them quite real. With modern economics and transportation we reach into
distant lands and into the lives of anonymous remote people and change
them, with little consciousness of the consequences.

This chapter connects the broken paths from our consumption and
lifestyle choices to environmental and health harms. From what kinds of
important things are we dissociated in our everyday experiences? How
do we affect them? How do our disconnections from them influence our
behavior? The environmental and health damages from the manufacture
and disposal of the high-tech electronic devices infusing our twenty-
first-century lives make a good entry point for studying the role of these
dissociations.

DIRTY CLEAN ROOMS, POISONED LANDSCAPES, AND E-WASTE

In 2010, about 40 million laptop and 23 million desktop computers were
sold in the United States. About 33 million flat-panel TVs and 236 million
mobile devices such as cell phones and smartphones were sold. Twenty-nine
million hardcopy devices such as faxes and printers went into homes and
offices.[15] Like most people around the world, Americans covet electronics
products, which shape how we do our work and interact with each other.
They let us order things from nearly anywhere for delivery to our homes
and offices; they're the stage on which much of our socializing happens.

Although electronics and computers are often thought of as "clean
technologies," the production of high-tech electronics components such
as circuit boards, semiconductor chips, LCD monitors, and disk drives
involves an astounding quantity of resources including hundreds of toxic
substances. Semiconductor microchips, the "brains" of high-tech electronic
devices, involve particularly large amounts of resources and hazardous

wastes. Microchips are inside most electronics products, including tele-phones, computers, televisions, cars, and fancy toasters.

The materials and parts that go into making microchips come from all over the globe and are thus extremely difficult to track. Nevertheless, researchers have estimated the weight of fossil fuel and chemical inputs necessary to produce and use a single 2-gram (0.07-ounce) microchip to be about 1,700 grams (about 3.7 pounds). Water and elemental gases such as nitrogen and argon add another 32,000 and 700 grams per chip, respectively, for a total of about 34,400 grams (76 pounds) of materials to produce a 2-gram microchip—over 17,200 times the weight of the end product![16] Computers and other complex electronics can contain dozens of microchips, among other high-tech parts. About 300 million electronic devices were produced last year for purchase in the United States alone, each with many microchips. In rough terms that's more than 33 million tons (30 billion kilograms) of materials for the manufacture of just their microchips, about 2.2 million tons (2 billion kilograms) of which is toxic. What industry can claim to be clean when it involves millions of tons of toxic chemicals for one year's production for the United States alone?

The great energy and material inputs required to produce microchips is a result of their extremely low-entropy (highly ordered) makeup. A large organizational structure for circuitry is encoded into an extremely small space on them to achieve high computational speeds. Because of the compressed space into which the circuitry is inscribed in the silicon wafer, any impurities introduced into the wafer during manufacturing create fatal (to the product) defects. In fact, a silicon wafer is the purest product manufactured on a commercial scale.[17] Consequently, there are extreme demands on the purity of the materials used in its construction. For instance, semiconductor-grade ammonia is 99.999–99.9995 percent pure, while industrial grades of ammonia are only about 90–99 percent pure. Distillation is an energy-intensive process, so the manufacturing of the chemicals used in microchip production requires a lot of energy. A typical semiconductor manufacturing plant also consumes two to three million gallons of water per day, and just one square centimeter of silicon wafer takes up to eight gallons (thirty liters) of water and 1.5 kilowatt-hours of electricity to produce.[18] These extreme low-entropy requirements mean that "the materials intensity of a microchip is orders of magnitude higher than that of 'traditional' goods."[19] A lot of stuff, some of it highly toxic, goes into making your TV, laptop, or smartphone.

Making High-Tech Electronics

On April 16, 2000, workers wearing the familiar white "bunny suits" of high-tech clean rooms emerged into the daylight from MMC Technology's CD-ROM manufacturing plant in San Jose, California, in the heart of Silicon Valley. A fifty-five-gallon drum containing chemicals had exploded, splashing nitric acid and producing a toxic cloud. Nitric acid destroys flesh, and inhaling the vaporized acid can instantly fill the lungs with fluid, causing immediate death. Seventeen employees were taken to emergency rooms of local hospitals, where they were treated and released. They were lucky. If significant amounts of the acid had splashed on them, the scene would have been horrific.[20] The bunny suits wouldn't have helped because they're not meant to protect workers—they protect the *products* from the workers. To be more precise, the bunny suits shield the high-tech components from the particles workers bring into the clean rooms: skin flakes, hair follicles, dust, cosmetics, and bacteria. Chemicals easily pass through their suits onto their skin and into their lungs.[21]

Three years earlier, on April 26, 1997, Jeffrey Saurman was less fortunate. Working at chip maker LG Epitaxy in Santa Clara, California, next door to San Jose, he accidentally mixed alcohol with nitric and hydrofluoric acids. The mixture exploded into a caustic red-yellow cloud, burning his arms, legs, and face. Coworkers found him frantically rinsing his badly burned body in an emergency shower and dragged him outside, away from the vaporized acids. Two weeks later, he died of multiple organ failure. Worker error was largely at fault, though the state also found safety problems at the company that may have contributed. The fine was a thousand dollars.[22]

The chemical intensity of high-tech electronics manufacture makes it one of the most hazardous industries for human and ecological health. Joseph LaDou, director of occupational and environmental health at the University of California, San Francisco, calls chip making "one of the most chemical-intensive industries ever conceived."[23] Many substances used in the production of microchips and other high-tech components are extremely toxic, he says. Others have unknown or partially understood toxicity. In the United States new chemicals are regularly produced and released in products or into the environment without ever undergoing thorough toxicity testing.[24] For microchip manufacture such releases are particularly salient because almost none of the materials used in the production of semiconductors end up in the final product (compare the 2 grams of a typical microchip with the 34,400 grams of materials used in its production). Instead, virtually all of the large quantities of materials used are recycled, processed, or released into the environment.

Manufacturers release toxic substances into the environment so routinely that the US Environmental Protection Agency (US EPA) records their annual releases in a database: the Toxics Release Inventory (TRI). Some experts believe that these figures are ten to one hundred times below the actual amounts of toxics released due to underreporting. The 522 chemicals listed in the TRI include elemental gases, deposition/dopant gases (impurities added to alter electrical or optical properties), etchants (chemicals that selectively dissolve wafer surfaces), acids/bases, and photolithographic chemicals used in wafer production. For most, even their most basic toxicity characteristics are unknown. Chemicals commonly used in disk drive coatings, chemical washes, microchip fabrication, and other high-tech production processes include poisons with foreboding names such as trichloroethane (TCA), acetone, xylene, epichlorohydrin, and bisphenol-A. High-tech electronics manufacturing employs various known or suspected carcinogens, genotoxins, mutagens, neurotoxins, or respiratory toxicants—that is, they cause cancer, genetic damage, or genetic mutations or disrupt neurological development or functioning or breathing. Even in relatively low doses, TCA, for example, can "damage the liver, nervous system, and circulatory system and has been associated with brain cancer in gerbils exposed through inhalation."[25] These are the invisible damages lurking behind our shiny, clean-looking computers.

Poisoning High-Tech Workers' Bodies

Clean rooms—the sealed-off chambers where the components of high-tech electronic devices are made and where futuristic-looking workers wear bunny suits—are clean of particles, not of chemicals. The tiniest speck of dust on a silicon wafer or on the surface of a magnetic hard drive can ruin the device. Alida Hernandez spent fourteen years at IBM's Cottle Road disk drive manufacturing facility in Silicon Valley, where among other tasks, she washed dust particles from disk drive platters using a soup of toxic chemicals that filled the air with fumes and often splashed onto her bunny suit, soaking it, her clothes, and her skin. These solvents and resins included known or suspected carcinogens and liver and nervous system toxins. She says that the company never told her that the chemicals could harm her. Two years after Hernandez left her IBM job, her doctor told her she had breast cancer. Together with former IBM employee James Moore, who had contracted non-Hodgkin lymphoma, Hernandez sued IBM for exposing her to cancer-causing chemicals in the workplace. In 2004, a jury decided that the technology giant was not responsible for causing the employees' illnesses.[26]

Another IBM worker never got the chance to sue his former company, though his family did. Having worked for years in a clean room at an IBM manufacturing plant in San Jose, Sherron Loanzan finally escaped the uncomfortable and toxic work for a job in the shipping department. But one day in March 1999, he began to realize he was losing control over his body. An MRI showed brain tumors. Surgeons removed them, but the tumors grew back, ultimately killing Loanzan. Two months before he died, he told journalists Christopher D. Cook and A. Clay Thompson that there had been no warnings from IBM about the cancer risks of the poisonous and potentially carcinogenic tubs of chemicals with which he worked. He was informed only of possible nausea and dizziness.[27] In 2004, a set of forty lawsuits against IBM for illnesses suffered by workers settled out of court.[28]

IBM's own "corporate mortality file" of worker deaths from 1975 to 1989 appears to show an inordinate number of brain and other cancers among its manufacturing workers, but the company has successfully stifled results of studies conducted with its data.[29] More recently, an extensive industry-sponsored study found no association between working in semi-conductor manufacturing and increased cancer mortality—overall or for any specific type of cancer.[30] But the authors of that study note that the young average age of these workers indicates that the study should be extended into the future. Also, it assesses only mortality and not illness or injury. A 1988 study remarkably found that 38 percent of pregnant women in high-exposure clean room processes suffered spontaneous miscarriages.[31] Glycol ethers were one of seven chemical agents linked to the problem.[32] Although manufacturers have begun to phase them out, various other chemicals linked to reproductive health problems are still in use.[33]

The uncertainty in these cases results from the fact that it's exceedingly difficult to establish clear and decisive links from these chemicals to particular cases of illness. Dr. Myron Harrison, a former medical director of IBM, contends with the slew of health problems of working in clean rooms in an article titled "Semiconductor Manufacturing Hazards." He writes, "Any large semiconductor facility uses several thousand chemicals. An attempt to review the toxicology of all of the thousands of chemicals in use at a typical fabrication plant is doomed to be superficial and of little value."[34] Useful toxicology assessments of chemicals are rarely done before they're introduced into manufacturing settings where workers are exposed. The extreme demands for particle-free air in clean rooms means the ventilation systems must introduce little fresh air. They recycle most of the air continuously throughout the day and thus leave in a lot of the airborne chemicals. Many clean-room workers thus breathe recycled toxic substances throughout their workday.

This intensive recycling of air, even though it's nominally filtered, allows the chemicals to react with one another and create whole new compounds with unknown health effects. Moreover, multiple chemicals entering the body can create additive effects—that is, one chemical can exacerbate the health effects of another. When workers are exposed to large numbers of chemicals simultaneously, the complexity of additive effects may overwhelm any attempt to diagnose them. Further complicating the ability to relate airborne and other workplace chemicals with particular illnesses is the fact that the processes and chemicals used in fabrication are under constant development. New chemicals and procedures are introduced frequently, changing the mixture to which workers are exposed.[35]

Although refinements to fabrication plants and processes have reduced worker exposure in recent years, skin absorption and inhalation are far from eliminated. Detecting and tracking these problems is complicated by the fact that the data that tells which workers are performing which tasks and are exposed to which chemicals is proprietary. Corporations resist releasing such data to protect themselves. The EPA and the US Occupational Safety and Health Administration (OSHA) collect data but allow for thresholds under which exposures don't need to be reported. These agencies don't link exposures with specific cases of cancer and other illnesses. Indeed, they're usually more focused on tracking injuries than analyzing illnesses. Even when people are known to be exposed, the complex, probabilistic etiologies and epidemiologies (causes and population patterns) of cancers and other serious diseases make it difficult to connect a particular disease to a particular exposure. The law reflects the statistical nature of such illnesses—in California it's legal to expose workers to many times greater concentrations of hazardous chemicals than the level the state has determined to cause one additional cancer per hundred thousand people.[36]

Nevertheless, injuries and illnesses resulting in work loss that involves "exposures to caustic, noxious, and allergenic substances" are estimated to be three to four times higher in the semiconductor industry than in manufacturing industries as a whole.[37] That's the case even though workplace statistics for semiconductor fabrication don't distinguish between actual clean-room workers, who account for only about 25 percent of the workforce in typical chip-making companies, and the remaining 75 percent of workers in fabrication companies.[38] So clean-room workers' exposures to these substances could be as much as sixteen times higher than in other manufacturing jobs. Regardless, there's compelling evidence that the hazards of microchip fabrication result in major health consequences for workers. According to the lawsuit filed by IBM workers, the company's own statistics recorded a death rate from brain cancer among its workers two and a half times that of the general public. Anecdotal data also suggest high incidences of cancers among the workers. As early as 1985, Gary Adams, a chemist for IBM in Silicon Valley, wrote a memo to company officials alerting them to a cluster of cancers in his building. Eight of fourteen of his immediate colleagues had contracted some form of cancer, including brain, lymphatic, blood-related, and gastric cancers.[39]

Clean-room work affects not only the body but also the psyche. "Clean room environments are notable for emotional sterility and dehumanizing surroundings," writes Dr. Harrison.[40] The identical bunny suits that everyone wears conceal individual identity—even name badges are avoided as a potential source of contamination. The hoods on the suits make it difficult to see facial expressions. To avoid producing contaminants, workers are discouraged from humming, singing, and laughing, and movements are supposed to be measured and deliberate—robot-like—to avoid stirring up dust. The main function of the workers is to supply and service automated machines, so the work is "repetitive, monotonous, and unchallenging": they may spend only 10 percent of the time actually working and the rest waiting for machines to finish a process step. The large number of processing steps, often up to five hundred, and the workers' lack of understanding of the overall process make each person's contribution to the whole microscopic device difficult to appreciate.[41] Clean-room workers are highly alienated from the products they make. They feel little pride and ownership for their work. Such conditions can lead to depression, anxiety, and other mental illness.

The continuous introduction of new and poorly tested chemicals into clean-room environments bodes poorly for worker well-being. The shift of high-tech manufacturing to China, where your computer and smartphone were most likely assembled and where semiconductors are increasingly being manufacturing, doesn't provide much hope, either. Worker rights, pay, and well-being are famously poor there. A rash of worker suicides at Foxconn in 2010, the country's largest high-tech manufacturer, led the company to install suicide-prevention netting on some of its buildings in the high-tech industrial city of Shenzhen.[42] In January 2012, 150 Foxconn workers in another town threatened mass suicide over low pay and poor working conditions.[43] The company's factories have been described as "labour camps,"[44] though they've improved due to pressure put on (and then by) Apple, the company of iPads® and iPhones® and a major Foxconn customer.[45] But in China and in the United States alike, the remote and unseen harms of high-tech manufacture go well beyond the workers themselves.

Poisoning the Landscape

In 1997, consumer and industrial electronics giant Hewlett-Packard (HP) completed construction of an EPA-monitored groundwater extraction and treatment system on Page Mill Road, a ten-acre site in Palo Alto, California, in the northern extent of Silicon Valley. Pipes emerge in various locations from below the grass, and steel disks cap numerous monitoring wells at intervals throughout the site. Behind a fence lies complex machinery that separates water from volatile organic compounds toxic to human and other life: arsenic, gallium, TCA, trichloroethene (TCE), 1,1dichloroethene (DCE), tetrachloroethene (PCE), 1,2,4-trichlorobenzene, and phenol. The purified water is pumped back into the ground. In the fifteen years preceding 2010, the system, a "pump-and-treat" operation, processed over a billion gallons of water from the aquifer, extracting thousands of pounds of these nasty chemicals from below ground to truck it away (to where?).[46] Hewlett-Packard discovered the leaking underground storage tank that caused most of the contamination at the site back in 1981 and promptly swung into action. Thirty years later, the mess is still being cleaned up and is still suspected of causing heightened levels of indoor air pollution in the area. Government agencies are trying to prevent people from ingesting the contaminated groundwater in the area.

Driving around in Silicon Valley, with its sunny weather, low-lying neighborhoods, wide boulevards, and unimposing office buildings, you might not suspect you're in one of America's most polluted places. The scene differs markedly from images of nineteenth-century industrial squalor ingrained in our imaginations with billowing smokestacks, furnaces, multi-story red-brick factories and row houses, dusty alleys, and filthy workers. But the HP groundwater cleanup site is just one of twenty-nine US EPA Superfund contamination sites on the National Priorities List (NPL) in Silicon Valley, which is home to more NPL sites than any other area in the country. NPL sites are severely or extensively contaminated and require major cleanup efforts and funds. In the Valley, they were created mainly by high-tech electronics manufacturing beginning in the mid-twentieth century. In the 1980s, dozens of high-tech manufacturers there, including IBM, Intel, National Semiconductor, Advanced Micro Devices, Fairchild Semiconductor, and Applied Materials, discovered that they had leaking chemical storage tanks.[47] Cleanup operations at many of these sites are still in progress decades after the original contaminations began. Unlike nineteenth-century pollution, the mess in the Valley is mostly invisible and underground.

One toxic contamination plume containing xylene, toluene, trichloro-ethane, and other volatile compounds from IBM's Cottle Road disk drive manufacturing facility lies almost two hundred feet (seventy meters) below ground and extends almost three miles (about five kilometers); the source was a leaking "Underground Tank Farm No. 1," where toxic chemicals were being stored.[48] A similar spill from underground storage tanks at Fairchild Semiconductor resulted in the contamination of underground aquifers by organic solvents; less than two thousand feet (610 meters) away, Great Oaks Well No. 13 drew water for the community of Los Paseos, where there was a spike in the number of miscarriages and birth defects.[49] Contamination plumes in the area are so numerous and large that they sometimes intersect and merge, and in the Valley you're rarely more than a few miles from a contamination cluster. Intel created one that mingled with others nearby to become over a mile (about 1.8 kilometers) long and five hundred feet (over 150 meters) deep. Samples taken from these sites reach four thousand times the levels deemed safe to avoid cancer and genetic mutations.[50] Contaminants from high-tech manufacture don't flow only into the air, ground, and groundwater. Some have passed into the sewer system and then into San Francisco Bay. The southern end of the bay, near Silicon Valley, is considered impaired due to industrial contamination by the heavy metals copper and nickel.[51]

In response to the environmental and health toll of the high-tech sector in Silicon Valley, Ted Smith founded the Silicon Valley Toxics Coalition (SVTC) in 1982. Smith had earned a law degree from nearby Stanford University, source of much of the brainpower of the high-tech firms in the area. SVTC has played a key role in disclosing, tracking, analyzing, and criticizing the burdens placed on the people and nature of Silicon Valley by the industry. According to its data, derived mostly from EPA reports, exposures by air alone to contaminants are calculated to contribute to somewhere between 47.6 and 1,544.75 additional cases of cancer per one million people per year in the area (while the 1990 Clean Air Act goal is one additional case per million people).[52] These and other data illustrate some of the health and environmental costs associated with the production of high-tech electronics, however dissociated they may be from consumers and high-tech decision makers. But their production is just one part of their full life cycle, which also includes disposal—and its deleterious effects.

Getting Rid of Our E-Waste

The ubiquity of electronic devices in our lives has a flip side: it all eventually must be discarded. Consider all the cell phones, TVs, computers, and GPS devices you've owned and thrown away over the years. Now add the ones you use but don't see: Internet servers, computers deep in your car, and data switching equipment. When talking about e-waste—discarded electronic devices and parts—it's helpful to use tonnage rather than number of items because some items such as cell phones are small while others such as TVs are big and heavy. About fifty million US tons (one hundred billion pounds or forty-five billion kilograms) of electronic devices such as computers, displays, TVs, peripherals, and mobile devices were sold in the United States from 1980 to 2009. That staggering amount is almost seven times the weight of the Great Pyramid of Giza. About eighteen million tons, more than a third, are estimated to still be in use. Five million or so tons sit around in garages unused. About twenty-seven million tons has headed for what the waste industry calls "end-of-life management." Three-quarters of that, twenty million tons, was put into the trash and headed to landfill, and one-quarter, seven million tons (a bit heavier than the Great Pyramid) was recycled. In 2009 alone, about 2.4 million tons of US electronics products went to end-of-life management, and of that 595,000 tons was sent for recycling.[53]

Where did it all go? As I discuss below, much of it went to villages and towns in Asia and Africa for disassembly and recycling with low-tech methods that release large amounts of toxic chemicals into an environment you'll likely never see and into the bodies of people you'll likely never meet.

The large amounts of toxic substances in high-tech electronics (*pounds*, in the case of some older desktop computers), combined with the vast numbers and fast obsolescence of these products, make the disposal of high-tech electronics commodities a significant problem for human and environmental health. Electronic products are put out of use for various reasons: products lose their basic functionality due to age or wear; changes in hardware or software make them obsolete; or the culture of "upgradism" and commodity fetishism entices people to replace things that work just fine. Most of us feel the continuous drive for ever-more functional, fast, and dazzling electronic devices. We use cell phones on average only about eighteen to thirty months before discarding them; for new laptop computers, the average lifespan is about two to three years and declining.[54]

Recycling and burying electronics can both create problems. When devices are deposited in landfills, their toxic substances eventually leach into the environment; the only question is how long it will take. Although much of the toxic material used in the fabrication of microchips is released or reclaimed during fabrication and not contained in microchips, e-waste includes far more than microchips. Circuit boards and CRTs (cathode-ray tubes) both contain lead and cadmium; switches, flat-screen monitors, and TVs contain mercury; computer batteries contain cadmium; some capacitors and transformers contain polychlorinated biphenyls (PCBs); and printed circuit boards contain brominated flame retardants (flame retardants, by the way, along with pesticides, regularly show up in American women's breast milk—where do they come from?). When high-tech products are processed for recovery or incinerated, additional toxic compounds are created. When burned to retrieve the copper of embedded wires, plastic casings, cables, and polyvinyl chloride (PVC), cable insulation releases highly toxic dioxins and furans.[55] Whether these materials are incinerated or placed in landfills, toxic substances ultimately escape into the air, soil, and water. Some of them must eventually enter humans, plants, animals, insects, and other living things.

What kinds of illnesses might be caused by these pollutants? CRT monitors and TVs that are still being disposed by the thousands each contain three to eight pounds (1.4 to 3.6 kilograms) of lead, which damages nervous systems, blood systems, kidneys, and reproductive systems. Mercury is contained in thermostats, sensors, relays, switches (including those on printed circuit boards and in measuring equipment), medical equipment, lamps, cell phones, and some batteries. It can damage various organs including the brain and kidneys and can disrupt development.[56] Plastics containing PVCs can account for several pounds of a typical desktop computer. Extremely toxic dioxin compounds are formed when PVCs are burned in disposal or recycling. Short-term exposure to barium, from the front panel of CRTs, can cause brain swelling and damage to the heart, liver, and spleen; chronic effects are not well known. Beryllium is commonly found on printed circuit boards and has recently been classified as a known carcinogen causing lung cancer.

While recycling may seem to solve the e-waste problem, it can also exacerbate it. First, the positive environmental image associated with the term *recycling* sometimes obscures the problem of the toxic content of e-waste and make people feel better about replacing devices that still work. It can stifle "innovation needed to actually solve the problem at its source—upstream at the point of design and manufacture."[57] Second, a lot of e-waste recycling shifts harms onto already disadvantaged populations, away from the consumer and her family to socially or geographically distant people. As much as 80 percent of US e-waste diverted to recycling is shipped to Asia (more than 70 percent to China alone),[58] and most of it goes to small towns where laborers work without protection, using recycling procedures that release large quantities of toxic substances directly into the air, soil, and water.[59] Some of the remaining approximately 20 percent recycled domestically is sent to US federal prisons to be processed under environmental and health regulations more lenient than those outside of prisons.[60] No one would knowingly dump toxic waste on someone else's family, but our efforts to do good by recycling can lead to this undisclosed, remote result.

Dissociating the Harms of High-Tech Electronics

In the early 2000s a team of investigators led by Jim Puckett of the Basel Action Network (BAN), a group working to "prevent the globalization of the toxic chemical crisis," and Ted Smith of SVTC followed the trail of electronic equipment recycling in the United States. The trail quickly led to Asia, where as many as ten million computer units were being shipped per year from the United States for recycling.[61] The authors reported their findings in the groundbreaking report "Exporting Harm: The High-Tech Trashing of Asia."[62] To visualize the number of computers recycled annually in poor villages in Asia, imagine a tightly packed stack of one horizontal acre that's 421 feet high (128 meters—higher than the Statue of Liberty), a total volume of 680,000 cubic yards (519,699 cubic meters). Economic pressures faced by e-waste collectors and businesses in the United States, together with poor export regulation, enable these high rates of export. Such businesses sell their e-waste to one of many highly competitive brokers, who in turn look for the best prices on the global market. Those often come from markets in Asia where labor costs are low and a lack of environmental and health regulations or enforcement enable the cheapest processing of materials.[63]

Puckett and colleagues uncovered shockingly unhealthy environmental conditions at overseas e-waste recycling operations: "The dirty little secret of the high-tech revolution." This cost is borne neither by Western consumers nor by the waste brokers who benefit from the trade but by low-wage workers and their local environments.[64] The investigators observed dirty recycling operations in Guiyu, China, and other poor communities in East and South Asia (they later investigated e-waste recycling in parts of Africa). Their report is especially valuable because in China it's often difficult for reporters to gain access to such controversial sites; police tend to usher investigators away. The cover of the report displays an image of a young Chinese child sitting atop a pile of e-waste recycling scrap, hands and feet blackened, perhaps by toner.

E-waste recycling is a major cottage industry in Guiyu. Chinese press accounts estimate that a hundred thousand people work in the industry in the area, an estimated 80 percent of the labor force.[65] In many cases people work without masks, gloves, controlled ventilation, or other protections, directly in contact with the toxic substances from inside computers as well as the vapors released when materials are burned. The investigators found some areas around work sites covered with black ash. Tools included only hammers, chisels, screwdrivers, bare hands, and open fires. Workers sweeping and processing printer toner, a suspected carcinogen, were found working in clouds of billowing toner with their skin and clothes blackened. In one area of town along the Lianjiang River, wires were being burned to recover the copper in them. Because wire insulation commonly contains PVC or brominated flame retardants, the airborne emissions and ash from burning wires contain significant levels of both brominated and chlorinated dioxins and furans—two of the most deadly persistent organic pollutants (POPs), which remain in the environment and in bodies indefinitely and accumulate. Carcinogenic polycyclic aromatic hydrocarbons (PAHs) are likely also present in the emissions and ash. Open fires release these materials into the air, soil, and water. Children play in the ash, and ash-contaminated surface waters are used for drinking, cooking, and washing.[66] In 2006, another group of researchers found elevated levels of lead in the blood of Guiyu's children.[67]

Figure 1.1. North American *technotrash* in Guiyu, China: Toxic land-scapes of work and play. (Image © Greenpeace/Natalie Behring.)

Puckett and his fellow reporters saw CRTs cracked open to recover the copper in their yokes, with the remainder of the CRT, containing lead and other toxic substances, discarded, dumped on open land, or thrown into rivers. Open irrigation canals were routinely filled with broken CRT glass and other un-recycled e-waste. Circuit boards were heated over coal fires to release the microchips, sometimes with fans to blow the toxic smoke away from the worker. The boards are sequentially stripped down to the bare fiberglass and then sent to burning or acid recovery operations along the riverbanks, which were full of charred circuit boards. This final burning process is estimated to release quantities of harmful heavy metals, dioxins, beryllium, and PAHs. Open acid baths, likely composed of 25 percent pure nitric acid and 75 percent pure hydrochloric acid, are used to salvage precious metals from chips removed from circuit boards. This process is carried out alongside rivers and other waterways, and plumes of steamy acid gases, often red, orange, or yellow, are emitted. Used acid baths with resinous materials are routinely dumped onto river banks, where a quick test showed a pH of 0—the strongest possible acid level. Efforts to recycle plastics from casings and other parts of computers involved heating up and melting them, which can release dioxins and furans. A lot of the plastic can't be recycled for various reasons, so piles of useless plastic could be seen dumped throughout the landscape and in nearby waterways.[68]

Figure 1.2. North American *technotrash* in Guiyu, China: Poisoned rivers. (Image © Greenpeace/Natalie Behring.)

Not just plastics but *most* of the imported materials and the process residues simply can't be recycled, so they accumulate in Guiyu. They can be found simply dumped along riverbanks and in open fields, ponds, irrigation ditches, wetlands, and rivers: leaded CRT glass; burned or acid-reduced circuit boards; mixed, dirty plastics including Mylar® and videotape; toner cartridges; and considerable amounts of other material apparently too difficult to separate. The landscape is marred by ashes from burning operations and spent acid baths and sludges.[69] The drinking water supply of Guiyu has become so contaminated from these recycling operations that beginning about a year after e-waste recycling operations commenced in the town, drinking water has had to be brought in by truck.[70] Along the Lianjiang River, in an area where circuit boards had been processed and dumped, investigators sampled sediment, water, and soil with alarmingly high levels of the heavy metals commonly found in computer equipment.

One water sample had lead levels 2,400 times higher than World Health Organization (WHO) Drinking Water Guidelines. The BAN and SVTC investigators measured barium in the soil at levels ten times higher than a US EPA environmental risk threshold. They found tin at levels 152 times the EPA threshold and chromium at 1,338 times the EPA threshold. Copper constituted 13.6 percent of one sample. Knowing precisely how the pollution has impacted the health of Guiyu's residents would require systematic epidemiological study that seems unlikely given the importance of the industry to the town's economy and the secrecy of the Chinese government.[71]

Locking the Problem Away

Like the residents of Guiyu, whose lack of other opportunities drives them to foul their environment and risk their health, inmates at the federal prison at Atwater, California, have found themselves literally locked in with the mess we on the outside created. "Exporting Harm" investigators Sheila Davis and Ted Smith observed practices in the prison recycling operations there that were "disturbingly similar to those found in developing nations."[72] Administered by the publicly subsidized firm UNICOR, the e-waste recycling business at Atwater paid inmates $0.20 to $1.26 per hour for electronics recycling work, including the primitive practice of manually smashing CRTs with hammers. Prohibited from speaking with the investigators, inmates later expressed concerns about health and safety conditions in letters. One worker reported, "Even when I wear the paper mask, I blow out black mucus from my nose every day. The black particles in my nose and throat look as if I am a heavy smoker. Cuts and abrasions happen all the time. Of these the open wounds are exposed to the dirt and dust and many do not heal as quickly as normal wounds."[73]

Inmates aren't considered employees and aren't protected against retaliatory acts by their employer (UNICOR) under the US Fair Labor Standards Act, nor are they allowed to unionize. State and local environmental and labor regulations that private sector recyclers normally must follow don't apply to the federal prison system. Conditions in these recycling operations would be considered substandard or illegal in US private industry (and the report describes a far safer and healthier private e-waste recycling operation in Roseville, California, run for Hewlett-Packard by Micro Metallics).

Workers in the prison weren't given representation on a health and safety committee overseeing the operation. An occupational health specialist from the California Department of Health Services wasn't allowed to accompany the investigators on their tour of the prison recycling facility. Prison and company officials denied requests for results of air monitoring tests.[74] In other words, they actively dissociated this information from people who could use it to make their decisions, such as inmates and consumers.

Why do such dissociations matter?

THE ANATOMY OF DISSOCIATIONS

In prisons; in rural villages in China, India, and Africa; and in clean rooms in the United States staffed by low-wage workers, high-tech electronics production and disposal take place outside the realm of the everyday experiences of the vast majority of those of us who buy and use these products. Exactly how are we consumers dissociated from these landscapes, processes, materials, and the people involved with them? How do these dissociations shape our decision making?

First, we remain separated from important knowledge. When I bought the computer I'm using to write this book, I knew nothing of the particular people creating its components, the toxic substances they were working with, or the actual pollution created in its manufacture. More important, we're dissociated from the actual consequences of our choices—we have no direct experience of them. Many people mediate between us and the workers nearest our consequences, including myriad corporate managers, salespeople, engineers, and bureaucrats and politicians who design and legislate regulations. They constitute a system over which we have little control. These dissociations combine and reinforce one another.

Dissociated Knowledge

The Frenchman Charles Pouliot (1628–1699) migrated from Saint-Cosme-de-Vair, France, to the outskirts of the village of Quebec, in the colony New France in the early 1650s.[75] That makes him one of my first ancestors to leave Europe for the New World. Still-existing contracts show that Charles, a master carpenter, built some of the first windmills on and near Île d'Orléans, an agricultural island with a climate moderated by the surrounding Saint Lawrence River.[76] The island needed windmills to turn the grains it grew into flour. Charles felled the trees, hewed them into beams and planks, and constructed most of his windmills and furniture from materials he molded himself from nature. He may not have made the metal tools he used, but he probably knew how blacksmithing worked. He certainly understood how the mill worked, and tending his own farm taught him how grains and other produce that supported residents were grown. In that small community, Charles recognized many of the faces of the people who ate grains milled at the windmills he built, and they likewise recognized him. He and most members of the community bartered and sold goods directly to each other.

Three and a half centuries later, a mere heartbeat in human history, my life is vastly different from that of my ancestor Charles. I occasionally make a point of buying things directly from the people who make them, but still, when I look around, most of the things in my immediate world are made by anonymous people in distant lands. The glass holding my drinking water comes from Russia. The laptop and keyboard in front of me were assembled in China of parts from countless unknown places. And someday they may be recycled there. For most of the other things around me, though, I don't even know the country of origin, never mind the people doing the work. Perhaps even more important, I really don't know how these objects were made, with their hundreds of chemicals and processes strewn far and wide. Charles knew how most things around him were made and how they worked. Even at the companies that make today's products, knowledge about their materials and construction is dispersed across myriad experts and executives, scientists and engineers, accountants and administrators.

Objects in our lives come from such obscure origins and embody such complex functioning that they're like aliens in our midst. Think of a microwave oven or flat-panel television. How are they made, and how do they work? A friend reading this chapter commented that she abhors toys with batteries and microchips because they alienate her toddler son from the cause-and-effect connection. He pushes the button, and the toy starts talking. "How, Mommy?" asks the son. She can only shrug. The uncanny complexity of our everyday items makes us all into children, in a sense, doesn't it? We can't really understand how they all work.

My forebear Charles also knew far more about nature than I do. He had to go out into the woods to select the best trees to cut, besides farming he probably hunted and trapped game and fished. Working directly with nature taught Charles a lot about it. He knew its rhythms and needs, and thus would've known more about how his actions affected the web of life. Perhaps he knew how to thin a stand of paper birch trees to promote timber growth for carpentry. Because being a successful master carpenter didn't require reading and writing, Charles could be, and was, illiterate. It's hard for me to imagine going through a single workday without reading or writing, interacting with this human artifact. He learned about nature with his senses and by talking to people. I've learned a bit about it in these ways, too, but more so by reading. So my knowledge of nature is far more abstract and fragmented. Most days my total involvement with wild nature is looking out the window to see if it will rain so I'll know whether to take an umbrella. Actually, I don't even need to do that—I can just check the forecast on the Web or radio. I don't know from which forest the oak trees that make my file cabinet were felled or how to manage that forest.

We can, however, know something about how our actions affect nature today. I outlined above how high-tech electronics result in harms to human and environmental health. But there are many other products I don't know much about and have to guess, even though I'm an environmental scholar. I know that cotton requires large quantities of irrigation water. Should I buy clothes made from another fiber instead? How much difference does it make if I buy organic cotton products? I guess it's a lot better because putting synthetic chemicals into the environment is usually worse, but I'm not certain in this case. From the lay perspective, our economy is a big mystery. We don't know basic things such as the ecology of the forests producing the wood we use or the agroecology of the fields and orchards feeding us. Through his direct engagement with the land and rivers, and those of other members of his community and the Native Americans who taught the newcomers, Charles would've had finer and more detailed knowledge both of how nature works and of how things such as trapping and logging affect natural communities.

But there's another kind of knowledge that Charles had and I don't. He could often see how his actual individual choices played out in the natural world. He could see the suffering of an animal he trapped, perhaps, so he would know the cost, even though that might not stop him. He could see how a crop of oats would yield less if not fertilized with enough cattle manure, and that might affect his family's well-being. Sure, Charles bought clothing, tools, and supplies from other towns and from France at market, but he had one foot anchored securely on the ground and could still directly see many of his actions play out in nature. Today, this knowledge of the consequences in nature of our individual actions, such as buying a smartphone, is almost completely absent and impossible to reconstruct. The economy distributes far and wide virtually all our effects on the natural world.

Of course, Charles's knowledge of nature and how he affected it was not complete—not quite as intimate, direct, and extensive as that of the indigenous people whom his society would eventually displace. And like many of my contemporaries, I'm not completely ignorant of these things. But the intervening generations between Charles and me have seen a progressive falling away of intimate knowledge of nature—particularly since the exploitation of fossil fuels in the nineteenth century vastly accelerated worldwide production and trade in goods. And this knowledge is important. If we were somehow able to trace the path of all the materials and processes associated with the production (and disposal) of an actual product when we're considering purchasing it, that information might give us pause. Knowing such things, we may purchase less or patronize different companies.

Knowledge dissociations don't affect only individual consumers. Consider all the hundreds of materials and thousands of processes in high-tech electronics. Information about them is vast and dispersed throughout organizations. The high-tech hardware CEO doesn't know much about 1,2,4-trichlorobenzene, and few or no chemical engineers in his company understand all the industrial properties *and* health and environmental effects of all the chemicals the company uses. Even if decision makers were to understand these materials and processes, other types of dissociation, outlined below, would impede that knowledge from guiding individual and corporate choices.

Many people have begun to learn more about the outcomes of their choices. They read writers such as the food and agriculture journalist Michael Pollan, who, among others, teaches about the consequences of industrial food production. Various scholars and organizations work to expose the environmental and social costs of our economy. Examples abound and can be found by anyone who wants to learn. The Silicon Valley Toxics Coalition educates about the consequences of high-tech electronics for people and the environment. The website goodguide.com provides health, environment, and society ratings for household and personal care products. These are good beginnings to countering dissociations of knowledge.

Dissociated Consequences

Even if we could somehow know everything about all these complex systems that sustain our lives and know fully how each of our individual actions affect nature and people, our consequences still happen far away. Our dissociation from consequences liberates us to continue creating harms while feeling fairly immune from them. It's strange indeed that in the modern world we're separated from *most* of the actual consequences of our actions. They're dispersed and dissipated throughout the globe by vast transportation networks and markets, aided by modern telecommunications. If I buy a new car, the materials and components may come from dozens of countries on several continents. If I dispose of a computer, landscape and people in a remote Asian town may become contaminated by the toxic materials contained in it. But I'm hardly affected, at least in any discernible way. Even eating's a global act in modern economies. Food travels thousands of miles from farms and ranches to supermarkets and restaurants. In response, "slow food" and "local-vore" movements aim to make eating personal, deliberate, and local again.

The economic and material networks of modern life together ensure that I directly experience relatively few of the consequences of my consumption choices. The economy is organized so that I feel mainly the positive effects of my actions while the negative ones are dispersed elsewhere, becoming what economists call "externalities." My direct experience is limited to the financial cost and the immediate benefits (occasionally harms) of actually owning and using what I buy. The damages of its production and disposal are manifest elsewhere.

Distance is, however, only one factor that mediates between people and their consequences. Between me and my e-waste contaminations lie thousands of people in high-tech companies and long, elaborate recycling networks with many middlemen. Companies and government agencies can rearrange harms and benefits geographically and across social classes. The people impacted are those with the least political and economic power; so e-waste gets sent to poor villages, not to wealthy Los Altos, California, where a lot of high-tech CEOs live. The stark reality of poverty and degraded environments is illustrated dramatically by an infamous December 1991 leaked internal memo authored by Lawrence Summers, then the chief economist of the World Bank: "Just between you and me, shouldn't the World Bank be encouraging MORE migration of the dirty industries to the LDCs [less developed countries]? . . . I think the economic logic behind dumping a load of toxic waste in the lowest wage country is impeccable and we should face up to that. . . . I've always thought that under-populated countries in Africa are vastly UNDER-polluted."[77]

Besides distance and organizations, time also mediates between people and their consequences in the modern world. Toxic chemicals or radioactive waste released into the environment, for example, remain harmful for many years, potentially affecting people long into the future. Carbon released into the atmosphere today will affect future weather patterns and contribute to coastal flooding, more severe storms, and loss of agricultural production for some future society. If Charles Pouliot had made major environmental mistakes, it might have cost him his life (and his descendants theirs!), but his environmental degradations probably wouldn't have lasted many generations. I, on the other hand, contribute to climate change and its effects, for instance, which will last a very long time: receding shorelines and vanishing glaciers. More dramatically, the 1986 nuclear accident at Chernobyl in Ukraine left a chunk of the planet irradiated and uninhabitable for centuries or longer, and in crowded Japan irradiated ghost towns surround the Fukushima Daiichi Nuclear Power Plant, which melted down in March 2011.

When consequences are shifted across time, they aggregate with other consequences to affect future generations and nature even more. Buying a computer may contribute to the accumulation of cadmium in the soil in a poor village where the computer is recycled; as the accumulation reaches a threshold concentration over time, it may then become hazardous. Sending a single computer to a poor village for low-tech recycling may not in itself cause illness or death, but combined with many others, it will do so over time.

Insulated from many or all of the direct health consequences of clean rooms and other aspects of production and disposal, we can purchase computers and other consumer electronics without fearing these harms. Brain cancers, miscarriages, birth defects—probably none of these will be experienced by individual consumers as a direct result of their consumption choices. Nor will they be experienced by the people responsible for the decisions about what goes into the products and the processes that make them. Although executives of high-tech electronics companies may have emotional and ethical responses when their companies hurt people or the environment, such responses usually don't outweigh the drive for profits as long as laws are not violated (and sometime then, too). Moreover, executives sometimes actively seek to suppress information about harms, as we've seen in the case of managers aborting a study of clean-room-related illnesses. Or they may act to lessen the harms, as Apple did in response to worker grievances at its supplier Foxconn (though problems remain).[78]

The dissociated relationship between high-tech executives and their workers merely echoes the one between most of us in wealthier countries and the poor people (and their environments) in less developed nations who bear the ill consequences of our acts. Think, for example, of the poor East Asians and Africans standing over acid baths, recycling our electronics, or people in "under-populated countries in Africa" living with our toxic wastes in their environments. The environmental scholar Rob Nixon coined a helpful phrase for this phenomenon: slow violence—a violence that happens "gradually and out of sight, a violence of delayed destruction that is dispersed across time and space."[79] Many environmental catastrophes such as deforestation, climate change, the melting of glaciers, and the rising of the oceans and their acidification due to more carbon dioxide in the atmosphere unfold so slowly and their effects are so dispersed that they fall under our radar. But the effects are real, particularly to the poor people most often impacted, such as the Bangladeshis (or New Orleans residents) whose homes and lives may be wiped out by increased coastal flooding driven by climate change. We usually think of violence as more immediate, visceral, and perhaps intentional, but when our actions result in illness, homelessness, and death, are we not carrying out a sort of violence?

Reflect on the novelty of these modern conditions of dissociated consequences. The past few centuries, during which people have become so removed from the consequences of their economic actions, constitute perhaps a quarter of 1 percent of human history (five hundred years, say, out of about two hundred thousand). For virtually all of human existence, slow transportation and communication meant that the consequences of almost all your actions were local. Your food, water, clothing, and other material goods were produced mostly within the village or region where you lived. Exceptions certainly applied. Wealthy, powerful people including royalty in various societies acquired resources and objects from large geographic areas. Trade networks have existed among most human societies, and even less wealthy people in many societies could acquire some goods, such as spices, from remote regions. Yet the proportions have reversed themselves: remotely sourced materials and products have become the rule rather than the exception. Now you have to go out of your way to buy local, and you may have to pay more. Even in California, a cornucopia of agricultural goods, it would take significant effort to obtain all my food from the local or regional landscape, and food is just one category (though certainly an important one) of the many goods produced, consumed, and discarded by Californians.

Dissociated Society

The social divisions running through modern societies give people power and influence over other people with whom they have little direct interaction or even knowledge. Corporate managers such as Intel's Paul S. Otellini or Apple's Timothy D. Cook make major decisions that affect the lives and health of low-wage production workers—those in clean rooms, for instance—while having little contact with them. Would the clean rooms be so toxic if the managers who decide about them had to work in them? Would the production processes of microchip fabrication be so toxic if the chip fabrication plants were situated in the affluent Los Altos hills alongside the homes of high-tech CEOs and managers? Clean-room workers are disproportionately ethnic minorities and immigrants and are generally not unionized.[80] They would have to traverse highly compartmentalized social structures, even within their own corporations, to reach the areas where the decisions about fabrication processes are made—even though they live with those processes throughout their workdays and can be profoundly affected by them.

Many US corporations have moved high-tech electronics production overseas to places where labor costs are less and, not coincidentally, worker and environmental protections are weaker. They realign the harms and benefits of production across national, class, and sometimes racial boundaries when unsafe, unhealthy labor is done by impoverished Chinese people working over acid baths in Guiyu, by Hispanic immigrants in American clean rooms, or by residents of American Samoa making garments under conditions that would not be tolerated in the States. Their work brings them into relationships with wealthier consumers, but it's a dissociated relationship because they're distant and often come from a very different social class and culture. There is, however, an even deeper sense in which dissociations corrupt our modern lives.

THWARTED ETHICS

Will a mother in Guiyu still get cancer from breathing the smoke of electronics recycling fires if I choose not to buy another computer? Am I responsible for her well-being anyway? It's unclear how much our individual choices matter. If I wait to replace my laptop computer, does that decision actually reduce the number of laptop computers manufactured and disposed? How much benefit is there to my choosing not to fly to another country for vacation? The flight will happen even if I don't take it. We modern actors are confronted by haunting questions: Does it matter what I do? Do my choices contribute substantially to harmful conditions, and if so, how much? I truly don't know; these problems don't seem to belong to me. Yet somehow the harms (and the benefits) ultimately depend on the decisions I make.

The computer that I may choose not to buy is already produced and will likely be purchased by someone else. Most commodities are manufactured long before anyone's decision to purchase and consume them, and their effects will persist in the environment long after the product is used.[81] Computers are an important part of the society in which I live, virtually essential to a productive work life for many people and, increasingly, most people's personal life. One person, even when possessing information about the consequences, can't change that fact single-handedly. It's difficult to know whether to follow the culture and purchase a computer, particularly when it could be used to mobilize against the use of toxic materials in manufacturing, or whether to avoid computers because of the harms associated with them. Participating in a society where common decisions are known to cause major problems for people and the environment, where real choice is reduced because many of the outcomes are dissociated from us, is apt to leave us with a sense of ineffectuality rather than a sense of being active, engaged actors in our world.

That sense of impotence, I argue, is one of the defining yet ironic characteristics of modern life. A typical modern person certainly wields more material power than one in almost any prior human society, except perhaps the most powerful elites throughout history. We mobilize vast material and social networks with our consumption and can communicate and travel over great distances. But because we're insulated from our consequences and are profoundly ignorant about them and the spheres of life on the planet where they play out, our ability to act as free ethical agents, choosing what effects we would like to create in the world, is hindered. We surrender a large proportion of our ethical freedom, responsibility, and agency to the political, legal, scientific, and other organizations and institutions that mediate between us and our consequences and between us and the spheres where they're manifest. And we're compelled to have confidence in institutions to continue to supply our needs and desires.[82]

Displacing our ethical subjectivity onto organizations thwarts ethical practice: it blocks us from fully appreciating the consequences of the actions we may choose and insulates us from actual outcomes. Worse, it opens us to manipulation. The more other parties, usually organizations such as corporations and government agencies, intervene between us and the consequences of our actions, the more information can be manipulated to influence our actions. Examples abound. Recent decades have seen myriad corporations "greenwashing" themselves, advertising that their practices are environmentally friendly. Some companies have indeed improved their environmental practices—sometimes dramatically. Others make only superficial changes and still manufacture harmful products while posturing themselves as green companies. A high-tech electronics company might refer to itself as green because it's using less packaging even while the products inside involve copious toxic chemicals.

Dissociated conditions make us particularly vulnerable to organizations whose sole purpose is the production and dissemination of information: media companies. Their manipulations and unintentional distortions can be damaging. Consider, for example, the conservative media outlets in the United States that give their audiences the impression that the main questions about global climate change—whether it's real and whether it's caused by human activity—are unsettled. They easily cast doubt by providing equal coverage and consideration (under the guise of "balance") to the extremely small minority of scientists who question whether it's happening or whether humans caused it. Or more cynically, they simply ridicule concerned scientists, politicians, and activists. On February 10, 2010, Fox News Channel showed images of US vice president Al Gore's book *An Inconvenient Truth: The Planetary Emergency of Global Warming and What We Can Do About It* planted in a snow bank during a snow storm. Actually, an increase in the number and severity of storms, including snow storms, is one of the expected outcomes of global climate change, so the storm was no proof against climate change and on the contrary might have been one of its effects. More broadly, in the United States large corporations insert themselves decidedly into the political process, occupying the gulf between citizens and their elected representatives.

It may be acceptable to surrender our ethical subjectivity to corporations, governments, and other organizations when they make choices aligned with our values. Maybe you shop at Whole Foods because it's known as a socially responsible business and you trust its decisions more than those of a competitor. But many organizations abdicate much of their ethical responsibility by following the lead of consumers and citizens, making particular products because people "demand" them. Oversized, inefficient vehicles come to mind at a time when the problems of burning fossil fuels are well known. Ethical concerns seem to fall between the cracks, to disappear between our fingers.

Organizations also sometimes actively seek to displace the ethical subjectivities, the ability to make choices, of their workers. They disempower workers by withholding information or decision-making power. A disk drive production worker at IBM in Silicon Valley identified by the pseudonym Alicia reported having to undergo occasional "highly secretive" physical examinations (apparently the company was concerned about regular worker exposure to certain chemicals). Her blood was drawn and she was x-rayed, but when she asked for the results, IBM officials responded, "If we find a problem, we will let your manager know." Another time, Alicia was required to wear a radiation exposure detection badge. When asked what would happen if she received an overdose of radiation, she was told, "If you don't hear from us, you're all right." In clean rooms, Alicia regularly had skin and lung exposure to various combinations of toxic resins and solvents. Her anxieties may have been validated later when she contracted breast cancer.[83]

Alicia and fellow workers were provided little information about the health and safety effects of the chemicals they worked with. Knowing too much about hazards to workers and the environment can be a liability to corporations. An Intel spokesperson said of a possible industry-government study on the health effects of clean room environments that it "would be like giving [legal] discovery to plaintiffs."[84] Another industry spokesperson declared that too little is known about the chemicals involved to point the finger at the industry: "There's not much scientific data out there" regarding the cancer risks to clean-room workers.[85] Thus, even while the electronics industry subsumes the ethical subjectivity of workers, planned ignorance prevents it from guarding their safety.

Ignorance and uncertainty abound in a dissociated world. The experts don't know all the potentially harmful consequences of the toxic substances that our industries create. How could anyone know all about the health and environmental effects of the tens of thousands of chemicals put into our environments every year by the chemical industry? And that's only one industry. What about the uncertainties of nuclear power? After the 2011 Fukushima Daiichi nuclear disaster, Germany, Japan, and other nations began reevaluating their use of commercial nuclear power. How much pollution do I avoid by buying an efficient hybrid vehicle? The answer is unclear, and it's possible I'm creating *more* pollution by doing so, if all the ecological costs of manufacturing and disposing a hybrid vehicle are taken into account.[86] What about all the electromagnetic radiation in our environments from the cell phones and other electronics we use? Humility is the appropriate response to uncertainty. The precautionary principle, which says that new technologies and policies must be shown *not* to be harmful before they're adopted, must replace techno-optimism, which gives new technologies the benefit of the doubt, as if they're persons with an inalienable right to exist.

High-tech electronic devices are but a small fraction of the material goods that support our contemporary lifestyles and simultaneously degrade our health and environments through their manufacture, use, and disposal. Cars, houses, energy, paper, food, and everything else we buy have environmental and health consequences, even if some are minor and acceptable. Think of one food, the delicious strawberry. Until recently, the toxic fumigant pesticide methyl iodide was used in California strawberry farming to sterilize the soil before planting (it's still used in Mexico). Known reactions to the chemical include eye irritation, nausea, vomiting, dizziness, slurred speech, and dermatitis, and it's a suspected occupational carcinogen.[87] Methyl iodide jeopardized the health of workers and farm neighbors exposed to it. It also wiped out the complex communities of micro-organisms that make soil fertile, thus requiring more artificial fertilizers. Given a choice, would people buying strawberries intentionally expose workers and the soil to this poison? Probably not. But we barely get the chance to engage our ethics when important information about our choices is complicated and technical and circulates mostly in the ethereal realm of corporations and governments who decide for us.

So our consumer desires march onward detached from consequences, many of which we'd find disagreeable. Powering it all is cheap energy. Our appetite for it results in a whole set of environmental, health, and safety problems such as the massive Deepwater Horizon disaster in the Gulf of Mexico beginning in April 2010 and the violence against indigenous people in the Niger Delta whose lands and waters are contaminated by spills and leaks from oil production.[88] Would we choose to pour millions of barrels of crude oil into the sea? To trample the rights of indigenous people to safe and clean access to their lands and waterways, as their ancestors had for centuries, when they don't share in the riches flowing to multinational oil companies? These are but a few examples of the health and environmental damages going on at our behest, in response to our "demands." Meanwhile, the poisons seep into the biosphere and bodies, the storm surges rise, species die off in droves, and forests fall.

CHAPTER 2

SENSE AND CONNECTION

The Texture of Dissociated Life

Have you seen nature today? Smelled nature?
Heard it, felt it, tasted it, touched it?

What's it like to be modern? What's it like to live with myriad dissociations ordering our lives, creating a wall between us and the larger world that supports us and that we in turn affect? Dissociations don't just damage the environment. They shift and color our experiences and shape our possibilities for growth and change. The structure of modern life determines what we have contact with every day, what we think about and engage with. These conditions make our lives qualitatively different from those of people throughout the vast majority of human history. Until now, people have always touched the soil, trees, wild game, and other natural things daily. The usual experience of modern life should be of great concern to us all, yet it's exceedingly hard to perceive and understand what it's really like. How can we see the fabric of the lives in which we're immersed? Fortunately, art, literature, and philosophy make good tools for rediscovering the world in which we live.

Let's begin with a glimpse at several of the many works of art and literature that lament modern experiences of dissociation. One of my favorites is the cover image for the November 27, 2000, issue of the *New Yorker* by the graphic artist Chris Ware. Produced in the aftermath of the first dot-com bubble, "Thanksgiving.com" portrays an Internet-commerce-inspired Thanksgiving meal in a futuristic world. A lone person is seated beside the sealed window of a barren, dull New York City apartment. Outside, other buildings tower over the Empire State and Chrysler Buildings, blocking out a bleak, gray sky. The diner sips liquefied dishes he (she? it?) selects by clicking on an image of a Thanksgiving meal on a monitor that, along with a large public-address-style loudspeaker, is set up before him, bringing information from the outside world. Pipes and cables deliver all the physical substances he needs. A small vent supplies air. Breast-like control knobs on the equipment and the loner's infantile hands and face point to his dependence on the nurturing yet anonymous techno-mother that dominates his life.

Figure 2.1. "Thanksgiving.com": Our alienated future? (Cover art for the *New Yorker* [November 27, 2000]. Illustration by Chris Ware.)[1]

There's no microphone or camera: the loner's a passive consumer of information. Connected to his full-body suit is a receptacle and hose to carry away bodily waste—he has no need to leave his chair (like the people in the 2008 Pixar film *WALL•E*, who must relearn to walk). In the distance we see hundreds of identical windows, implying that this scene is replicated millions of times over throughout the city and perhaps the planet. The image conveys a feeling of sublime alienation, a recurrent theme in Ware's work. Who would want to live in this world?

At first, Thanksgiving.com seems to simply portray the relentless dot-com commodification and commercialization infusing contemporary life—a commentary on the availability of everything on the Internet. But the issue of the *New Yorker* that it fronts is dedicated to "the digital age," the new era in which computers are truly reshaping daily experience. It has articles on GPS systems that track our whereabouts, the question of whether computers truly enhance productivity, the rush for Internet jobs, and the dazzling potential future of the Internet's infrastructure (which hasn't quite played out). Ware's cover doesn't simply say that we're now all busy with our technology. Its commentary runs deeper, suggesting that rampant commercialization and virtual living through electronic devices have social and psychological consequences, too. Rather than showing a family sitting around a computer on Thanksgiving, ordering traditional dishes that have become commercialized, the image portrays a diner who's hauntingly alone on this special day—disengaged, passive, and at the mercy of the systems supplying him with nutrients and a holiday celebration.[2] Thanksgiving comes through a tube and involves no social interactions. We (and presumably the loner) know little or nothing about the meal's origins. By amplifying aspects of contemporary life, as so much art and literature do, Thanksgiving.com becomes a haunting indictment of the ways we've become detached from the origins of our sustenance and the destinations of our wastes—and potentially from community—to become yielding, infantile consumers of the latest the great techno-mother offers up to us.

As dark as it is, Thanksgiving.com is hardly the bleakest glimpse of contemporary dissociated life. It looks like paradise compared, for instance, to the more explicitly dystopian world presented in the Wachowski siblings' 1999–2003 film trilogy *The Matrix*. In the deep future a society of machines has forcibly alienated humans from their bodies and from experiences of the physical world to make them into their instruments. In a reversal of the human-machine relationship in which people use machines as their tools, humans are installed in vast arrays of pods feeding a colossal power plant that supplies the energy driving the machine world. The *matrix* refers to the artificial reality wired into humans' brains to make them think everything's normal while their bodies are exploited.

The human-machine role reversal in *The Matrix* may already be happening in our world, though in less vivid terms. Our mechanized economy makes nature into an instrument, a tool, a means to economic ends. And we are, of course, part of nature. *The Matrix* thus nicely illustrates the reversal that the German Frankfurt School sociologist-philosophers Max Horkheimer and Theodor Adorno described in the 1940s. By disenchanting the world, removing the magic and rejecting the mythic in favor of a rationalist and mathematical view, they said, the Enlightenment made human thought an automatic and self-activating process. The human mind then impersonates a machine, and that opens the possibility for machines to do the thinking. By following a course of rationality that makes nature into an instrument of human will, humanity indirectly makes itself into a mere instrument as well.[3]

The Wachowskis and Chris Ware aren't the only artists reacting to the various alienations of modern life. Consider George Orwell's *Nineteen Eighty-Four*, Mary Shelley's *Frankenstein*, Aldous Huxley's *Brave New World*, or recent films such as *WALL·E*.[4] Orwell and Huxley decried the soulless, dominating administrative imperium ruling lives from afar. Shelley showed science objectifying people by using human bodies for spare parts. In *WALL·E* the last remaining humans roam the galaxy in a ship, cut off from nature, which survives in only a single precious seedling. Many authors and artists seem to believe there's something lamentable and perhaps even tragic about being alienated from nature and society in the ways we are. Environmental philosophers concur.

SENSE AND ENVIRONMENT

> For it is only at the scale of our direct, sensory interactions with the land around us that we can

appropriately notice and respond to the immediate
needs of the living world.
—David Abram, *The Spell of the Sensuous*, 1996*

The environmental philosopher David Abram argues passionately and
eloquently for the vital role of the senses in being human and being animal.
He says we all must have sensual engagement with wild nature to be able
to care for it. But direct, daily sensory experience with the landscape,
particularly the lands used to sustain our lives, is uncommon for most
people today. How many times a week do you get your hands dirty with soil
or sweat from working the land? How often do you explore untamed, wild
spaces? Are your children able to play, hide, and get lost in natural areas
beyond the safety of the playground? For most of us, nature is a distant
object, mediated by layers of bureaucracy and industry.

We observe and touch nature remotely, through the machines of science
and industry—microscopes, bulldozers, combine harvesters—and by proxy,
with the hands of low-wage agricultural workers. On the occasions we
experience it up close, nature's usually framed by the context of leisure in
which we encounter it: we're not there to gather berries for winter canning
but rather to camp on our few precious days off. In our leisure time, nature
touches us only in limited, controlled ways in the form of parks, gardens,
pets, aquariums, and zoos. Nature's a constant presence in our daily lives
only in its most highly modified, processed, and rarified forms—plastic
pens, paper, computers, automobiles, houses—that seem far removed from
the raw materials from which we make them. Under such conditions, it's
difficult to see how the needs of the living world could be met.

The voluminous artifice in our immediate surroundings makes us
forget our alienation from rich, wild, productive nature. Nature remains
"out there," beyond the horizon of our artifice, but we don't fully realize
we're disconnected. Our dependence on and participation in the productivity of nature nevertheless persists. It's merely obscured by the many
layers of corporations and government organizations that produce and
deliver our essentials—food, water, clothing, shelter, and energy—and take
away our wastes. According to scholars in differing fields, such as the eco-
feminist environmental philosopher Val Plumwood (1939–2008) and the
German social theorist Niklas Luhmann (1927–1998), the problem with

having institutions mediate our relations with nature so intensively is that large-scale institutions of modernity fail to adequately register ecological needs and thus can't respond to them.[5] Ecological needs that aren't known can't be met; ecological degradation results.

Abram says it's our proper human, animal, living role in the world to be directly and sensuously engaged with nature.[6] Dissociated from it, we can't appropriately respond to it, nor can we adequately register the consequences on it of our actions, both individual and collective. Our modern life experiences fail to deliver multisensory knowledge and understanding—a *feeling* of the natural world—gained through direct, sensory encounters in which we meet nonhuman others—shrubs, animals, bugs, mountains—not as objects but rather as interacting and worthy subjects in their own right. In other words, without our senses fully engaged in nature, we fail to see trees as actors shaping the world, though they certainly are. Experiences are important because they mold our thinking and our choices. But how do they do so in our modern lives? To know that, we must take a look at typically modern experiences—what fills our awareness and attention, what we deal with on a daily basis.

A PHENOMENOLOGY OF MODERN LIVING

Phenomenology, simply put, is the study of experience. It's a field that looks at how we perceive things in the physical world and how we become conscious of these perceptions. The French philosopher Maurice Merleau-Ponty (1908–1961) focused his philosophy of phenomenology on the body and its role in producing consciousness and knowledge of the world.[7] Phenomenologists address questions such as, How does knowledge first enter our minds through perception? They study how the human mind encounters the world in the body; how it is that objects, with their multiple potential facets of perception, their perceivable and hidden sides, end up being represented in our minds. A phenomenologist might ask how our direct observations of a tree lead to the representation or consciousness of a tree in our minds, and what is the role of our perceptions of trees in our lives. Phenomenology differs greatly from the analytical study of ideas that are only indirectly associated with experience, such as propositions and logical principles, which is the concern of much modern, academic philosophy.[8] I use the term *phenomenology* in a broad and less technical sense to examine "what it feels like to live in the world of modernity."[9]

I want to challenge people to examine more deeply the experiences we all take for granted. What do we most commonly perceive, interact with, reflect on, and engage with? Modern life-worlds—the totality of experiences in modern life—differ dramatically from those of hunter-gatherers, ancient agriculturalists, or even people living today in small, less modern cultures. On the other hand, they're also unlike those of the hyper-modern loner of Thanksgiving.com or the human body pods of *The Matrix*.[10]

This chapter examines experiences that are typical in modern life to better understand our phenomenal dissociations—our losses of direct experience and sensory engagement with important aspects of our world. We'll see that two kinds of phenomenal dissociation are common. First, all the built, artificial things in our personal and work environments and experiences—cars, houses, computers—cut us off physically from wild nature. Second, our attention and involvement has shifted from the outer, surrounding material world (even artificial things) to our inner, conceptual worlds. Before proceeding, I'd like to clarify what I mean by two vitally important terms: *nature* and *artifice*.

Nature versus Artifice

On the face of it, the distinction between nature and artifice is clear. A tree in the woods is nature. An iPhone® is artifice. But in reality the distinction is a difficult one that environmental scholars grapple with all the time (though the more postmodern ones seem to prefer to abandon the distinction altogether). What about potato plants, which have been domesticated over many millennia? Aren't they simultaneously the product of wild nature and of human thought and labor, a technological creation just like a car? And isn't the iPhone also a product of nature in the sense that all its component materials and the energy going into its manufacture are mined from mountains and synthesized from petroleum oil? After all, the building blocks for our products don't spring directly from our minds; they come from nature.

Drawing a too-sharp distinction between nature and artifice is a legacy of Western dualistic, dissociating thought, which seeks to hyper-separate everything human from everything natural. We then see the human mind and its productions as a separate and higher order of existence from the human body and the rest of nature. This kind of human-nature dualism underlies all forms of human exceptionalism (the notion that humans are unlike the rest of nature and thus have special moral standing), including Plato's and Descartes's claims that cognition is a uniquely human capacity. (The scientific field of animal cognition demonstrates repeatedly that many other animals besides humans think. Some New Caledonian crows, for instance, can figure out without training the proper sequence of tools to use in a laboratory to access food.[11]) Religious traditions that oppose "man" (particularly "his" spirit) to nature likewise foster this conceptual hyper-separation of humanity from nature. This opposition can be seen in the biblical tradition when "man" is introduced into a preexisting nature (the Garden of Eden) or is given the role of dominion over nature (not partnership with it), which to many people legitimates subduing and controlling it.[12]

Even with such pitfalls, it's both possible and necessary to maintain a distinction between nature and artifice. In an age when virtually no part of the planet escapes the imprint of humanity, it's dangerous indeed to think of the entire physical world simply as artifice or as a hybrid between nature and artifice. That view opens the whole physical world to unlimited transformation. And things such as the iPhone are clearly different in many ways from other things, such as trees growing in forests—and entire forests, for that matter—in important ways that matter to the health of our environments. We don't have to insist that any substance touched or molded by humans is plainly artifice and that nature is simply the untouched. We can maintain a "productive tension" (as the British philosopher Kate Soper calls it) between the two, remembering that humans already and always are a part of nature and that the nonhuman world exhibits its own productive powers—the growing of grasslands and fisheries or the purification of water in hydrological cycles—on which humanity depends.[13] Nature in a rainforest or a coral reef, even with the marks of humanity, reveals elements of complexity and dynamism not seen in plastic pink flamingoes and jumbo jets. The *negentropic* processes of life—the processes that work against the thermodynamic law of entropy (which says that disorder tends to increase) to produce the material order that is life—mark an activity and an agency in the world *beyond* humanity. Let's not imagine ourselves as gods, creating everything that sustains us *ex nihilo* (from nothing), without nature's help.[14] Let's instead acknowledge that there's nature in everything we create.

For any particular thing, then, the question will always be, What's the unique balance between artifice and nature in it? Our modern experiences are filled with hybrids that combine human agency and natural productivity. Artifice and nature are the ends of a spectrum, with plastic pink flamingoes and jets near one end (artifice) and coral reefs and rainforests near the other (nature). Intercropped farms and gardens lie somewhere in between. If there are any ecosystems remaining on Earth untouched by human agency, perhaps buried under the Antarctic ice cap, they lie wholly at the nature end of the spectrum. Even if few things are purely artifice or purely nature, the two terms are indispensable for distinguishing between things that are more one or the other. In describing modern life, the distinction becomes crucial.

The Feel of Modern Life

Consider a typical office worker in the United States named Joe. Joe lives in a suburb and commutes by car to his office in an industrial park. He wakes up each morning in his family's house, which he mortgages from a large bank. His home is furnished and supplied with things purchased from stores stocked with goods made at locations around the world. On a typical weekday, Joe engages with his wife and children while preparing to leave for work. He rides to work alone in his car. On the way, he sees flowers and trees and sometimes squirrels and pigeons. After parking in a space under the building in which his office is located, Joe takes an elevator up to his office. Having gotten into his car in his home garage and exiting it in his work garage, Joe doesn't have to go outside during his workday. At work he comes across acquaintances and "work friends." He may have lunch with several of them. Lunch is served by people with whom he has only casual acquaintance. It's produced and prepared by people he's never seen.

At work, Joe spends most of his time with symbols, concepts, and figures that are abstractions of material-world things. Most of his interactions are with a computer and printed documents. He's an accountant whose job is to decide how much forest his company should log in the next quarter, where the timber should be processed, and so on, though he has no particular knowledge of the relevant ecological implications or of the people involved in the logging and lumber manufacture. Joe doesn't work directly with trees or forests, though his number crunching helps determine what will happen to them. He does work directly with some people, his colleagues with whom he meets to discuss the numbers representing the financial outcomes of different production plans for their leased forest lands and lumber mills.

After eight or nine hours at the office, Joe returns home tired but happy to be with his family. Usually separately but occasionally together, they eat dinner, watch television, shop on the Internet, or seek some other form of entertainment. Activities such as bowling or Little League punctuate the rhythms of the evenings. On weekends, there are trips to parks, shopping malls, sporting events, restaurants for dinners with friends, or movies, or the family just stays home to watch television or movies. Joe likes to watch the game with friends on his giant flat-screen TV. He has an Android smartphone that delivers news, sports scores, emails, and text messages to the palm of his hand at any time, making his complicated life easier to manage. He's both fascinated by it and proud of it—it makes him feel modern and accomplished. It's part of his identity. Joe's life is by design the normal life of a middle-class, suburban, white-collar resident of the United States.

Although Joe is fictional, his basic daily routine and everyday experiences are recognizable to most people living in the United States and, increasingly, beyond. As a model for contemporary phenomenological experience, Joe works well enough. His life articulates many contemporary ideals, particularly that of the opportunity for financial success and material well-being—and it exhibits many forms of modern dissociation. Perhaps your life is similar; perhaps it's quite different. The point here isn't to capture every modern life but to draw a sketch that comprises features we recognize as common to contemporary life, particularly in the Western modern world. Individual counterexamples can't debunk this portrayal of the average Joe because it's meant as a composite. The goal isn't to criticize Joe but to paint a picture of a typical life.

Joe's Phenomenology

What does Joe deal with in his everyday life? What does he interact with, perceive, and think about? Clearly, his regular experiences—his contacts, knowledge, and interactions—involve chiefly a stream of material and conceptual artifice (the latter meaning ideas, numbers, concepts, and so on). On a daily basis, wild nature barely enters into Joe's direct experience. Human-made things instead capture his attention: supermarkets; bank accounts; shopping malls; e-mail; nuclear-family interactions, as wonderful as they are; and television, including perhaps smart, meaningful programs. Joe hardly even has contact with the food he and his family eat beyond selecting it in a restaurant or supermarket and perhaps sometimes cooking it. So he doesn't know much about what was involved in its production or about the processes, people, and landscapes involved in growing it.

Sure, Joe often sees birds on the way to work or in the backyard in the morning. During lunchtime he walks by lawns and gardens. He feeds, walks, and plays with the dog and maybe even tends a backyard vegetable garden. Occasionally, he enjoys wilderness excursions in distant parks and preserves. He likes to hike and would do so more often if he had the time. But these and most of his other nature experiences are framed entirely by the narrow context of leisure activities in which he encounters them (except maybe the gardening). Joe's dissociation from nonhuman nature is reflected in his limited knowledge of these things. He can't name more than a few trees or other plants, animals, or rocks in his region or their distinguishing characteristics, ecology, or natural history. But Joe's life ultimately depends on these things—weather patterns, soil, forests—so his dissociations from them are entirely relevant.

Consider also Joe's work life, which constitutes about half of his waking experience. At his job, Joe attends mostly to abstractions of nature—accounting figures that represent trees and other materials derived from nature—and abstract economic details such as labor costs. Various human constructions such as meeting schedules, deadlines, names of contacts, and the performance of subordinates fill his consciousness. Even with the focus of his work entirely on the forest products industry, Joe has little overall knowledge of the social and ecological context of the institutions making up that industry. What are the blue-collar workers' lives like? What kinds of animals live in his company's leased forests? In doing his work, Joe uses a range of artifacts such as computers, telephones, and photocopiers. His expertise is in accounting; mathematics; general office skills; word-processing and accounting software; communication, time management, and interpersonal skills; and his company's business. In sum, Joe's work life involves little experience or contact with nature—or with most of the rest of the human economy, for that matter.

From the long scope of human history, what might seem missing from Joe's experiential world? Perhaps the sight of the rings of a tree his firm has cut down or the pungent aroma of its newly exposed sap, its lifeblood. The peaty smell of the damp soil tilled to grow his produce. The wetness and steam of the blood from the chicken butchered for his table. The sound of a pig squealing as it's led to the killing floor or the glistening of the sweat on the foreheads of the low-wage immigrant laborers carving the animal's flesh into commodities. Joe doesn't see the billowing black clouds of toner dust blowing up in the poor town where his old laser printer cartridges are recycled or the acrid yellow clouds above the acid bath where his computer motherboard is being stripped. Missing from Joe's experiences are the feel of wood in his hands as he works it to build a tool or a house; the ache in his shoulder from carrying water for his family to cook and drink; the sight of the misty horizon when the sun rises, finally bringing warmth; the taste of morning dew on a fresh-picked apple. Joe misses the sound of an engulfing nature—singing birds, creaking trees, chirping insects, gurgling creeks, and splattering rain—that's alive, vibrant, and productive, always working and supplying much of what he needs, never resting. Without these things, Joe's life is impoverished and, perhaps more important, he can't fully realize how his living and choices affect the world.

Joe's Fractured Living

On the one hand, Joe has a lot of artifice, material and conceptual, in his experiences, and on the other, he's quite removed from the productive nature that his life depends on. The two are related, of course. The more our work lives involve artifice—facts, figures, machines, computers—the less they involve direct experience of nature. In a life dominated by artifice, there's less room for nature. Nature appears in Joe's life in the distilled, controlled forms of monoculture backyard lawns, gardens, pets, and parks. The preponderance of artifice and the scarcity of nature in Joe's life skew his perceptual and cognitive (relating to conscious mental activity) experiences and his knowledge toward the artificial end of the nature-artifice spectrum.

Another remarkable characteristic about Joe's life is that it's radically dispersed through space, across the whole planet. The material conditions that both affect and are affected by Joe are spread over a far greater spatial extent than could ever be the case in a non-modern historic moment in, say, medieval England. Joe's work in accounting sometimes involves the clear-cutting of forests thousands of miles away. The products he and his family use come from virtually everywhere on the planet, and the effects of his purchases are similarly scattered. His construction of a new addition to his home may help propel the clearing of virgin rainforest in Indonesia for plywood, though he doesn't know that. Joe's social universe is also spread around the globe to a degree that was impossible before modern telecommunications, air travel, and other technologies that both enable and foster relationships across large distances.

A final remarkable feature of Joe's hypothetical life is that it's highly compartmentalized—despite its great reach. During the workday Joe's social interactions involve mainly other workers in his company, mostly other people in his department and the vice president to whom he reports; his community and family members aren't really included. Compartmentalization orders Joe's personal life, too. His nuclear family lives separated from other generations and extended family members in a single-family home, a situation that would seem strange to people living during most of human history. Neighbors almost never drop by unannounced; it's considered impolite. Play dates are scheduled, and children are discouraged from playing outdoors where they may casually run into each other. Even though Joe and his family are friendly with their neighbors, their level of interaction doesn't quite match that of folks in less modern or non-modern places. In Balinese villages, for example, many neighbors see each other daily, know each other's business, wander into each other's homes, and play essential roles in each other's elaborate life-cycle ceremonies. The lack of social structure in Joe's suburban neighborhood—no real leaders, no neighborhood roles for residents, no walkable town square for festivals and performances—mirrors a lack of organization, participation, and community engagement.

Joe's time is also divided in a way that's new in human history. The compartmentalization of time has roots in the rationalization of labor in the industrial revolution—its measurement, accounting, and systematic management.[15] In most traditional societies, work and social time are mixed throughout the day, but in Joe's life, they remain separate. Modern divisions of time and labor break apart various relationships. During the eighteenth-century parliamentary enclosures in England, lands that had been communally managed for pasture and horticulture for centuries were transferred to private ownership and fenced off, usually with hedgerows. The enclosures compartmentalized space in the name of national strength and private rights.[16] Around the same period mechanical clocks were adopted, giving industrialists heightened control over the labor force by allowing finer regimentation of time and its allotment to tasks.[17] And spaces became further divided into production zones and leisure areas.

In the industrial revolution time and space were transformed together in a historical shift that optimized both for the market. Linear, evenly divided, and divorced from natural processes, abstract industrial time conforms to the ideology of mechanism, the early-modern idea that the physical universe is like a clock that "man" is supposed to maintain. Linear, rationalized time is a practical tool of mechanized, disciplined labor—an industrialist must be able to perform arithmetic operations on time, to count it precisely in order to fully control labor. The rationalization of labor and time elevates numbers and divisions—pieces of work accomplished and wages earned—and divorces work life from personal life. Time became alienated from the sensuous world of the physical, ecological reality of the seasons, daylight, moon cycles, human bodily cycles, weather and climate patterns, and harvest cycles. Ask yourself: Is the moon waxing or waning today? What fruits and vegetables are currently in season in your region? For most of human history people would be able to answer these questions immediately; they had their finger on the pulse of the skies and land around them.

Joe experiences the compartmentalization and rationalization of time in various ways. The forty-hour work week (which often expands into his private time by ten or more hours) means that large chunks of time (Monday through Friday, nine to five) are almost totally committed to work. During those long hours Joe puts personal relationships on hold or barely attends to them. Most of Joe's ancestors, by contrast, mixed work and leisure—

singing or chatting while working together in the fields, perhaps, or social-izing while preparing a community festival. When leisure time becomes subordinated to work time, as in the rhythm of Joe's life, it becomes pre-cious and therefore must be regimented just like work time. We schedule all sorts of recreational activities and time with friends in advance; today, kids must have "play dates." The pressure on leisure time may also provoke anxiety in Joe when he's just sitting around, relaxing, and doing nothing—something people in less industrial cultures do much better.[18]

Should Joe Care?

Setting aside for a moment real-world problems beyond his own life, Joe may care about all this dissociation business because it could have an emotional toll on him. Joe may feel like he's at the top of his game, successful in his career and family life. But his separation from the nature and production processes on which he depends can be a source of anxiety and stress. American Psychological Association surveys find that stress levels in the United States are high (22 percent of Americans report extreme stress). Researchers conclude that the nation may be "on the verge of a stress-induced public health crisis."[19] Various factors contribute to high stress levels, but lack of outdoor time and a feeling of being at the mercy of complex, inscrutable modern systems likely add to the burden.

The British sociologist Anthony Giddens noted a "lack of control which many of us feel about some of the circumstances of our lives."[20] To stay alive, Joe quite literally depends on anonymous systems such as agri-culture, but he has almost no say in them and doesn't quite understand them. It's easy to imagine the Thanksgiving.com loner being a bit nervous about whether the next meal will come down the tube. In that sense vul-nerability to wild nature in non-modern conditions—diseases or animals that may attack, or deadly weather—has been replaced by vulnerability to far-reaching institutions that we have little control over. They make up the "juggernaut" of modernity.[21] We must all ride the juggernaut because we're all laypersons in virtually all the "expert systems"—finance, automo-bile manufacture, the oil industry, bridge building—that make up modern institutions.[22] Joe can't steer the juggernaut or slow it down.

The juggernaut comprises cultural, political, and economic institutions that produce the symbolic tokens, such as money, and expert systems, such as nuclear engineering, that Anthony Giddens calls "abstract systems."[23] Giddens's theories elucidate Joe's dissociations and illustrate how they propagate through space and time. Symbolic tokens and expert systems remove social relations from the immediacy of context—they dissociate (Giddens calls them *disembedding* mechanisms). Expert systems are "systems of technical accomplishment or professional expertise that organize large areas of the material and social environments in which we live today."[24] They carry out the work of modern life and possess our most vital knowledge. Joe relies on expert systems such as law and medicine to sustain him. But they mediate between him and nature, producing the human-nature alienation and other dissociations that run through his life-world. Joe maintains confidence in these systems, both because their breakdown is rare and because in truth he doesn't know what else to do.[25]

Joe must bring that confidence to work, too. Because of functional specialization in the workplace—Joe crunches numbers, someone else cuts trees, a third person interprets environmental protection laws—he must be confident that other parts of his firm, industry, and government will attend to the vital things outside his expertise. So Joe at work is much like Joe at home—surrounded by expert systems in which he's no expert. He may assume that upper management or the government will prevent his company from perpetrating major environmental harms, but the problem is that everyone else, just as dissociated from vital things as Joe is, holds similar assumptions. And most workers in these expert systems are specialists only in abstractions such as forests and logging rather than in particulars such as the Stanislaus National Forest and its giant sequoias. Joe might care that various expert systems and institutions act on his behalf and might want to know what they're doing.

Like every organism, Joe changes his environment. As a modern person, particularly an American, Joe affects the world more than virtually all organisms or people who have ever lived, but he remains unaware of most of the changes that he directly or indirectly causes. They happen outside of his phenomenal range, so he's sheltered from having to confront them. But Joe might care to know about these things so he can incorporate them into his decision making, since they are, after all, connected to his choices. How can he make an informed decision when he doesn't fully know what the outcomes will be? Being phenomenally cut off from nature prevents Joe from having full awareness of how he shapes nature and what living nature needs to be nurtured and healthy. How could he possibly know about nature's complexities and vulnerabilities, and respond appropriately to them? As David Abram explains, adequate knowledge of the natural world requires the full capacity of the human senses opened directly to wild nature in an immediate mode of being.

Perhaps the most important aspect of being dissociated from his outcomes is that Joe doesn't directly experience the harms he helps create. When he drives his gas-guzzling car, he harms people and nature: global warming, flooded coastal zones, farmers trying to survive on newly arid lands, and so on. But he doesn't experience any of that as directly connected to his actions. He mainly experiences the financial costs of owning and driving the car. Waste products from his car's manufacture probably contaminate an economically disadvantaged area where Joe doesn't live and where toxins are dispersed widely. Wars to maintain access to resources are fought mostly by poorer people unrelated to Joe. The pollution from driving his gas guzzler spreads throughout the atmosphere and is shared by people around the globe, not just by Joe and his family. He vaguely knows about some of this, and that knowledge takes its toll.

Joe may understand these problems and his role in them, but his understanding remains abstract, and abstractions can't substitute for direct, sensuous experience. The immediate experiences of his everyday life powerfully shape Joe's life choices. Culture, custom, and trust in the institutions that support modern life determine Joe's decisions more than he may realize. Shouldn't Joe want to fully understand the conditions of his own decision making? He bought his gas guzzler because he saw ads that convinced him of how happy he'd be with it; he liked its style and features, and he saw other people driving that model (and greenwashing diverted any guilt: "best gas mileage in its class"). Moreover, Joe lives in a culture that associates consumer products with happiness. Abstract knowledge of problems such as climate-change-induced floods and droughts carries little weight against such powerful forces.

Of course, Joe's dissociations don't hold complete sway over his choices. He does respond to some of the knowledge about environmental damages he hears, and he does recycle, for instance. Many people choose to drive less or buy less stuff they don't really need. Maybe by taking political action,

Joe could help reshape the institutions that determine the choices available to him. His awareness of deforestation may someday lead him to buy less paper or fewer wood products even while his work may increase deforestation. But abstract knowledge can only go so far. Creating a healthier planet needs more. In the meantime, we continue to ride the juggernaut, and our dissociations make the consequences seem unreal. These conditions may not simply be an accident of history. They may be encoded in our modern values, as expressed in philosophy and popular culture.

Living Plato's Dream

> No wonder most fleshers had stampeded into the polises, once they had the chance: if disease and aging weren't reason enough, there was gravity, friction, and inertia. The physical world was one vast, tangled obstacle course of pointless, arbitrary restrictions.
> —The orphan Yatima in
> Greg Egan's *Diaspora*, 1999†

Greg Egan's 1998 science fiction novel *Diaspora* takes place in the thirtieth century CE, a thousand years from now.[26] Prefigured by ideas of disembodied beings in *Star Trek* and other science fiction, the humans in *Diaspora* no longer need bodies to live—in fact they shun them. Instead, their minds live on inside *polises*, computer networks that comprise the total life environment for communities of conscious but incorporeal post-humans.[27] Presenting a world of maximum mind-body dissociation, *Diaspora* takes phenomenal dissociation to its limit. Yatima's an orphan, conceived from a mind seed in the elaborate conceptory of Konishi polis without replication of codes from any existing "citizen." *Ve* (the gender-neutral third-person pronoun in *Diaspora*—Yatima has no gender) is born within a virtual-machine "womb" and grows freely in a process of "psychogenesis," becoming a full "citizen" a few seconds after conception. The "fleshers" are a minority population: the small number of people who've remained in organic human bodies and live in enclaves on Earth.

In the scene excerpted above, young Yatima is experiencing *vis* (the possessive of *ve*) first moments of embodiment and physicality, having cross-translated into an abandoned *gleisner* (a flesher-shaped robot body) to satisfy *vis* curiosity about the flesher enclaves by exploring the one at Atlanta, Earth. *Ve* is a "truth miner," living until now in the midst of an almost purely mathematical, simulated world, with input on demand from remote sensors on and around Earth. In a clumsy and constrained humanoid body moving through a landscape of dense undergrowth, yearning for *vis* usual quantum leaps through *scapes* of simulated space-time, Yatima has suddenly come to a fuller realization of the persistent limitations of embodiment and materiality.

The haunting truth is that Yatima's contempt for material nature isn't an arbitrary musing born of the aesthetic of an alien, distant-future society. Rather, it's a reflection of modern Western values and the moment in history in which the author lives. Science fiction's literary power derives from the truths it reveals about today's world. As I discuss in chapters 5 and 6, the fantasy of transcending material existence has been an important aspect of Western thought at least since Plato's efforts to free the human mind from the human body. Plato wrote, "Every seeker after wisdom knows that . . . his soul is a helpless prisoner, chained hand and foot in the body."[28] The starkly dissociated life-worlds of *Diaspora* serve as a useful foil against which to view both the dissociated phenomenology of modern life and Plato's disparagement of the body.

In *Diaspora* having a body is optional, and "citizens" in its computerized polises experience few material constraints. They leap instantly from one scape to another at will. Time is malleable; individuals can accelerate or slow it according to their own wishes. They can clone themselves at will (the term *diaspora* refers to a thousand clones of an entire polis that disperse throughout the universe)—a capability that projects forward current interests in human cloning.[29] Life in the polises provides protection from the vulnerabilities of living in bodies and in nature. Nature remains outside the polises, where it's usually only a hindrance or an object of curiosity, not an active and productive agent or something humans should partner with in a relationship of respect. Reminiscent of Joe at work, *Diaspora* citizens spend virtually all their consciousness within the immaterial realm of concepts, ideas, models, and representations that can be manipulated without limit or ill effect. The messy business of a material life immersed in nature has been eliminated. That may appeal to many people.

Even while raw, messy nature is held at a distance, citizens and the polises exercise virtually unlimited control over it. Models of reality and of fantasy, mathematical models and theoretical models, are created at will by the bodiless citizens, who can also create physical objects by directing nanomachines to modify matter according to any arbitrary design. Citizens create a thin machine billions of kilometers long in the solar system (sources of its materials include asteroids and other celestial bodies). There's talk of nudging planets into new orbits to protect them from a cosmic catastrophe. In the end at least one polis in the diaspora has discovered a way to travel to other universes. This disembodied citizen of the polis is a futuristic analog of the contemporary office worker at a desk, in front of a computer, working on concepts and other intangibles for her entire work life, with little direct contact with nature, yet ultimately impacting nature.

The life-worlds of *Diaspora* may seem extreme, but in truth they express ideals taken from ancient and modern Western philosophies. Early-modern efforts to conceptually separate mind from body and reason from animality—and to bring all of material nature into the unconstrained service of the human mind—advanced an agenda of liberating our minds from the constraints and limitations of a complicated, difficult nature. The latter was to be brought under control and dominated by people. Our modern life-world lies on a trajectory between two points: the traditional life-worlds of humans through most of history, working directly with nature, and one that, like *Diaspora*, realizes Plato's vision of minds fully independent of bodies. In the transition from sensuous engagement with the other-than-human world of plants, animals, human bodies, and soil, our modern preoccupation (and occupation) is increasingly with conceptual and material artifice: ideas, interior spaces, mathematics, programs, cars, computers, emails, television programs, and movies.

The world of *Diaspora* reflects modern sentiments and history in other ways as well. On *vis* expedition to an enclave of fleshers in a "gleisner" body, Yatima expresses condescending pity for these earthbound people in their fleshy, organic bodies when *ve* first comes (artificial-) face-to-face with them. *Vis* disdain recalls early European encounters with the "savages" of the New World and their "miserable," poor material conditions (never mind that the "savages" thought the Europeans could use a bath).[30] Yatima and these Europeans both observe the locals living their more intensively corporeal lives, in more direct contact with Earth, and both consider the locals inferior because of their sublunary conditions. Both arrive at their destinations using "superior" technology, and both seek to return home with goods—materials, knowledge, or people.

In our world and in that of *Diaspora*, remote-sensing technologies, satellites, and other apparatuses acquire information about remote environments, which is then gathered, accumulated, calculated, and acted on much like the knowledge brought back to the colonial centers in London and Lisbon in the European Age of Discovery in the fifteenth through seventeenth centuries. Today's remote-sensing technologies monitor myriad aspects of Earth and the cosmos, from Earth's weather to ecological conditions including forest cover, to deep-space images of celestial bodies. While many of these images are vital for our daily lives, disaster preparedness, and understanding how we're transforming the planet, some are just for war planning. In *Diaspora* artificial sensors provide all the sensory input received and acted on in the polises. Nothing is sensed directly by these bodiless people, leaving open the possibility that they could be manipulated. Just as a virus can take over a computer in our world, perhaps one citizen in *Diaspora* could gain control over that society's artificial sensory systems, an ironic possibility, given the novel's moral schema of freedom and personal empowerment.

This imaginary scenario reminds me of today's intensive monitoring and control of natural resources and human activity from afar, in control rooms and "situation" rooms. One such room in Sacramento, California, houses the equipment and managers that watch and regulate the flow of both natural and artificial rivers all over the state for agriculture, industry, and homes. Like a giant central nervous system, elaborate systems of sensors deliver flow, level, and other data to the control room, where human managers serve as the brain. Electrical cables imitate nerves through which signals pass to operate gates and pumps that respond like muscles to the commands of the remote brain. Vast reaches of nature are controlled from one central point. Similarly, in the White House Situation Room, the president and his aides control the deployment of troops and resources for wars and other strategic events like a single mind operating a globally dispersed body. Such cybernetic systems, mimicked in *Diaspora*, phenomenally dissociate mind from consequences by extending the human reach across space. Dissociations extend our bodies across the globe.

The remote sensing and control of *Diaspora* likewise reverberate in our modern daily experiences. Televisions, radios, smartphones, and computers deliver information acquired from a variety of sources around and beyond the globe. Sitting before a computer, processing information in emails and other forms, is redolent of both Plato's vision of the contemplation of the Forms and *Diaspora*'s disembodied citizens "mining truth." We control remote computers, cameras, and weapons (as do the US Air Force personnel operating deadly drone aircraft) with minimal physical interaction, much as *Diaspora* citizens control remote machines. The expansion of human control over the material world envisioned in *Diaspora* is virtually limitless: a few citizens avert a cosmological-scale catastrophe that would have destroyed Earth. But that control comes at the price of a dearth of unmediated sensory experience in the natural world. The combination of control and lack of sensory knowledge could be disastrous, as could our increased involvement with material and conceptual artifice.

The Ascendance of Artifice

Plato dreamt of lives spent living in deep cerebral engagement with conceptual artifice, the Ideas. What does it mean that today we spend so much of our waking time interacting with artifice—the ideas and physical objects that we ourselves create? Although every interaction with artifice doesn't necessarily replace one with nature or people, increased engagement with it certainly crowds out time for the outdoors or friends. Our preoccupation with artifice not only physically blocks our perception of nature; it also renders a heightened self-reflexivity: we're becoming more involved with our own creations and less involved with the world beyond our stuff. Abram points to this "self-reflexivity" as a unique and prevalent characteristic of modern life.[31] We're involved with the productions of our own species, as contrasted with those of nonhuman nature, to a radical degree. Nature's moving into the background of our consciousness.

One problem with this shift is that the artifice we generate and infuse into our phenomenal lives is full of reductive characteristics: rectilinear forms such as the straight lines that make up most buildings and many American city layouts; the spatial (and logical) linearity of text; the temporal linearity of films; and the smoothness and regularity of plastics, metals, and glass. Even the most sophisticated items we produce, such as passenger jets, televisions, and abstract art, don't exhibit the same degree of complexity, dynamism, irregularity, and unpredictability found in the natural world. Compare a passenger jet with a bird. Although the jet is indeed a (perhaps beautifully) complex piece of machinery, with thousands of interacting subsystems (a complexity I worry about too often when I fly), much of its complexity substitutes for a lack of natural sophistication and efficiency. The burning of jet fuel in combustion engines can't compare with the elegant biochemistry in which birds transform organic matter into chemical energy and then mechanical energy. Picture the physical form of the bird, with wings that move through an infinity of positions, pushing air in multiple directions; feathers that adjust through a continuum of positions; a body that grows, reproduces, and heals itself; and flight paths that exhibit mesmerizing agility. The jet airliner seems crude in comparison. Its stationary wings and linear thrust cut much simpler, smooth, geometric curves in the air (the bumpy shapes and movements of turbulence are, of course, a consequence of the irregularities and chaos of nature).

The reductiveness of our artifice is a problem because it impoverishes our senses, says Abram. Caring for nature requires a deeply felt, sensuous, subject-to-subject engagement with the natural world (one in which we meet natural others as subjects, not objects). Equally important, sustaining our sensory abilities requires sensuous engagement, too. We lose the acuity and immediacy of our senses when we don't practice sensing wild nature. The journalist Richard Louv, author of *Last Child in the Woods: Saving Our Children from Nature-Deficit Disorder*, argues that it's an essential ingredient of healthy intellectual and psychic development as well.[32] As our senses get accustomed to interacting with the artificial world of rectilinear buildings and streets and uniform plastics, they lose their sharpness. Subject-to-subject relations with dynamic, complex, and productive nature—Abram's more-than-human world—are necessary to maintain our most refined and sensitive human capacities, both mental and physical. This way of relating *with* nature gives us a deeper, embodied sense of nature's workings and its responses to human activity. The institutions of modernity, all our expert systems in all their complexity and sophistication, fail to establish this kind of sensitivity to nature's workings.[33] When we're not out in nature, we start to lose some of our humanness, and institutions can't make up for that loss.

That said, we don't have to completely shun artifice and "return to nature." Doing so would only re-inscribe the old misleading dualism between nature and culture. If we instead refocus our phenomenal worlds to include more wild nature, we'll find that the quality of our relations with *both* nature and artifice has shifted. Reducing artifice in our phenomenological lives can open up our energy and time for more intentional, thoughtful, and caring relations with both nature and the artifice that remains.[34] At the same time, we can recognize the mutual human and natural productivities that merge in the "nature-culture" hybrids inhabiting our world, including cars and smartphones. Caring for and about our artifice more intentionally, in relationships based on attention, quality, and caring rather than simply quantity, will reduce our environmental impacts.

A more intentional and considered mode of engagement with nature, artifice, and hybrids of the two would create greater understandings that reengage us with the spheres that we touch remotely and thereby foster less destructive choices. The simple act of turning on a light switch can be deepened by reflection on the switch, the electricity, the light, and on all their sources in the material world—and thus all the implications of flipping the switch. For instance you might contemplate, however briefly, the long networks of labor, energy, and materials that converge to make the light possible. You could consider the coal-fired or nuclear power plant generating acid rain, mercury, or radioactive waste, contaminating freshwater bodies and fish, or laborers working to maintain the distribution network or the processes and pollution associated with the manufacture of the bulb. Opening the drawer of a wooden desk, you might momentarily contemplate the trees and forest harvested to make the desk, together with the smoothness of the desktop, the grain of the wood, and the sound of the drawer rolling out as it's amplified by the interior space created by the shape of the desk. You may feel the smoothness of the disposable plastic pen you hold while imagining the accumulation in landfills of all the plastic pens you will have used during your lifetime. These thoughts, feelings, and phenomenological experiences build a deeper appreciation of the artifice in our lives—an appreciation that can inform our decisions and actions to be more environmentally beneficial. With more awareness you might turn the lights off, buy fewer pens, and think twice about replacing that old desk.

THE INDUSTRIALIZATION OF SENSUALITY

At work and at home, Joe still uses all his senses but not nearly to their full potential. Our human sensuous capabilities evolved over the millions of years that we and our ancestors were immersed day and night in the landscape, not walled off from the rich, vibrating, flowing realm of nature by our artifice. Although we risk losing our sensory acuity in the long run if we don't use our senses to their fullest, a few centuries of engulfment in the regularities of the modern human-made world are not nearly enough to grossly diminish either our ability or our need to take in the rich fullness that wild nature has to offer. Joe is still an animal. He hasn't lost his senses or the desire to use them.

Joe's desire hasn't gone unnoticed by modern industry, which is ever on the lookout for wants and needs—or busy producing new ones through ads: faster computers, thinner smartphones, bigger cars. Industry has rushed in to fulfill our longing for sensuous engagement and enjoyment, something fundamental to being human that until now was always free. Suppressed from everyday experience, sensuality has become ripe for commodification, and we've been made into its consumers. When industry sells nature back to us, sensuality takes center stage.

But commercial reconnections with nature contain an irony: as a commodity, nature remains alien to us because the opening of the senses in this way is a controlled endeavor, part of the industrial mode of life. It's nature in a voyeuristic, leisure form we get, not nature as an active, engaged partner in the sustenance of life. Certainly not all our experiences with nature are mediated by industry—we can have many moments of intimate engagement with nature beyond the reach of any market. Yet much of the time nature is served to us as a product we encounter only in leisure situations in which we interact with it mostly superficially.

Related industries falling under the broad umbrellas of travel, leisure, hospitality, and recreation sell products and services designed to reunite us with nature's visceral presence. Their seductive images target sensually impoverished people. And those of us who can afford to do so quite reasonably take up their offers because we feel the loss of firsthand experience with nature in our lives. We buy nature and sensuality as various commodities, from equipment for a do-it-yourself stay in the wilderness to all-inclusive packages that house vacationers close to nature, often in remote, exotic locations, where nature looks different from the way it looks in our backyards. Selling an "antidote to civilization," vacation and travel brochures lure us with something modern, civilized living leaves behind: vivid sensory engagements with the natural world.[35]

Sinking Your Feet into Warm Sand

Pick up vacation-oriented literature and be prepared to have your senses enticed by images and words. *Islands* magazine, for instance, is a treasure trove of the sensuous in advertising.[36] Within its pages companies sell "paradise" with detailed evocations of sensual experiences in lilting prose and colorful photos. Ads and articles alike engage all the senses: the sights of beautiful sunsets, lush tropical scenery, or white sand beaches; the feel of balmy breezes, fine sand under foot, the sun on your skin, the refreshing water of the sea or a pool, the relaxing touch of a massage; the smells of air pure and fresh, perfumed with the scents of spices, or fragrant with massage oils; the tastes of sumptuous fine dining or spicy food; the sounds of leaves rustling in the balmy breeze or soothing tropical music wafting through the air. These images invite us to return to the neglected body. They're hard to resist.

Hotel and resort accommodations cater to sensory- and nature-deprived people by providing rooms as close as possible to nature, the most desirable ones usually facing the beach. Sun and sea air embrace you. Many resorts offer bungalows situated among gardens, rice fields, or tropical rain-forests, immersing guests directly in landscapes and seascapes. No corridors or lobbies intervene between visitors and these delights. Some resorts push this idea to the limit by placing pampered, air-conditioned, fully serviced patrons directly above turquoise waters, sand, and coral in personal bunga-lows suspended on stilts—paying no mind to the fact that coral communi-ties can be killed off this way.

Ads for warm-weather destinations sell the most direct, sensuous contact with nature. Bare-footed vacationers bypass the shoes, asphalt, and build-ings that separate skin from Earth. Beachwear or, at some resorts, nudity allow for unmediated experiences of the wind, water, waves, sun, sand, and friendly others, and can add to the enjoyment of other vacationers. Images of swimsuit-clad couples frolicking in the water or on the beach attract

Figure 2.2. The Grenada Getaway: "This idyllic island . . . will seduce you. . . . From luxurious lodging, cultural treasures, and magnificent architecture in the south to lush rain forests, lakes, and waterfalls to the north. From white sand beaches and unspoiled coral reefs. . . . Even the air is perfumed with the enticing scents of nutmeg, cloves, cinnamon, and vanilla." (Advertisement from *Islands* magazine [July/August 2003].)[39]

weary office workers with sexual innuendo. Some advertisements even commercialize sex with organized "sex tours," which often take advantage of poor human rights situations in less developed countries.[37] When there's no time to take a vacation out of town, spas bring the senses to the city with massage, aromatherapy, salt baths, herbal wraps, mud baths, floatation tanks, Jacuzzis®, and saunas. Spa ads sell sensual pleasures with the same kinds of imagery and descriptions used in brochures for vacation packages.[38]

Figure 2.3. Paradise Awaits: A sensuous immersion. (Advertisement from *Islands* magazine [July/August 2003].)[40]

Farther from the beach, vacations on working ranches have become popular for office-weary people. Places such as Badger Creek Ranch in Colorado and Estancia Ranquilco in Argentina provide a taste of manual labor, immersion in a rural landscape, and a heavy dose of interaction with powerful animals. You can help groom horses and ride with ranchers rounding up cattle. Even work can be a vacation if it involves using your whole body, which has been neglected during long days at a desk. The senses come alive with the sights of sprawling landscapes and the smells of animals, manure, hay, leather, and sweat. But vacationers on working ranches and farms encounter "working nature" contingently—nothing is necessary about the occasion. This type of phenomenal engagement, however real, enjoyable, and compelling, is limited and temporary.

Figure 2.4. British Virgin Islands: "Turquoise sea, white sandy beaches, hidden coves, swaying palm trees, tropical fragrances, pastel colours . . . you'll receive a warm welcome from the islanders delighted to introduce you to their particular brand of Caribbean music, food, and drink." (Advertisement from *Islands* magazine [December 2012].) [41]

Encountering the Wild

Wilderness adventure tours and the array of outdoor equipment available to modern consumers offer other possibilities for nature-deprived people to reengage with the wild. Even the most extreme wilderness experiences are now marketed. Nearly any person of good basic health and financial means can experience an ascent (or an attempted one) of Mount Everest. Guided by professional adventurers, customers climb from highly populated—and polluted—base camps and leave loads of modern litter such as oxygen bottles on the mountain.[42] Wilderness adventures provide an "antidote" to daily civilized life by removing all but the most essential artifice (that which you carry on your back and in your head), affording an experience of immersion in wild nature. The senses feast on sun-warmed skin, the cool water of the river, the slippery ice of the mountain, and the smells of the forests.

One of the pleasures of being in the wilderness is the feeling of independence and self-sufficiency. Hikers might carry all their food, clothing, and shelter, gather water and firewood, fish or hunt, and prepare food under conditions that are primitive compared with those of daily life. But the idea of self-sufficiency is at least partly an illusion. For most people such trips would be impossible without modern materials and equipment (and the knowledge of how to use them). Moreover, manufacture of outdoor gear, mostly from synthetic materials, creates pollution and uses resources, but those problems remain out of sight, beyond the wilderness, in "working nature." The apparently pristine state of the leisure nature that we encounter lets us forget about all that for a moment.[43] But we need these moments in our sensually deprived lives. The visceral way we're sold these things and our craving for nature only highlight our daily lack of sensuous engagement.

Drive-Up Bears

The abundance of national, state, and regional parks in the United States makes it fun and convenient to immerse ourselves in nature. Bears visible from the car, geysers, mountains, and cliffs, in all their visual magnificence, can be enjoyed by most of the American public because beautiful national and state parks are close and inexpensive. Millions of people, including many foreign visitors, drive into Yosemite Valley to gaze at its granite walls, an experience that can't be reproduced in media. (The first time I ever got out of a car in the valley, I was amazed at how far up my eye had to follow the walls to reach the top.)

Yellowstone Park visitors can feel the spray of Old Faithful, a tangible experience of the living Earth. Deer, buffalo, elk, and wolves are there to see, too, in all their majesty. In-park lodging allows visitors to remain among the natural splendor. The spectacle of animals and landscapes in the great parks punctuates the quotidian void between us and nature. Native Americans, too, must have felt a certain void in the nineteenth century when the US government removed them from those landscapes to create the parks, driven by the destructive idea that real nature, wilderness, can't have people in it.[44] The irony is poignant: Native Americans' working nature was taken from them and turned into leisure nature for people who need a respite from their own degraded working nature. Today, we can nevertheless be grateful for the parks, and even these more commercial ventures, as powerful ways of experiencing nature and regaining, however momentarily, a fuller experience of our senses.

LIVING PHENOMENALLY DISSOCIATED LIVES

As our civilization becomes increasingly preoccupied with its own creations—text messages, stock prices, houses, tablet computers—nature recedes into the distance and so, too, does our awareness of it. It becomes a distant backdrop, an occasional friend. We end up living as if nature doesn't matter, hoping that modern institutions—government agencies, corporations—will do what's necessary to protect the ground on which we stand, the ecologies in which we're enmeshed. But politicians, bureaucrats, engineers, and corporate executives are also sensually cut off from natural phenomena and the outcomes of their choices. Although they have statistics and sophisticated scientific knowledge of the material world, few of them ever see firsthand the nature their policies and products remold for human use. Modern institutions can't adequately sense, communicate, and account for nature's rhythms, forms, and vital processes.

Joe's not an immoral person. He's simply going about his life following the moral precepts of his society, in accord with the consequences of his choices that he can see. He rides the juggernaut of modernity, and it seems perfectly fine with him, even if the ride is vaguely anxiety producing. He sees himself and his family well fed, physically healthy, and content. All their material needs are spectacularly met, and they're protected from the whims and limits of nature, as far as Joe can see. But his belief that his lifestyle can be sustained without hurting others is an illusion created by his phenomenal disconnect from nature. Like the curtain initially shielding the Wizard of Oz from view, the distances, technologies, and institutions that lie between Joe and both nature and his consequences create this illusion.

Joe can pull back the curtain to regain some engagement with the flesh of sensually rich and complex wild nature and to learn more about how his life remolds and shapes it. So that the forests his company logs can be less of an abstraction, Joe might visit them. He might study where waste from his community goes, how it's handled, and who does the work. He might learn more about the problems with making and disposing of high-tech electronics. These efforts will put Joe more in touch with nature and his consequences. All the outcomes of our phenomenally dissociated life-worlds may never become clear and present personally to us in our own lifetimes, but we can make immediate improvements and begin to move toward a society in which people are more phenomenally connected with these important things. In the next chapter I explain the psychology that makes these phenomenal connections so vital.

CHAPTER 3

FROM DISSOCIATION TO DESTRUCTION THROUGH THE PSYCHE

C onsider for a moment a seemingly absurd but instructive scenario. You have to get someone to commit an act of destruction. How would you set things up to make him follow through, if he knows he's going to be hurting someone or something? First, you'd make it a physically simple task, like pushing a button. Then, you'd put him in an enclosed room far from the target of harm. Next, you'd tell him as little as possible about the person or animal he'll be hurting or the forest he'll be burning down. Then, you'd show him that many other people will be pressing buttons just like his. If possible, you'd make this act of destruction just one in a long series that starts with very minor ones. You'd tell him that the destruction won't happen right away—it'll be delayed for some random amount of time, and he won't know when it takes place. The final step would be for someone to pose as an authority figure, perhaps wearing a white lab coat or business suit, and tell the button pusher something like "You must do it now. If you don't, the whole system you're part of will fail." If you do all these things, he'll be more likely to comply.

This situation may sound bizarre, but it's not so different from what happens in our everyday lives. The authority figures in our world—the corporations, the politicians—tell us to buy more to make ourselves happier and to stimulate the economy for the common good. They communicate these things through persuasive advertisements. We see everyone else doing the same thing. And it keeps getting easier: the click of a mouse button delivers a new cookware set to our door. The damages happen far away, to nature and people we never see: pollution from the mining waste, global warming gasses released into the atmosphere from all the energy spent mining and smelting the metal, flooding and drought intensified by climate change.

Even the part about you being put into this position isn't particularly unusual. Military trainers dissociate their trainees from the objects of their harm as a well-worn tactic to facilitate violence. To get a soldier to kill or an "operative" to torture someone, their superior creates distance between the killer and the victim—any kind of distance: physical, psychic, social. Soldiers are taught to create social distance by dehumanizing their victim, referring to them with derogative or impersonal terms—such as "the enemy"—and thinking of them as radically different from themselves. Physical distance is perhaps the easiest dissociation to create. It's emotionally easier to drop a bomb that might kill several people from a jet high in the sky than to shoot someone with a gun while you look into her eyes. It's easier still to drop bombs from a drone aircraft operated via a video-game-like console in a remote location.

But why do these tactics work? If I know that I'm burning down a forest, for example, why isn't knowledge of the consequence enough? Why would it deter me to witness firsthand the caustic fumes from acid baths used in high-tech electronics recycling? If I know I'm hurting someone, why does it make it easier to not see him? The short answer is that, like other animals, we best grasp—and respond to—the world directly through our senses. A more complete answer comes from a groundbreaking and controversial series of psychology experiments and related analyses that show, among other things, that physical and psychological distance facilitates violence. This chapter looks at these to explain why knowledge of our environmental damages is necessary but isn't enough to fully rein in our destructive choices. First, however, it'll help to explore the concept of dissociation in psychology and how my concept of dissociation differs from it.

DISSOCIATION IN PSYCHES

Psychologists have related disconnectedness to dysfunction for centuries. In the early nineteenth century the term *dissociation* began to be used to mean the pathological (dysfunctional) disconnection or alienation of parts of a person's consciousness, memory, and experience internally in the psyche.[1] Due to a work-related trauma, for instance, your mind might block out all consciousness and memory related to your work life. Suddenly, you can no longer do your job. In the late nineteenth century, the French psychiatrist and philosopher Pierre Janet (1859–1947) used the term dissociation (*désagrégation*) to describe how aspects of personality and consciousness become split off from the dominant personality and lead to psychopathologies such as multiple personalities, disconnection from normal stream of consciousness, and physical illness.[2]

The two Austrian inventers of psychoanalysis, Josef Breuer (1842–1925) and Sigmund Freud (1856–1939), later conceived of dissociation as a healthy phenomenon—a natural psychic process that allows complex tasks to be pursued without disrupting a person's stream of consciousness.[3] But this normal defense mechanism becomes pathological when intolerable conflicts arise between two streams of consciousness, which then can't be rejoined to establish a unified experience of identity. The psyche seems to handle wishes, perceptions, and memories that contradict a person's overt values and self-concept by splitting them off, only to have them become seemingly unrelated symptoms, such as mysterious illnesses or facial tics.[4] Do we similarly disconnect from knowledge of our environmentally destructive actions, causing them to become strange and unrecognizable psychological maladies?

Psychological dissociations can also result from major emotional problems or traumas such as physical or sexual abuse.[5] Today, clinical psychologists work with a whole class of disorders related to dissociations taking place within the psyche. The *Diagnostic and Statistical Manual of Mental Disorders (DSM-IV)* defines dissociation-related disorders as "a disruption in the usually integrated functions of consciousness, memory, identity, or perception of the environment."[6] Psychic dissociations are often accompanied by somatic (bodily) problems such as eating disorders and insensitivity to pain.[7] People with dissociative disorders can even experience distorted memory, identity, and perception of the body and the surrounding environment.[8] Certainly, there's a parallel between dysfunctions arising from dissociations within the psyche and problems resulting from the kinds of dissociations from nature and consequences this book is concerned with. Both are breakdowns among organically related parts, and both lead to abnormal functioning.

In the worst-case scenario, psychic dissociations can lead to self-destructiveness. The clinical psychologist Israel Orbach has tied dissociations in the psyche with experiences of dissociation from the body (feeling out of touch with one's physical self) and self-destructive, even suicidal, behavior. Psychic dissociations, Orbach believes, result from distorted or minimal parental caring behaviors during childhood, lack of attunement to a child's needs, and inappropriate parental responses to a child's bodily experiences. They can produce a "sense of detachment, numbness, and bodily estrangement."[9] If your mind isn't working in a normally interconnected way, you can start to feel disconnected from your body, and that feeling of disconnectedness can prompt you to harm your physical self.

Do we feel similarly estranged from nature because of our experiences of dissociation from it? Does our estrangement drive us to be destructive toward nature? It's not far-fetched to think of nature as our extended body. It's the material realm from which we emerge and with which we merge again after death. During our lives, materials and energy constantly flow into and out of us, from and to the environment. If nature is our collective body, or an extension of our bodies, then perhaps this jump from the individual to the collective might make perfect sense, and so might the parallels between intrapsychic dissociations and those from nature. But my concept of dissociation is much broader than the ones coming out of psychology: I'm concerned with all sorts of breakdowns of connections between elements that normally work together to form a coherent and healthy whole, not just psychic ones. Nevertheless, the path from psychic dissociations to destructiveness provides initial yet tentative support for the dissociation and destruction connection. Fortunately, much stronger, more direct evidence is available to show how the phenomenal dissociations that concern me lead to destructiveness.

FROM PHENOMENAL DISSOCIATIONS TO DESTRUCTION

> Our spatial relations shift from one situation to the next, and the fact that we are near or remote may have a powerful effect on the psychological processes that mediate our behavior toward others.
> —Stanley Milgram, *Obedience to Authority*, 1974*

In the early 1960s a young Yale University psychologist, Stanley Milgram, wanted to understand the destructive obedience of soldiers in the German death camps from 1933 to 1945 when "millions of innocent people were systematically slaughtered on command."[10] To that end, he performed some of the most surprising, controversial, influential, and famous experiments in the history of psychology: at least twenty-one variations on an experiment designed to measure destructive obedience to authority.[11] The studies made a deep and lasting impression on our understandings of human behavior. In 1975, the Turkish social psychologist Muzafer Sherif, a founder of the field, said, "Milgram's obedience experiment is the single greatest contribution to human knowledge ever made by the field of social psychology, perhaps psychology in general."[12] More recently, researchers performing a comprehensive review of the studies declared, "Stanley Milgram's experiments on obedience to authority are surely among the most celebrated in the history of psychology. . . . The Milgram experiments . . . have stimulated thought as has perhaps no other single research program."[13] Among other results, some of Milgram's variations showed that phenomenal dissociations make people more destructive.

In Milgram's obedience experiments, a naive subject (a volunteer), playing the role of teacher in a staged learning experiment, is instructed to administer increasingly severe electrical shocks to a supposed learner.[14] The shocks allegedly serve as punishment for incorrect answers and thus improve learning. But the learner isn't a volunteer as the subject is told; rather, he's an actor who only pretends to receive shocks. If the subject (teacher) protests against moving to the next higher shock level, an authority figure (supposedly the experimenter but actually another accomplice) instructs the subject to continue delivering shocks and thus fulfill his role. The instructions follow a preset script that rises in level of insistence to match increasing levels of resistance from the subject.

The switch array used by the subject to deliver fake shocks runs in 15-volt increments from 15 to 450 volts. Labels on the switches range from "SLIGHT SHOCK" at the lower end to "DANGER-SEVERE SHOCK" at the higher end. Before the experiment begins, the subject receives an actual shock of 45 volts, applied by pressing the third switch on the shock generator, to convince him or her that the generator is real.[15] The subject then helps the experimenter strap the learner into the chair.

The following describes the baseline version of the experiment, the basic procedure of which many variations were carried out. The subject hears the learner mention that he has a "slight heart condition." During the experiment, the subject and learner remain in adjacent rooms, and the subject can hear but can't see the learner. The learner's audible responses to the supposed shocks begin at 75 volts, when he grunts. The responses increase to a verbal complaint at 120 volts and a demand to be released from the experiment at 150 volts, when the learner also mentions that his heart is starting to bother him. At 270 volts he begins to produce an "agonized scream." The learner's shrieks and demands to be released become increasingly vehement and emotional as the voltage is increased. Forty subjects were tested in this baseline version of the experiment.[16]

The psychologist Thomas Blass, a foremost scholar on the obedience studies, argues that they've remained not only controversial but also relevant for many reasons. First, there's the "unexpected enormity of the basic findings."[17] Sixty-five percent of the subjects (twenty-six out of forty), all American adult men, complied with the experimenter's instructions fully, shocking the learner all the way to the maximum level of 450 volts. Eighty percent of the subjects (thirty-two out of forty) continued past the point when the learner said his heart was bothering him and demanded to be freed from the experiment.[18] These results defied predictions by groups of Yale University seniors and professional psychiatrists, who predicted total obedience rates of 1.2 percent and 1.25 percent, respectively—a far cry from 65 percent.[19]

A second reason for the salience of the obedience studies is that together they make up one of the largest integrated research programs in psychology. Milgram conducted extensive variations on the baseline experiment and developed an integrated analysis of the aggregate findings. Third, psychologists have fervently debated the obedience studies in print, both praising and criticizing them.[20] A "storm of controversy" grew around the experiments because they actively deceived naive volunteer subjects, they seemed to show how easy it is to get people to do harm, and they exposed participants to emotionally grueling conditions.[21] Although the subjects were told after each experiment that no shocks had actually been administered, some experienced severe psychological reactions during the experiments, including signs of extreme tension while delivering the most powerful shocks, ranging from sweats and trembling to nervous laughter and uncontrollable seizures (experienced by a remarkable number of the subjects).[22] Some volunteers believed they could be seriously injuring the "victim."

Fourth, Milgram's obedience research is relevant to and has been used in fields outside of psychology, from communications research to philosophy, political science, education, and Holocaust studies. Finally, the research remains significant because it revealed a "fundamental and far-reaching" implication about human nature: that situations can override personal dispositions in determining behavior.[23] In other words, even kind, sympathetic people can, under certain circumstances, be influenced to harm others, even when their own well-being isn't at risk.

Context is crucial, the experiments confirmed. People don't simply act according to their own predispositions, which may help explain why some people who care greatly about the environment consume and create pollution just as much as others. Without the setting created in the obedience experiments—an authoritarian figure continually demanding that the subject deliver shocks—few subjects would have "hurt" the learner, having been told what to do only at the start. A man dressed in a white lab coat posing as a scientist with serious-looking equipment could get people to inflict pain on others. A variation on the experiment showed that the subjects wouldn't have delivered painful shocks of their own volition, that they weren't acting out their latent aggressions: when told they could choose any shock level to be administered, most of the subjects delivered shocks in the lowest range. Only two ventured into the "danger" zone.[24]

In the long-standing debate between psychologists who are situationists (seeing behavior as arising from context or situation) and those who are dispositionists (seeing behavior as resulting from personality or disposition), the obedience experiments come down firmly on the situationists' side. Situation prevailed over disposition for most of the volunteers. The experiments provided ample evidence of subjects forcing themselves to act against their own personal dispositions to do no harm. The conflict expressed itself as sweats, trembling, nervous laughter, and seizures for many subjects. Strangely enough, these symptoms are a hopeful sign that people are predisposed not to harm others.

But neither situation alone nor disposition alone determines behavior. Most subjects in the baseline experiment succumbed to the situation, but fourteen out of forty (35 percent) disobeyed.[25] Interestingly, the ones who refused to hurt the learner scored higher in "social responsibility" on a well-known personality test, the Minnesota Multiphasic Personality Inventory.[26] Nevertheless, from the start, just by demonstrating the power of situations to influence behavior—particularly to do harm—the obedience experiments already begin to support the idea that dissociations can be harmful, because dissociations are about contexts—ones that lead people to make destructive choices seemingly against their personal values.

You can see situations at work all the time. Perhaps you know about global climate change and even about some of the many problems it causes, and maybe you want to reduce your burden on the planet. Meanwhile, you're bombarded with ads to buy a new car (and just the manufacture of a car makes a large global-warming impact). In the United States car companies for decades have been pushing large, consumptive vehicles because they return the highest profit margins; today, SUVs and muscle trucks dominate many American parking lots and streets. The two main candidates in the 2012 US presidential election at times seemed to be competing over who would drill for more oil, implying there's no urgent climate change problem. Commercials, peer choices, government officials—all seem to condone the choice that conflicts with your values. You buy. Your environmental values take a back seat in your roomy new vehicle.

Starting with the compelling power of situations, the obedience experiments provide a solid foundation to begin to understand the links between dissociation and destruction despite the criticism they've received, particularly around the ethics of experimental deception.[27] Milgram studied what was approaching a thousand American adults in the series.[28] Psychologists have carried out and published at least twenty experiments modeled on Milgram's obedience paradigm worldwide, with overall results confirming the original findings.[29] The validity of the results hasn't diminished in time.[30] A similar experiment with an authentic victim—a puppy receiving actual shocks to the point of becoming animated and howling, sad to say—yielded similar results for male subjects (curiously, *all* thirteen female subjects obeyed fully in delivering shocks to the puppy).[31] Milgram's results and those from the puppy experiment closely match those from an experiment conducted in 1924 in which subjects were instructed to manually cut off the head of a live rat.[32] The relevance of the obedience results outside of psychology, their remarkable demonstration of the power of situational, contextual factors, and their central concern with human destructiveness all suggest they can shed light on our harmful choices.

The strongest confirmation of the link between dissociation and destruction comes from a set of variations on the obedience experiments: the "proximity series," which manipulated the closeness of the supposed victim to the subject across four different setups. The first variation, "Remote-Feedback," differs from the baseline experiment in that the victim (learner) makes no vocal complaint and the subject can't see him. But at 300 volts, the "laboratory walls resound as he pounds in protest." After 315 volts the victim no longer answers questions, and the pounding ceases. In the next experiment, "Voice-Feedback," the victim is in a separate room, but his complaints can be clearly heard through the walls. The "Proximity" variation puts the victim in the same room as the subject, a few feet away, thus making him visible as well as audible. In the final variation, "Touch-Proximity," the victim gets a shock only when his hand rests on a shock plate. At the 150-volt level, the victim demands to be set free and refuses to put his hand on the plate. The experimenter then orders the subject to force the victim's hand onto the plate, requiring the subject to have physical contact with the victim beyond the 150-volt level.[33]

Forty subjects were studied in each of these four experimental variations. Obedience rates (the percentage of subjects who obeyed the experimenter fully and delivered all shocks up to the highest level) fell as the subject became more proximate to the victim: 65 percent in the Remote-Feedback condition, 62.5 percent in the Voice-Feedback condition, 40 percent in the Proximity condition, and 30 percent in the Touch-Proximity condition.[34] In the most dissociated condition, Remote-Feedback, no subject stopped before administering the 300-volt shock, at which point the victim kicked the wall and no longer answered the subject's multiple-choice questions.[35] Overall, the subjects became less willing to inflict harm as the victim (and his suffering) became more immediate and salient to them.[36] Results from earlier pilot studies support this relationship. In the pilot the victim gives no protests, verbal or pounding, but is still dimly visible through a mirror to the next room. In this setup "virtually all subjects, once commanded, went blithely to the end of the board, seemingly indifferent to the verbal designations [on the shock generator before them]

('Extreme Shock' and 'Danger: Severe Shock')."[37] Likewise, if we can't see and hear nature, if we can't witness our degradations, there may be no limits to our destructiveness. Proximity is the complement of phenomenal dissociation, so the proximity series shows that people are more likely to make harmful choices when they're more dissociated from the consequences and the others they're affecting, even when they know they're causing harm.[38]

Milgram's View

Although the obedience results provide ammunition for the argument that there's something inherently destructive about our dissociations, they don't quite tell us why. This question concerned Milgram. Why should it be relevant to actually witness a harm you're creating if you know about it? Moreover, when you're able to witness it, why should it matter how close you are? With the results in hand he sought to develop a theory to explain the relationship between distance—phenomenal dissociation—and destruction, among the other findings. He presented his elaborate framework in a paper, "Some Conditions of Obedience and Disobedience to Authority," in which he identifies six factors that make people more harmful under conditions of dissociation.[39] Most of them, described below, also seem to apply to situations in which what's harmed isn't human.

Empathic Cues: In the more remote conditions, the victim's suffering has an abstract, remote quality for the subject. "He is aware, but only in a conceptual sense, that his actions cause pain to another person; the fact is apprehended but not felt." A conceptual engagement doesn't necessarily lead to an emotional response. Milgram noted that this is a common enough phenomenon and gives the example of a bombardier who knows his weapons will inflict suffering and death, yet his knowledge is "divested of affect and does not arouse in him an emotional response to the suffering that he causes." Visual cues of the victim's suffering may trigger empathic responses in the subject and give him a more complete grasp of the victim's situation. The empathic responses themselves may be unpleasant and thus curb destructive behavior. You might not enjoy looking up close into the eyes of the pig being slaughtered for your dinner.

Retired US Army lieutenant colonel Dave Grossman studies the conditions that enable killing in war and everyday society and its psychological costs. He writes, "At close range, the resistance to killing a person is tremendous. When one looks an opponent in the eye, and knows that he is young or old, scared or angry, it is not possible to deny that the individual about to be killed is much like oneself."[40] Grossman quotes a Vietnam Special Forces veteran as saying, "When you get up close and personal, where you can hear 'em scream and see 'em die, it's a bitch." Proximity to a source of authority and distance from a victim facilitate killing, Grossman says.[41] Empathic cues work, of course, with nonhuman others as well. It's easy to imagine feeling empathy for a pet or perhaps even a "head" of cattle. I've experienced this with inanimate things, such as trees, and others do, too. Proximity and face-to-face encounters encourage empathy and provide the context for all sorts of genuine emotional connections to arise, including those that lead to caring and nurturing choices.

Denial and narrowing of the cognitive field: Milgram writes, "When the victim is close, it is more difficult to exclude him phenomenologically."[42] In the more remote conditions of the proximity series, it's easier to exclude the victim and his suffering from thought. In the two most remote situations, feedback is sporadic and discontinuous, while in the two most proximate conditions, inclusion in the immediate visual field renders the victim continuously salient and harder to ignore.[43] Tellingly, in the conditions in which subjects could see the victim, they often averted their eyes.[44] Clearly, when our victims lie outside our cognitive fields, when we don't even know they exist, their well-being can't be considered, regardless of whether they're humans. It's easier to not think about high-tech workers' bodies becoming contaminated with chemicals when you can't see them.

Reciprocal fields: In proximity conditions, not only can the subject observe the victim but the actions of the subject are now under scrutiny by the victim. When the victim witnesses the subject's actions, it may bring about shame or guilt in the subject, an emotional response that can block harmful action. Blindfolding the victim of a firing squad may cause less stress not only for the victim but also for the executioner, which may be why executioners wear hoods. Being part of the victim's field of awareness may make subjects more self-conscious, embarrassed, and inhibited in causing destructive violence against the victim. Reciprocal fields are defeated by dissociations that prevent the victim from seeing the actor. Perhaps this factor is most relevant when the victim is human. But consider also the powerful effect of the gaze of nonhuman animals. I'll never forget the frightened, desperate look in the eyes of a very sick dog I'd taken to the vet as the doctors approached her to draw blood. It was hard to resist the urge to stop them.

Experienced unity of act: Under dissociated conditions, it's more difficult for the subject to be aware of the connection between his actions and the consequences for the victim. The act and its repercussions are physically separated. The two events of pressing a lever and protests in another room are in correlation but "lack a compelling unity." In the proximity conditions, this unity is more fully achieved.[45] The experienced unity of an act is disrupted when actors are dissociated from consequences in space or time. There's little experienced unity between buying a ream of paper and the felling of the trees that went into it.

Incipient group-formation: Putting the victim in another room affects the social relations of the situation. It draws the victim farther away from the subject while the subject and experimenter remain closer together. A group begins to form between subject and experimenter, and the victim is excluded. In the remote condition, the victim is truly an outsider who stands alone, physically and psychologically, like nature, which stands alone, away from our daily lives. When the victim is brought closer to the subject in the proximity conditions, it's easier to form an alliance with him against the experimenter. The subject now has an ally against the experimenter. Alliances shift with changing spatial relations. Another of Milgram's experimental variations, the "closeness of authority" variation, further supports the notion of incipient group formation. When the experimenter is physically removed from the room where the subject sits and the two communicate by telephone (with all other conditions remaining equal), obedience drops sharply.[46] Alliances are of course more relevant to situations involving other humans being harmed, but perhaps they occur when animals are victims, too.

Another experimental variation confirms the importance of group formation: the "bring-a-friend" condition. Milgram never published this condition, possibly because he used an unethical procedure in it.[47] Subjects were instructed to bring an acquaintance of at least two years. This person became the learner and was secretly instructed by Milgram in how to deceive the subject into thinking the shocks were real. In this condition, in which a relationship already existed between the subject and the learner, only 15 percent agreed to administer every shock. For most people, the existing relationship outweighed the one between the subject and the experimenter, though it's strange to think that even 15 percent of people would agree to give dangerous shocks to a friend or acquaintance. Nevertheless, it's clear that having a preexisting relationship with someone—or some thing, perhaps—reduces harm, a result we can apply to reducing destruction in the real world by establishing more relationships with the others our actions can affect.

Acquired behavior dispositions: People and other social animals learn not to harm others mostly in contexts of the proximal relations in everyday life, dealing with people in face-to-face interactions at home, in the neighborhood, or at the grocery store. In the past, aggressive actions against physically close others may have resulted in retaliatory punishment, while aggression against physically more distant others may rarely have led to reprisal. In the obedience experiments "the concrete, visible, and proximal presence of the victim acted in an important way to counteract the experimenter's power and to generate disobedience" against destructive orders.[48] We may in effect be taught to respect and protect others who are physically closer to us. Perhaps that's why so many indigenous cultures living in close contact with nature exhibit great respect for the natural world.

Milgram's concepts of *agentic state* and *strain* help in understanding the tension we feel between our environmental values and our participation in actions that harm nature.[49] Milgram believed that the obedience-experiment subjects were drawn into an agentic state—they became, in part, agents of the experimenter, carrying out his wishes. Several factors drew them into this state: the experimenter's elevated authority in the institutional setting, the existing agreement between the subject and experimenter, the relative ignorance of the subject compared with the experimenter in the setting, and the subject's loss of responsibility. Strain is the internal force that drives subjects to want to stop their harmful acts.[50]

Doesn't Milgram's concept of the agentic state aptly describe our abdication of responsibility to corporations and institutions that decide how our food is grown, what drugs are safe, and how the environment will be used? And doesn't his idea of strain describe how many of us want to stop the environmental damages to which we contribute and end up feeling disempowered to do so? Most of us do have choices, just as the obedience subjects could stop obeying. But I think we relinquish much of our responsibility to institutions to tell us what is right and proper. And they tell us, for instance, that driving is completely acceptable in almost all cases—it's our right as free individuals—even while driving is well known to contribute to some of our worst environmental problems.

Several aspects of the situation kept subjects in the agentic state. Milgram called them "binding factors." The *sequential nature of the action* makes it hard to give up at any particular point once you've started giving shocks because doing so might imply that your prior actions were bad.[51] The subjects also had *situational obligations*: they made a promise to help the experimenter and felt obliged to keep that promise.[52] Because we're socialized to follow rules set down by authority figures, the subjects experienced *anxiety* when they considered not following the experimenter's instructions. Their anxiety over violating the rules took the form of nervous laughter and trembling. These symptoms disappeared as soon as subjects chose to disobey, resolving the tension of the situation.[53]

It's always easier and less anxiety provoking to go along with social norms and rules, which bind us into maintaining the sequence of harmful acts we participate in daily. Why not just buy one more plastic bottle of water or one more smartphone? Many aspects of daily life place us in situational obligations to do things we know result in harms. To maintain friendships often means driving long distances and flying. To many Americans, being a good mother means buying plenty of gifts for the children at Christmas, even knowing many will be used only briefly before being discarded.[54] In what other ways does Milgram's analysis of destructive obedience play out in everyday life?

Cogs in the Machine?

When people know they're inflicting harm but continue to do so because situational factors compel or entice them, they feel strain. Some people might feel slightly guilty about driving instead of taking public transit, even if riding the bus would be inconvenient. But factors that increase dissociations between the subject and action on the one hand and the object and consequence on the other alleviate the strain. The ill consequences of driving are so remote, and there are so many layers of industry and government between us and the consequences, that it's usually easy for most of us to just drive without considering them at all. Milgram summarizes this effect:

> Any force or event that is placed between the subject and the conse-quences of shocking the victim, any factor that will create distance between the subject and the victim, will lead to a reduction of strain on the participant and thus less disobedience [to the demand to inflict harm]. In modern society others often stand between us and the final destructive act to which we contribute. . . . Indeed, it is typical of modern bureaucracy, even when it is designed for destructive purposes, that most people involved in its organization do not directly carry out any destruc-tive actions. They shuffle papers or load ammunition or perform some other act which, though it contributes to the final destructive effect, is remote from it in the eyes and mind of the functionary.[55]

Think of Joe and his accounting work at the logging company, detached from the logging but nevertheless involved. Milgram envisioned fields of force that diminish in effectiveness with increasing psychological distance from their source. They can either inhibit or promote certain types of behavior.[56] Fields of force emanating from the experimenter promote compliance with his instructions, whereas those emanating from the victim inhibit compliance. The more we're inundated with ads to buy new smartphones, for example, the more we may be under the influence of the field of force of the companies selling them. Conversely, the more we hear about, see, and maybe even feel the toxic effects of their creation, the more we may be influenced by a competing field of force.

In some versions of the obedience experiments, the field of force ema-nating from the learner was muted. One obedience subject said, "It's funny how you really begin to forget that there's a guy out there, even though you can hear him."[57] His comment echoes modern humanity's relation-ship with nature. Most of us know about our society's abuses of nature. We can hear nature at a distance. But institutions of science, government, and industry hold authority in our lives and can put us in an agentic state. We comply, yielding to their authority, carrying out acts we know to be destructive toward nature and other people, like Milgram's subjects who yielded to the experimenter and delivered (what they believed were) painful, possibly damaging shocks to the victim, even when hearing his screams. Milgram observed, "Any competent manager of a destructive bureaucratic system can arrange his personnel so that only the most callous and obtuse are directly involved in the violence."[58] Relatively few people are needed for the most socially and environmentally destructive tasks, and corporations can usually find people sufficiently obedient, tolerant, ignorant, or unconcerned. Many lack other opportunities.

As I discuss in more depth below, our continued abuses of invisible others can have effects on us, not just on them—just as Milgram's subjects found themselves trembling, laughing, or worse. Ecopsychologists talk about environmentally related despair in ways that recall the emotional effects of strain experienced by obedience experiment subjects.[59] The authority of modern industrial and other institutions prescribes behaviors—purchasing inefficient vehicles and other features of the "good life"—that contribute to results people find distressful, such as global climate change and deforestation. Most meat eaters are appalled when they finally find out about the conditions on the factory farms that their purchases support. The dissonance between acting in accord with authority and the knowledge of undesired consequences of such actions may harm our mental health in ways we're only beginning to appreciate.

The Cycle of Destruction

A long time passed from when I realized how wasteful it is to bag my groceries in disposable paper or plastic bags until I actually did something about it. Staying in Germany in the late 1980s, I found that I was expected to bring bags to the grocery store or pay for heavy-duty plastic ones at the register. That experience made me more aware of the problem. When I returned home to Northern California, I could've brought my own bags to the grocery store without raising any eyebrows. Why did it take years for me, with all my concern about the environment, to change this habit that consumes trees and produces unnecessary waste?

The answer may lie partly in one of Milgram's explanations for continued obedience in the experiments. To change a routine, to give up a practice that may be harmful runs the risk of implicitly condemning our own past behavior. The new behavior would create a self-critical stance toward the old one, which would then conflict with our positive self-image and thus create cognitive dissonance: the discomfort or anxiety of holding two conflicting ideas or beliefs. In the obedience experiments the step-wise progression and the gradually increasing nature of the harms probably helped launch and propel the sequence forward. At lower shock levels, it's easier to obey the authority figure and deliver shocks because the effects are much smaller. As the shock level increases, it becomes emotionally more difficult for subjects to continue to obey, yet to break off the pattern recriminates oneself for delivering the previous shocks. Subjects may feel compelled to continue to the end to justify the shocks they've already given.[60] Because the shock levels increase gradually and uniformly, there's no obvious dividing line at which the subject can justify stopping without condemning her own previous behavior. The overall effect is similar to the "foot-in-the-door" technique: a person is more likely to comply with a major request after carrying out a more minor one.[61]

Similarly, environmental destruction has increased in scale and scope continuously over the five or so centuries of the modern period, with some exceptions. Modern technology and wealth, bolstered by the exploitation of new energy sources, particularly fossil fuels, have increased each person's individual potential for destruction. Continuing on the same path may be a way of avoiding the cognitive dissonance that could be created by tacitly condemning past behaviors with a change in direction. Put another way, we may have a need to validate past behaviors that degrade the environment by repeating them or changing them only gradually.[62] People who seek major social change would do well to account for cognitive dissonance, besides comfort, greed, and the like, as inhibitors of change.

Other researchers have come to similar conclusions. In *The Roots of Evil* the psychologist Ervin Staub, a pioneer in the psychology of peace and violence, sought to understand how destructive practices such as genocide and group violence are perpetuated. Staub notes that the further destruction has progressed, the more difficult it is to stop. He builds on the psychologist Kurt Lewin's (1890–1947) conception of the "goal gradient . . . the closer you are to a goal, the stronger the motivation to reach it."[63] Interrupting goal-directed behavior creates tension, Staub found, and people are motivated to reach closure, to resolve psychic tensions. One of Milgram's subjects said to himself, obviously loud enough for the experimenter to hear, "It's *got* to go on. It's *got* to go on,"[64] as if the goal of completing the experiment were paramount. A goal brings the promise of closure.

What goal might lead to environmental destruction? The modern story of progress drives much of the change that we see. We believe in progress—a linear historical trajectory in which human welfare continually improves through better science, technology, and social organization. According to this story, we're growing less dependent on and less vulnerable to nature. We're becoming nature's masters, as envisioned by the early modern philosophers I discuss in chapter 6. Perhaps the thought of getting closer to the goal of true independence of and mastery over nature, as expressed in science fiction, drives us ever onward. As Lewin said, as we get closer to the goal (or believe we're doing so), our motivation becomes stronger to reach it. And this drive may persist in full light of the damages we cause along the way. So far, better technology and management haven't delivered us from environmental ruin, in spite of developments such as wind power and hybrid vehicles.

In his work in Burundi, Rwanda, and elsewhere to promote caring, nonaggressive peoples and societies, Staub applies "just-world" thinking to understand human destructiveness and the absence of helping behavior. Harm is self-perpetuating. People's naive beliefs in a just world lead them to devalue victims of harm. At some level most of us believe that people get their just deserts, that victims have earned their suffering by their actions or character. Staub observes that genocidal conflict is fueled by an intense devaluation along class or other group lines.[65] Milgram also saw this effect in his subjects, many of whom grimly devalued the victim after hurting him. A typical comment was "He was so stupid and stubborn he deserved to get shocked." Milgram writes, "Once having acted against the victim, these subjects found it necessary to view him as an unworthy individual, whose punishment was made inevitable by his own deficiencies of intellect and character."[66]

People, moreover, can devalue entire classes of other people who've been victimized, a phenomenon not foreign to the United States, with its long history of slavery and racial discrimination. Just-world thinking may carry over to social institutions, to lead us to believe that our institutions of industry, science, and governance are operated by mostly righteous people performing mostly righteous acts, so that those who are harmed are justly so. We can ask, Does this phenomenon apply at some level not just to other people we may harm with our choices—such as the coastal populations in Bangladesh who will be inundated as the seas rise with global climate change—but also to nonhuman nature? Do we begin to think that, having suffered our abuses, nature, including other animals, somehow deserves such treatment?

Social Division and Destructiveness

> Evil deeds are rarely the product of evil people acting from evil motives, but are the product of good bureaucrats simply doing their job.
> —Philip G. Zimbardo,
> "On 'Obedience to Authority,'" 1974†

Is it possible to test the idea that participation in a bureaucracy can perpetrate harms? Another variation of the obedience experiments showed that it is. The "peer administers shocks" variation removed the subject one additional step from the victim in the experiment's social hierarchy by placing another person between the subject and the victim. The subject in this variation doesn't press the shock levers but tells someone else to do so. The person sitting at the shock-machine controls is actually an accomplice of the experimenter, though the subject is told he's another subject. The subsidiary act of ordering another person to administer the shock remains vital to the overall progress of the experiment in which the subject believes he's participating.[67]

In this condition, when the experimental subject doesn't press shock levers but rather orders someone else to do so, only three subjects out of forty (7.5 percent) refused to continue to the highest shock level.[68] Recall that in the baseline experiment, fourteen out of forty subjects (35 percent) defied the experimenter. That means that introducing only one level of social intermediation into the situation, and no other changes, dramatically increased the likelihood of subjects to be destructive, from 65 percent to 92.5 percent. Similar experiments verified these results.[69] These findings cast a shadow on the bureaucratic structure of modern institutions of industry and governance, in which decisions traverse uncountable layers of intermediaries. Adding only one layer in the experiment—one person—between deciders and their material-world consequences increased the chances that the deciders would make a harmful choice by almost 30 percent. How many intermediaries might there be between a high-tech executive and a high-tech factory worker? Researching bureaucratic destruction, the psychologists Nestar J. C. Russell and Robert J. Gregory concluded that bureaucracies actively seek to broaden the "zones of indifference" enveloping their members so they can complete the inhuman tasks of the organization as efficiently and smoothly as possible.[70] Feeling and emoting people only disrupt a well-oiled administrative apparatus. Perhaps bureaucracies *are* inherently destructive.

Bureaucratic harm motivated the "Utrecht Studies," a series of nineteen experiments exploring the willingness of intermediaries to carry out harmful acts. Modern bureaucracies are full of intermediaries, noted the experimenters, Wim H. J. Meeus and Quinten A. W. Raaijmakers. Participants were instructed to disturb a job applicant undergoing a test that supposedly would determine qualification for a job (the applicant was actually an accomplice of the experimenters). The participants were told to say fifteen negative "stress remarks" cleverly designed to hurt applicants' performance (and supposed job prospects) and to affect performance cumulatively. In the basic setup over 90 percent of participants complied. "Obedience is extremely high when the violence to be exerted is a contemporary form of mediated violence," the experimenters concluded.[71] Positive attitudes toward social institutions and distant relationships with fellow citizens lead to the high levels of "psychological-administrative violence" found in the experiments and in modern society more broadly.[72] Granted, institutional authorities in contemporary society don't necessarily have "violence" toward people or nature as their goal. Nevertheless, it's certainly one outcome.

Hostility and aggression are surprisingly easy to induce in experimental settings and, by extension, in real life. The psychologist Philip G. Zimbardo and colleagues devised the famous prison experiment at Stanford University in 1971 as a companion to Milgram's obedience experiments.[73] The experimenters created an artificial hierarchy in a mock prison by assigning each student participant to act as either prisoner or guard and by using uniforms, rules, and physical enclosures to simulate prison life.

The perpetrators and victims weren't separated by distance, by a wall, or by human intermediaries but rather by power differences. Putting people who were formerly cordial peers into a hierarchy of power, a kind of social dissociation, made them more aggressive.

Interactions between these normal people quickly became hostile, affrontive, and dehumanizing. "Prisoners" became more passive; "guards" became more active. Aggressive behavior (mostly indirect, due to experimental constraints) became more frequent, particularly on the part of the guards. Verbal affronts became common. Some guards went beyond their roles to engage in "creative cruelty and harassment." Emotional reactions to the situation were so severe that the experiment had to be terminated prematurely—after six days instead of two weeks. Five prisoners had to be released because of extreme depression, crying fits, rage, and acute anxiety. One participant developed a psychosomatic rash on his body. Interestingly, on termination of the experiment, prisoners were delighted, but most of the guards seemed dejected, perhaps because they'd lost the heightened control and power they'd temporarily enjoyed.[74] Or maybe their prior aggression now made them feel guilty. Either way, these artificially produced social dissociations clearly elevated the destructive tendency of those on the side of power and distressed the disempowered ones.

Torturing Nature?

Social division based on social *difference* has long been used to make people violent. In a 1991 article, "The Education of a Torturer," the psychologists Janice T. Gibson and Mika Haritos-Fatouros explain that US military training seeks to induce a strong sense of connection among soldiers while actively alienating them from both their own civilian society and "the enemy." In basic training soldiers are isolated from non-military life, introduced to new (often arbitrary) rules and values, and allowed little spare time for clear thinking. Enemies are dehumanized and distanced with derogatory names or by being called "the target." Beyond standard military aggression, more severe forms of violence, including torture, can be taught, with social division as an underlying method. Studying torturers in the Greek military police and elsewhere, Gibson and Haritos-Fatouros learned how psychologically normal adults can be trained to carry out horrific acts against others. Torture trainers use initiation rites that isolate trainees from society, elitist attitudes and "in-group" language, social modeling that includes watching other group members commit violent acts, and blaming and dehumanizing victims.[75]

It's not only people doing the dirty work who are made more violent by hierarchies and other social divisions. Top executives and political leaders also are susceptible to becoming more prone to order aggressive actions through mediating social structures. Studying group violence and genocide, Ervin Staub noted that leaders who commit atrocities are often surrounded by a small group of decision makers and isolated from those they affect. Dissociation from the victims of their violence makes the violence easier. Milgram maintained that conscience, which regulates impulsive aggressive action, is diminished when one enters a hierarchy.

According to Staub, compartmentalization in organizations and society enables us to focus on goals that may conflict with our own strongly held values. Perpetrators of violence may concentrate on the execution of immediate tasks, ignoring the larger picture of ethics and consequences. When Himmler and Hitler accepted responsibility for the Nazi death camps, it allowed the personal responsibilities of lower-level soldiers and others to percolate up the hierarchy and become diffused. But Himmler and Hitler didn't carry out the program of genocide against European Jews by themselves. In modern society, functional specialization—our highly specialized modern labor force—makes it easy to disregard one's responsibility for final outcomes. Consider for example the workers who did nothing but schedule trains to the Nazi death camps.[76] Consider also Joe from chapter 2 crunching numbers in his spreadsheets, playing his small part in deforestation. Felling trees and making wood certainly isn't equivalent to genocide, nor is it harmful in all cases. The point is that Joe may not be able to tell when it's excessive and damaging and when it's not.

Being in a social hierarchy helps people to evade responsibility for their actions. Actions are fragmented when you participate in complex organizations filled with hierarchies, divided responsibilities, and highly specialized separated duties.[77] Although we must in the end consider ourselves responsible for all our conscious and freely chosen actions, we may only feel responsible when the consequences are the ones we intend. Bureaucracies take advantage of this cleavage by arranging that everyone need only intend to follow the rules, thereby feeling they're acting responsibly. Bureaucracies are extremely effective at organizing evil.[78]

The shift of responsibility in destructive acts can be observed. In Milgram's experiments, subjects who thought the experimenter had more responsibility for the shocks were more obedient.[79] Subjects clearly disavowed their responsibility, as demonstrated by statements such as "If it were up to me, I would not have administered shocks to the learner." And the "Subject Free to Choose Shock Level" variation,[80] in which most chose very low shock levels, suggests that they were telling the truth. A loss of identity can accompany release from responsibility. One of Zimbardo's prison guards said he felt he'd lost his identity and was now (number) 416. Under such circumstances, he noted, it's easy to forget that others are human.[81]

Responsibility is diffused constantly in our everyday lives. Is the pollution coming out of the tailpipe of my car my responsibility, the responsibility of the car company, the engineers who designed the car, or the EPA regulators who regulate its emissions? Milgram believed that when one acts in the context of institutions, the functions of the superego shift "from an evaluation of the goodness or badness of the acts to an assessment of how well or poorly one is functioning in the authority system."[82] So I may care less about the pollution coming out of my car's tailpipe than whether I'm abiding by laws, earning a good living, and generally doing what I'm supposed to do to be a good citizen, which seems to include driving.

Responsibility doesn't shift only up and down in a hierarchy; it can also shift laterally to peers in any social structure. The psychologists Bibb Latané and John M. Darley wanted to know why thirty-eight bystanders would watch a young woman being brutally stabbed to death without intervening or telephoning the police—the famously tragic 1964 murder of Kitty Genovese in a New York suburb. The perception that others are witnessing a crime or other emergency markedly decreases the likelihood that an individual will intervene. Conversely, if a person is alone when faced with an emergency, she's more likely to feel responsible and to respond. The passivity of other witnesses can make you feel like nothing needs to be done.[83] Are we like Kitty's bystanders, watching, dazed and stunned, as the destruction of our global environment unfolds? Not quite. Many have begun to respond to our slow-motion disaster. But perhaps many others linger, feeling that it's not necessary to respond since so few are doing so. If my friend drives this huge car, it must be fine. If golf is a popular sport, the water problem (golf courses consume large quantities of it) mustn't be so bad.

WOE IS PSYCHE

> Our current educational, political, economic, and religious institutions . . . were apparently designed by a mad architect whose deliberate plan was to alienate us from one another and to wrench us from the soil that would nourish humanness and self-fulfillment.
> —Louise J. Kaplan,
> *Oneness and Separateness*, 1978‡

If modern dissociations are harming the material world around us, leading us to damage nature and other people, are they also affecting our inner worlds, our emotional lives? Surely we have feelings about the ongoing destruction we see in the world. I believe most of us feel some amount of despair or sadness when we hear about our environmental devastations—the loss of nearly all the great old-growth forests of North America; the demise of entire species such as the baiji river dolphin of the Yangtze River in China, declared effectively extinct in the past decade; the dwindling of glaciers and their life-giving summer melts; the radioactive contamination of a swath of sea and landscape around the Fukushima Daiichi nuclear power plants in 2011. We might get angry seeing intensified storms inundate another coastal city. We might feel regret when the hydraulic fracturing technique used to mine for natural gas permanently contaminates water supplies with poisons. Mostly, we block these feelings to get on with our lives: make a living, come home to cook dinner, get the kids ready for bed, and do it all over again. But however suppressed these feelings may be, we would do well to remember the connection between harmful behavior and strain vividly felt by the subjects in the obedience experiments. A mature and poised businessman was reduced in twenty minutes to a twitching, stuttering wreck on the verge of nervous collapse due to the internal conflict between his values and his role in hurting someone.[84]

We likely feel inner tension when we know our everyday actions contradict our ideals of natural and human health, but the dissociations that enable such harmful actions may also in themselves hurt us psychologically. They can arouse unhappy feelings and even lead to dysfunctions. Our lack of contact with wild nature gives us ailments such as stress, depression, and anxiety, and in children it seems to contribute to obesity and hyperactivity disorder, according to research brought together by Richard Louv in *Last Child in the Woods*.[85]

A field called ecopsychology has emerged in recent decades to study the relationships between our psyches and both nature and environmental crisis. Some ecopsychologists believe that chronic emotional trauma, pain, and anger result from knowing that we're degrading nature.[86] These emotional responses can make us more destructive, too.[87] Ecopsychologists propose various remedies to human-nature alienation: ecologically attuned psychotherapy; deep-ecology-inspired environmental protection actions as worship; wilderness encounter sessions; "councils of all beings" in which nonhuman others are given representation; and re-Earthing rituals that seek to foster joy, commitment, and inspiration through reconnection with Earth. The historian Theodore Roszak, a pioneer of ecopsychology, writes that its goal is to "bridge our culture's long-standing, historical gulf between the psychological and the ecological, to see the needs of the planet and the person as a continuum."[88]

In his 1992 book that propelled the field forward, *The Voice of the Earth: An Exploration of Ecopsychology*, Roszak brings together psychological insight, speculative cosmology (theory of the natural order of the universe), cosmological history, and a reenvisioning of the concept of ecology to understand the troubled modern human-environment relation and to envision a future humanity more highly attuned to the "deep systems of nature, from which our psyche, our culture and science ultimately derive."[89]

Another ecopsychology researcher, Laura Sewall, tells how being removed from sensual engagement with nature is rewiring our brains and making us less able perceive and respond to the natural world. What we pay attention to actually determines the synaptic connections forged in our brains, a phenomenon known as *neuroplasticity*. That is, we shape who we are—our very identities—by how we attend to our physical and social environments. In everyday life we're bombarded with images and messages from the advertising industry that make us into ideal consumers. And the ubiquity of flat screens is rendering our "'screen-dazzled eyes' incapable of perceiving the depth and ambiguity of the 'earthly terrain' around us."[90] Modern people's sensory capacities are atrophying with our lack of contact with the pulsing, infinitely detailed natural world beyond our artifice. A loss of perceptual abilities does not bode well for the human species because "all organisms must receive accurate signals from their environment to adapt to changing conditions."[91] Sewall explains from a neurological perspective why directed, focused attention to both nature and our natural degradations is necessary to properly grasp and respond to our environmental damages.[92]

Our dissociations within the social sphere also take their toll. The social alienations induced in the prison experiments of Zimbardo and colleagues quickly led to suffering. Five prisoners were released because they experienced depression, bouts of crying, rage, acute anxiety, and psychosomatic rash. The hostility and aggression in the situation, created by the artificial hierarchy and power divisions, probably directly caused many of these responses, but the alienations and animosity are better seen as a mutually producing complex: alienation leads to hostility, which leads to more alienation, and so on. Both guards and prisoners experienced these unpleasant emotions; the bad feelings grew directly from the experiment's social dissociations.[93] Conditions of power that dissociate classes of people in daily direct contact with one another may make both groups less happy.

Social alienation has long been linked to depression and related ailments. In the modern Western world, alienation and depression are common, particularly in urban settings and among the elderly, who often have little or no extended family structure to fall back on when their immediate social circles of peers thin out.[94] Communication technologies such as the Internet, Facebook®, and Twitter® certainly connect people in a particular way, but they can also make people more depressed or anxious, depending on their individual psychological and social makeup.[95] And they certainly can't counteract nature deficit disorder—such technologies usually only increase it. Electronically mediated relationships can displace face-to-face, fleshly ones and make people feel connected. But there's no true substitute for a life lived in direct contact with communities of people and nature.

LETTING PSYCHE SEE

Our psyches need to see. They need to witness the world and our impacts on it, with all our senses active, without the blindfold created by modern arrangements. That's the lesson of the experiments discussed in this chapter. When our psyches are masked, when other people or distance or anything else mediates between us the people and the places we affect, other forces can take over, and we're no longer operating on our own accord. We enter an agentic state, fulfilling the mandate of corporations and other institutions and perhaps the culture at large rather than following our own values and morals. And we sense this discrepancy. Information trickles in from out there, the places we touch. The conflict creates strain that we may just be starting to recognize. Letting our psyches see will let us align our outcomes with our values.

What does the psychopathology of dissociation mean for Joe from chapter 2? It plays out in his life in many ways. For instance, he has little chance to feel empathy for those others whom his choices affect. When he calculates the clearing of a forest, he can receive no empathic cues from the forest or its creatures, for he never witnesses them. Nor can he receive empathic cues from the cattle suffering in the factory farms producing the meat he buys. The situation of Joe's work life narrows his cognitive field: for the forests, it's "out of sight, out of mind." Reciprocal fields aren't activated either. Joe isn't seen by the creatures in the forest being felled, by the pig being slaughtered for his dinner, or by any workers who've suffered cancer caused by chemicals they used in making his flat-panel TV. These concepts don't even enter Joe's world. He experiences no unity between his actions and their outcomes. Buying a laptop is wholly separate from any related consequences out there in the world beyond his laptop screen. And if Joe decides to do something good for the environment, such as buying organic fruit instead of fruit grown with synthetic pesticides, the benefits of that choice are invisible, too. Positive change comes harder when both positive and negative results are hidden.

Throughout his life, Joe has acquired behavior dispositions to act based on everything that's proximate in his life: family, job, church, coworkers, consumables. The situational obligations in his life emerge out of his ingrained duties to be a normal, consuming person who does his job well and takes care of his family and perhaps his immediate community members. Those other parts of the world out there affected by his choices are not part of his obligations according to the society that has shaped him from birth. The sequential nature of his everyday choices and actions—he drives every day, he regularly buys things that look attractive in advertisements, he and his family drink from plastic water bottles, his family slowly increases their consumption each year—creates a self-propelling and gradually increasing cascade that's difficult to stop. The dissonance between his positive self-image and the implicitly criticized, guilt-laden former self that would be created by breaking off the cycle is too great.

In a similar vein but on a larger scale, this modern society drives ever onward, toward the enticing goal of total mastery of nature. Perhaps one day, we'll be able to build a "Dyson sphere" as suggested by the physicist Freeman Dyson: some type of shell completely enclosing the sun (or another star), capturing all its energy for human use. Perhaps medical science will vanquish illness. Or maybe we'll be able to download our minds into computer hardware and live body-free forever. On the way to these developments, however, if we proceed in that direction, it seems clear that there will be winners and losers, the latter suffering disproportionately the costs. Techno-scientific mastery has so far come at the price of organized and ordered neglect—shoving many consequences out where they can't be seen by the psyches of the winners, whose responsibility becomes diffused throughout the system. Only from the corner of their eyes can they see, barely, the suffering and destruction on which their accomplishments are built.

As he analyzed the obedience experiments, Milgram wrote, "Proximity as a variable in psychological research has received far less attention than it deserves."[96] We need to know more about the psychopathology of dissociation. Little work in this area has been done since Milgram's studies. And we have virtually no empirical psychology that directly tries to understand what happens to our decision making when our consequences are diffused and distributed throughout the natural world. It would be helpful to know, for instance, how much we can gain by simple knowledge of particular environmental problems versus witnessing them firsthand. It would be fascinating to carry out obedience experiments in which the target isn't a simulated human victim but a plant, for example.

The experiments discussed in this chapter are inspired by the desire to better understand the origins of major human problems. Milgram wanted to know more about the mass violence in the Nazi concentration camps. The "Utrecht Studies" sought better understanding of the mediated, administrative types of violence that occur in everyday bureaucratic life. Staub's concern lies in inter-group violence, particularly genocide, which has unfortunately risen in modernity. But what about perhaps the most encompassing complex of problems challenging the future of humanity—the global environmental crisis? One problem alone in this category, global climate change, threatens to bring on major challenges including mass destitution, wars, and political chaos. Psychologists could contribute greatly by giving us a fuller understanding of the psychopathology of phenomenal dissociations and how we might best address it.

CHAPTER 4

DISSOCIATIONS IN WESTERN PSYCHES

Only after venturing beyond North America and Europe did I finally begin to grasp the full range of human cultures. Arriving in Istanbul in the late 1980s in the middle of Ramadan, I wandered out around midnight to find the city alive with worshippers and peddlers in places such as the spectacular Blue Mosque. I stood transfixed by the whirlwind of activity. Years later, I discovered the intense, almost intrusive friendliness of people in rural villages in Southeast Asia. As I walked through the countryside in Lombok, Indonesia, children followed me and my traveling companion for long distances, shouting "Hello!" repeatedly. In Aceh, Indonesia, a band of college students escorted me through the city for days, asking me about America and telling me about their proud beliefs. Through the fine art of bowing in Japan I began to learn just how much respect people might pay to strangers. Staying with a hill tribe in Thailand, I sat on the floor with a family, eating searingly spicy rice dishes with our hands out of communal pots and later relieved myself among the dogs and pigs. Life feels drastically different in different places.

I came to realize that differences among cultures go beyond the mere material conditions of everyday life. People's experiences of their worlds and understandings of what it is to be a person vary tremendously. When I saw a ticket clerk at a train station in Jaipur, India, just shrug and grin when I asked him what time the next train to Udaipur would run, I realized I was in a place where the fluidity of time refused to be disciplined by mechanical clocks. At a three-day wedding celebration at a longhouse seven hours up a river in Borneo, where families seemed like they never wanted to leave each other's presence, I realized that community means something different there than in the United States. After Balinese friends reminded me many times to honk the motorbike horn when crossing rivers to scare off of the malevolent spirits living there, I was certain that some people still see souls infused throughout the material world, not just in humans. The spirits also live on in more modern cultures such as Japan, where I accompanied a friend's father late on New Year's Eve to their family business to place *mochi* (rice cake) offerings for the *kami* (spirits).

Many people experience their worlds as intensely interconnected and fluid. For them, time might be less regimented, community ever present, and the world alive and interwoven. If people of different cultures see and experience their worlds so differently, are these differences reflected in their habits of mind—*how* they observe and think about things (not *what* they think about)? Psychologists have long thought that basic mental processes don't differ meaningfully among healthy people, but the new field of cultural psychology, combined with ethnographies of personhood (ways different cultures understand and enact what it is to be a person), prove them wrong.

Perception, cognition (thought), and other mental processes vary, sometimes dramatically, among cultures. Shown an image of nature, groups from different cultures remember aspects of the scene in distinctly different ways, as I describe below. Asked to explain the causes of an event such as a mass murder or the outcome of a sporting match, some people tend to focus on the attributes (characteristics) of the perpetrator or the players, such as the evil intentions of the murderer or the skill of a key team member, while others are more apt to talk about the context of the event, such as the family members who've let the person down or the friendly home audience of the match. Moreover, the very idea and experience of being a person differs starkly among cultures. Some people think of themselves and others as self-made, independent, and autonomous. Strangely enough to common Western perspectives, in some cultures people see themselves as social beings defined almost entirely by the social and natural worlds around them. Many studies show that, although there's great variation among cultures, dissociations run through Western mental processes all the way down to our most intimate experiences of self. We tend to think of ourselves as separate from nature and other people and to see and think of the world as more divided.

Cross-cultural comparisons of mental processes are fraught with hazards. Some scholars point out that the differences within any particular culture are greater than those between any two cultures. One Japanese person may have a thought style more like the average American, for instance. But I'm less interested in describing individuals (or applying these conclusions to any one person) than in understanding the tendencies of groups or cultures in their entirety, something that can be known only by observing many people and their cultural milieu as a whole. There's no need to give up trying to distinguish common characteristics of people in one culture from those of another simply because there's overlap between the two. But we should remain wary of agendas to make one group seem inferior and avoid tendencies to explain or predict how any particular individual thinks or acts. Such efforts usually distort the facts to their liking.

None of the contrasts drawn in this chapter reveal people of one culture to be intellectually inferior to those of another. Although we have some habits of perception and thought that create problems, as I explain, Westerners aren't, of course, on the whole less intelligent than anyone else (nor are we more intelligent, as some presume). The goal is to see whether and how characteristically Western mental habits tend to overlook relationships. Indeed they do, and this realization is important because dissociating tendencies of thought cause us to establish a world full of disconnections— we mold society and nature according to our mental constructions. And our disconnections from nature and consequences lead to environmental degradation, as chapters 1 through 3 discuss. So the link between dissociated Western mental processes and environmental deterioration is indirect, but it's nonetheless momentous. These habits of mind, together with the dissociating intellectual legacy that I discuss in chapters 5 through 7, lead us to create a life-world with a *de*-structive structure. Moreover, a mental style that fragments everything may inhibit us from perceiving the sum of our environmental damages as an environmental crisis that's whole, interrelated, and real.

The more holistic and interconnected thought styles prevalent in many Eastern cultures are no guarantee of environmental benevolence. Indeed, some of the most polluted, ravaged nature on the planet today is in East Asia, particularly in China, where industrialization has taken on a carefree exuberance, speed, and scope new in history. Choking air pollution engulfs cities, and some rivers are so contaminated they're almost completely dead. I've already outlined some of the harms of crude e-waste recycling in China, and releases from electronics manufacture are most likely polluting air and landscapes as well, though it's difficult to be certain with the national government's control of information. Although the country's laws forbid many such abuses, enforcement is weak due to rampant corruption and the fact that some of the most polluting companies are owned by regional governments.[1] China has become the world's factory, so many of us not living there nevertheless bear some responsibility for its environmental damages.

But China's environmental damages arise mainly from its entry into the global market economy and its eager, unchecked adoption of modern industry, both born in the crucible of the Western worldview and mindset. These aspects of modern culture layer on top of local cultures and mix with them to form unique hybrids. Long-standing Chinese styles of perception and thought exist alongside Western influences even while the former may slowly erode. Many of the millions of Chinese moving to cities are rejecting old connections to landscape, community, and tradition in favor of making easy money. Imagery from the West beamed throughout the world via movies, television, and news media stimulates desires everywhere for the material life. Modern dissociating thought fragments human-nature relationships worldwide in the process of globalization. Environmental damage from modern industry in the East only makes it more urgent to understand how perception and cognition have become dissociated.

THINKING THE WORLD FRACTURED

> In his strivings for order, Western man has created chaos by denying that part of his self that integrates while enshrining the parts that fragment experience.
> —Edward T. Hall, *Beyond Culture*, 1981*

Two related yet divergent experiences, psychoanalysis and anthropological research, led the comparative anthropologist Edward T. Hall (1914–2009) to see how distinctly dissociated Western thought and experience are.[2] In the three decades since Hall wrote the above comment, groundbreaking work in cultural psychology has bolstered his assertion about Western people's fragmentation of experience, if not the creation of chaos. Western psyches divide the world, dissolving essential relationships. Our fragmented experiences seem to both reflect and reinforce the peculiarly Western fragmentations in our thought processes, and they separate modern Western cultures from most others. Dissociations inhabit even this most intimate of the spheres of life—the psyche—guiding how Westerners view, conceive, reason about, and respond to the world around us.

Cultural psychologists study differences and similarities in perception, cognition, attention, and other psychological processes among people from different cultures. Findings in the field are sometimes striking, and they contradict long-held assumptions claiming that basic human thought processes, the style and method of thought, are universal—the same for all normal, healthy people. Only the *content* of thought is different, it was believed. Seemingly basic processes such as inductive reasoning (creating general statements from specific examples) and deductive reasoning (creating specific conclusions from general statements) have, until recently, been expected to be the same among normal humans everywhere. Psychologists working around the globe now devise and replicate ingenious cross-cultural experiments to assess *how* it is that people perceive and think about their worlds at the most basic levels.

The psychologist Richard E. Nisbett brought together many cultural psychology findings in his 2003 book *The Geography of Thought: How Asians and Westerners Think Differently—and Why*.[3] He and other psychologists make compelling arguments for "deep and broad differences" between East and West generally and between specific cultures.[4] Not only do belief systems vary greatly by culture, but so do people's naive metaphysics (their most basic assumptions about the makeup of the world, such as whether nature is more like a web to which people belong or more like a hierarchy with humans at the top), tacit epistemologies (how people can know the world, such as whether everything is always changing and therefore not entirely knowable or static and thus quite knowable), and cognitive processes (how observations and accepted facts are combined to create new knowledge)—in sum, the ways people perceive and *think* the world differ significantly by culture.[5]

But in the radically globalizing world we live in, you may be thinking, people migrate widely between countries, homogenizing their cultures. How can we designate an American when the United States is a nation of

immigrants? True, cultures are becoming increasingly mixed and hybrid-
ized, but major differences remain among them, and cultural psychology
researchers take pains to assess distinct groups. In the studies I discuss
in this chapter, participants considered "Japanese" are all native Japanese
who've lived most of their lives in Japan and are acculturated to Japanese
ways; "Chinese Americans" are people of Chinese ancestry born and raised
in the United States; "European Americans" are people of predominantly
European ancestry born and raised stateside; and "Americans" are people
born and raised in the United States. In the results of cultural psychology
experiments, first- and second-generation Asian Americans tend to lie on
a spectrum between Asians and European Americans. With increasing
generations, differences in mental habits between Asian Americans and
Americans recede.

Some of the studies I cite below show that Western habits of thought
are prone to certain kinds of errors, but that doesn't mean that Eastern
ones are superior on the whole. Characteristically Eastern habits also have
their pitfalls, as we'll see. It would be misguided to think that Westerners
simply need to mimic more holistic and integrative Eastern styles of
thought. Instead, we can seek to understand the differences and respond to
them, getting to know the weaknesses and how they might be remedied.
Most important, given how fast modern industry is changing the face of
the planet, is to uncover the fragmenting mental processes on which it's
built—the templates for our destructive phenomenal dissociations.

Fragmented Perception

In one experiment American and Japanese volunteers were shown an
animated underwater scene with fish and other typical aquatic objects and
were then asked to describe it. Americans tended to talk about the big
fish, the most obvious focal point. Japanese participants recalled just as
many details of the focal fish, but they made more than twice as many
statements about the background and contextual aspects of the scene, and
twice as many statements relating inanimate things to animate ones, such
as a fish swimming past some particular seaweed. The Americans' attention
to the qualities of the big fish paid off in one sense because they were
better able to recognize them when presented against other backgrounds,

but they missed many other aspects of the scenes that the Japanese caught.[6] In general, East Asians tend to be better than Westerners at perceiving contexts and relationships.

In another experiment, native Chinese, Chinese Americans, and European Americans were shown Rorschach cards and asked to describe what they saw. The native Chinese were more likely to give "whole-card" responses, in which a card's entire composition formed the basis of the response. European Americans tended to focus on individual parts of the image, for instance, two things that look like legs sticking out. Chinese Americans fell between the European Americans and native Chinese in their attention to details versus the image as a whole. Seeing an image in its undivided entirety is an integrative, connective way of looking at it.[7]

Other experiments directly study attention to relationships. One tested the ability of Americans and Chinese to detect the amount of "covariation" between objects—a test of ability to perceive a fundamental relationship. Participants were asked to judge the degree of association—functional (having related functions, such as two vehicles), formal (having similar shapes), or other—between pairs of arbitrary figures presented sequentially on a computer screen. A car and a bicycle have a higher degree of association than a car and a flower, for instance. Chinese participants more successfully detected the covariation between objects than Americans did; they better recognized relationships between objects, such as airplanes and trains, not trucks and kites. The psychologists concluded that Asians' dialectical epistemology, a form of reasoning that focuses on relationships and contradiction (think yin-yang), may make them particularly sensitive to relations between stimuli and better able to judge the strength of associations.[8] These tests showed that Americans were less attuned to relationships.

In another experiment, Americans' greater tendency to dissociate objects from the visual fields they're embedded in was tested in a situation in which doing so gives them an advantage. Volunteers were asked to look into a rectangular box framing a rod and judge when the rod was vertical. Americans performed better than East Asians. They exhibited less field dependence, or context dependence, in perception, and as a result could see objects more independently of their contexts. Whether it's helping them or not, Americans tend to be less aware of or concerned with relationships among things than East Asians.

These comparative studies show that Westerners tend to perceive things torn from their full contexts. They perceive various objects, even less vital ones, more prominently than contexts, focus on those objects, and see them disconnected from other objects in a scene. This perceptual bias is pronounced when a scene contains strong focal objects that stand out due to size, motion, or other dominating features. Non-Westerners are more likely than Westerners to attend to the field in which focal objects are embedded and to take in wholes or gestalts of a scene. Westerners tend to see the world as being fractured.

Dissociated Explanations

Until recently, one of the habits of thought considered universal across cultures was the "fundamental attribution error" (FAE—also called "correspondence bias"). The word *fundamental* suggests that psychologists thought it applied to everyone. According to the FAE, people tend to see behavior "as a product of the actor's dispositions and to ignore important situational determinants of behavior."[9] For example, a person might more readily see a criminal as morally deficient rather than lacking in positive alternatives. Thanks to cultural psychology, we now know that the FAE isn't quite fundamental: "The fundamental attribution error," wrote social psychologist Ziva Kunda in 1999, "may be fundamental only in Western cultures, where the person is viewed as autonomous, independent, and separate from the surrounding environment."[10]

Experimenters compared Americans with Hindu Indians by asking subjects to choose and explain actual events from their lives: the actions of other people. Americans tended to describe people's behavior in terms of traits (such as recklessness or kindness), Hindu Indians made more references to social roles, obligations, the physical environment, and other contextual factors when explaining the behaviors of other people. In another study, Americans tended to relate events such as mass murders almost entirely to a presumed mental instability and other negative dispositions of murderers. Chinese accounts of the same events tended to speculate on situational, contextual, and even societal factors, such as a lack of educational opportunity or the failure of the person's family to provide sufficient socialization. In a study of sports writers in Hong Kong and America, those in Hong Kong focused on contextual explanations of game results, such as having had a good preseason. The Americans talked more about the characteristics of individual team members, including talent.[11]

This Western habit of considering actors' dispositions before situational factors is so strong that we keep doing it even when given strong cues that should defeat this tendency and the errors it can cause. In one set of experiments participants were assigned positions on an issue, either pro or con (without regard to their actual positions), and were asked to write a short essay expressing their assigned position (for example, in favor of Fidel Castro's Cuba). Then they had to read an essay written by another participant, whom they'd been told went through the same procedure. Americans, but not Koreans, believed that the essay writers were expressing views they actually held, even though they'd been told otherwise and knew from their own experience in the experiment that the writer had been assigned the position.[12] Results for Japanese participants were similar to those for Koreans.[13] Apparently, when we Westerners (Americans, at least) try to explain a person's behavior, we feel so compelled to exclude the context of the action that we end up denying obvious facts.

East Asians depend on context to explain not only human behaviors but also events in the physical world. Various studies find that when they explain physical events, Chinese are more likely than Americans to refer to the field, or context. For example, they might say "The ball is lighter than the water" rather than "The ball is light, so it floats"—also one of Aristotle's misconceptions, or "The glass broke because it fell from high up onto a hard surface" rather than "The glass broke because it's fragile."[14] So Chinese and other East Asians tend to apply context-focused explanations in *both* social domains—judging whether someone favors Castro's Cuba—and physical domains—explaining why a glass broke, while Westerners tend not to do so in either. The Eastern habit of considering context in all sorts of explanations points to a distinct metaphysical difference between East Asians and Westerners. The Easterners are not simply learning to apply certain aspects of particular contexts, such as the fluid medium of something floating; they're identifying relevant facets of every type of context. We can surmise that they hold a distinctly different view of the world as being more interconnected, and they apply that view across dissimilar situations to explain events.[15]

From a psychological perspective, it's no wonder that Western world-views tend to see humans as separate from and independent of nature, the very ground and context of our existence. The separation is encoded in our psyches. And if that's the case, it's also no surprise that modern industry treats nature as a backdrop or exploitable object rather than as an active player in human welfare.

What Makes a Group?

Westerners also seem to discount relationships when mentally assembling items into groups. One experiment presented American and Chinese children with images of objects from various categories, such as vehicles, furniture, and food, and asked them to choose items that went together and state the reason for their choices. The Chinese children tended to use "relational-contextual" reasoning for their pairings. For example, when shown a man, a woman, and a baby, a Chinese child might group the woman and the baby together because "the mother takes care of the baby." American children, in contrast, were much more likely to group objects based on category membership or shared qualities, for instance, grouping the man and the woman together because "they are both adults."[16] These results echo the findings discussed above, in which Westerners tend to focus more on the attributes of objects when making causal inferences (reasoning about the causes of events), and Easterners tend to be more "context-inclusive."[17]

Another way Westerners de-emphasize contexts is by relying more heavily on rules than on relationships when grouping things and assigning them to categories. Again, the Western focus is on isolated objects rather than on objects enmeshed in a web of relationships. Working with Chinese and American adult volunteers, experimenters described things that shared both a relationship (for example, a pencil and a notebook: one writes on the other) and a feature (a blue house and a blue car). The Chinese were more likely to group items on the basis of relationships, and the Americans to group them according to features. The latter approach uses an implicit rule stating that similar features qualify for membership in the group.[18] If we wish to see ourselves dependent on and immersed in nature, Westerners must struggle against our tendency to focus on objects at the expense of connections and the fuller context of life.

Noncontradiction or Dialectics?

If the West's styles of reasoning tend to dissociate events and objects from their contexts, it shouldn't be surprising that its formal logic does so, too. Eastern formal reasoning tends to embrace interconnection, fluidity, and change. Western logic by comparison breaks things apart and prefers to think of them as existing in unchanging isolation. Consider the three rules at the core of Western-style reasoning, bequeathed to us by the fourth-century BCE Greek philosopher, Aristotle:

1. *The law of identity*: A thing is identical to itself.
2. *The law of noncontradiction*: No statement can be both true and false.
3. *The law of the excluded middle*: Every statement is either true or false.

It's striking how contrary these principles are to assumptions and methods prevalent in Eastern philosophies, which tend to use a dialectical approach—but one that diverges from Western dialecticism. In the latter, a thesis (a statement about anything) spontaneously gives rise to its opposite, an antithesis. The contradiction between the two is resolved only by the creation of a synthesis. Consider the following example (meant only to illustrate the Western dialectic, not to advance a new argument about the environment). Thesis: "Economic growth is good." Antithesis: "Economic growth harms society by harming nature." The synthesis then could be a new economic paradigm that doesn't harm nature.

In contrast with this basic outline of Western-style dialecticism, the style prevalent in the East often transcends, accepts, or even insists on contradictions rather than seeking to resolve them. Kaiping Peng and Richard Nisbett describe this Eastern "naive dialecticism" in terms of three principles:

1. *The principle of change*: Reality is a process that isn't static but rather is dynamic and changeable. A thing need not be identical to itself at all because of the fluid nature of reality.
2. *The principle of contradiction*: Partly because change is constant, contradiction is constant. Thus old and new, good and bad, exist in the same object or event and indeed depend on one another for their existence.

3. *The principle of relationship or holism*: Because of constant change and contradiction, nothing either in human life or in nature is isolated and independent, but instead everything is related. It follows that attempting to isolate elements of some larger whole can only be misleading.[19]

The Eastern principles contextualize and integrate, while the corresponding Western ones dissociate. Eastern dialecticism incorporates aspects of relationships and contexts. The principle of change connects present to past and future and refuses any disjuncture in the flux of nature or society. The principle of contradiction creates a union of opposites, connecting them rather than seeking to separate them. The principle of holism envisions a fully interwoven, interdependent universe, a nature that includes humans.

By contrast, Western laws of logic create sharp conceptual cleavages and assume that the world is highly divisible. The law of identity forces us to conceive objects plucked from the ravages of time so they can maintain a stable identity. But an apple, for instance, is constantly and relentlessly changing, as it grows on the tree, is harvested, and immediately begins the slow process of decay. The law of noncontradiction requires that no statement can be both true and false, forcing severe divisions in the realm of truth. I want to say that the apple is red because it mostly is, but it's also not red because there are other colors—a bit of green, brown, yellow, and white—on the outside and in. Since the apple's mostly red, I believe it's valid and practical to call it red, but the principle of noncontradiction gives me little leeway. The law of the excluded middle likewise confounds gradations of knowledge and understanding. It says that the statement "the apple is red" must be either true or false. No in-between state is allowed. When it comes to my direct observations of a material world that's nuanced and changing, Western formal reasoning seems to insist on an impractical purity of thought. It appears more attuned to ideal forms, such as an imaginary red apple, perfectly uniform and eternal.

Aristotle's principles of logic divorce people, things, and events from their contexts to divide the world into disconnected, more tractable quantities of truth and identity. The result is fragmented fields of being, both in the physical sphere and in the realm of truth. Qualities such as reddish-green can't be quantized into discontinuous, discrete values, so they're discounted.

It seems the ancient Greeks sought to remove the messy complicatedness of nature to make it easier to represent and manipulate.

Methods of dialectical thought that embrace change, contradiction, and interdependence are of course not alien to Western thought. More nuanced "post-formal" principles are learned late in adolescence and early adulthood, and wisdom depends on supplementing formal logic operations with a more holistic and dialectical approach. We Westerners know intuitively that the sky around dusk isn't a single, nameable color and that as individuals we change constantly throughout our lives. Yet in coming to conclusions and making choices, Westerners don't rely on these approaches as much as Easterners do. We depend more on our formal logical principles, particularly the principle of noncontradiction, which can be observed in action. When experimenters presented volunteers with vignettes of social contradictions, such as a conflict between a mother and a daughter over going to school versus staying home and having fun, American responses tended to come down more strongly on the side of either the mother or the daughter. Chinese responses were more likely to find a "Middle Way" that involved a contradiction. That is, they looked for merit and fault on both sides and attempted to reconcile the contradiction rather than eradicate it, as in the statement "both the mothers and the daughters have failed to understand each other."[20]

Western aversions to contradiction can produce strange results. In one experiment, volunteers were presented with either two arguments of unequal plausibility or just the more plausible one (Chinese and American participants agreed on which of the two arguments was more believable). When presented with both arguments, Americans tended to judge the more plausible one to be even more plausible than they did when presented with it alone. In other words, Americans actually found a contradicted assertion more believable than the same one not contradicted. They apparently felt so much pressure to resolve the contradiction that they clung more strongly to their favorite argument when it was contradicted. Environmentalists would do well to consider this phenomenon. Climate change deniers presented with a weak or seemingly implausible argument about the reality and urgency of climate change may just dig in their heels more.

In contrast, Chinese participants presented with two unequally plausible arguments resolved the contradiction by finding both propositions equally credible. In fact, their responses were just as questionable as the Americans': they saw a less plausible proposition as more plausible when contradicted by a stronger argument. In a similar experiment Americans increased their confidence in their initial position on an issue when presented with an additional weak argument against their position. Koreans, by contrast, decreased their confidence and became more unfavorable toward the initial proposition when presented with an additional weak counterargument. The Western taste for dissociating logic—one that leads to a clear and decisive outcome and utterly rejects other possibilities—again resulted in bizarre outcomes.[21]

In sum, Easterners are less concerned than Westerners with contradictions. On the whole Easterners don't draw sharp lines between truth and falsehood. They seem to have greater preference for practical compromise solutions and holistic perspectives and are more willing to endorse contradictory positions and shift their belief in the direction of a new argument, even when it's weak. When asked to justify their choices, they move to a compromise, a "Middle Way" rationale. The West's greater adherence to noncontradiction is clearly no guarantee against questionable conclusions.[22] Western styles of reasoning divide the truth into discrete packages that seem incompatible with a complex, interconnected material world full of change. The law of noncontradiction adds to the divisiveness of Western mental habits. Yet it's not necessary to simply judge Western or Eastern thought as superior to the other; both styles of thought have their problems, as we've seen. Moreover, the categories of Westerner and Easterner each include a huge range of cultures and people. The comparisons are nevertheless striking and instructive.

◆━◆◈◆━◆

What do these differences mean for the environment? Eastern habits of perceiving and thinking recognize a dynamic, flowing, interdependent world of nature and humans, so they're more consistent with ecology, which sees connection as paramount. They have more potential to be environmentally positive. Unfortunately, in much of the East, traditional styles of thought are being sidelined as the region rushes to modernize and industrialize. In China alone, a large measure of the world's resources are being used to power consumption and construction, which has proceeded at such a breakneck pace that several ghost cities looking for sufficiently wealthy inhabitants have sprouted.[23] Western thought styles, practiced in the West or East, apply a "divide and conquer" approach grounded in a drive to exercise power over the natural world, as I discuss further in chapter 6. They shun connection in favor of absolute and cleanly conceived conceptions of lived reality that sanction decisive action. But more important, Western divisive thought supplies the cognitive foundation for the dissociations infusing our lives. We think of the world in a fractured way and then shape it and live it that way. Providing the substructure for our removal from nature and our consequences is perhaps the most damaging legacy of dissociating Western mental habits.

Social Origins of Systems of Thought

Why should such strikingly divergent styles of thought arise among cultures? And why do so many differences seem to line up along the broad categories of Western and Eastern mental habits? Nisbett and his colleagues speculate that some of these asymmetries originate in historical and geographic differences in the ancient world. They suggest that the intensely agricultural society of ancient China required much more cooperation than the ancient Greek economy, which depended more on herding and fishing in accord with the region's ecology. And these two ancient societies would have deep and lasting influences on thought throughout their respective regions.

As I elaborate in the next chapter, ancient Greek city-states were highly autonomous; people could easily leave one city for another; and the sea provided an escape route for dissidents. Situated at one of the crossroads of the world, the Greeks' extensive participation in trade may have stimulated their curiosity and debate—and their individualism. These turbulent conditions led to social relations that were highly contingent (dependent on various, possibly changing factors) and changing, so debate posed fewer personal risks in ancient Greece and was an integral part of social life and the political system.[24] Competition, including combative winner-take-all public debates, was a staple of Greek life. Greek history, topology, and society may have contributed to Western dissociating mental habits.

This speculative view is bolstered by the history of the Renaissance.

Leading up to the Renaissance, fewer people were participating in agriculture, and independent city-states such as Florence rose up, driven by economies based on crafts and trade. Ancient Greek social forms and intellectual traditions, including a rediscovery of science, became highly influential again in the West. Thus conditions increasingly resembled those of ancient Greece and its individualist society. The idea that stronger social networks of interdependence among agriculturalists (versus currency-based trade among craftspeople) foster more holistic metaphysical assumptions is supported by empirical evidence that farmers in various societies exhibit more field-dependent perception and cognition than hunters, herders, or industrialized peoples.[25]

But social practices don't simply produce systems of thought. Rather, the two produce one another. Cognitive styles are given stability by being embedded in larger systems of belief and social practice.[26] Balance and trade-off are more pronounced in the communitarian cultures of the East, for example. Decision processes in Japanese boardrooms avoid conflict and favor compromise. Balinese neighborhood organizations make decisions by consensus, in which everyone eventually agrees or submits to a decision after extensive discussion, rather than by majority rule, in which the majority imposes its choice on the minority. In the West contracts are virtually set in stone. In the East people continue to negotiate contracts long after they're signed.[27] Although it's frustrating to Westerners, the fluidity of contracts in the East takes into account the fact that the world keeps changing after a contract is signed, and the situations of the parties involved change, too, as do the relationships among them.

Low-Context Cultures

Empirical studies support the old formula that Eastern cultures are more collectivist and holistic and Western cultures are more individualist and analytical, but what are the limits of this idea? The philosopher Donald J. Munro and colleagues present a more nuanced understanding of this notion in *Individualism and Holism: Studies in Confucian and Taoist Values*. One counterexample can be seen at the end of the Han dynasty (approximately 206 BCE–220 CE), when a cry emerged in China for diversity and there was a flowering of individualism.[28] Conversely, Western culture and life have hardly been monotonously individualistic and analytic. Aspects of Plato's metaphysics, in the ideal, are holistic, and "the holistic vision dominated Europe from the decline of Epicureanism until well into the Renaissance," writes Munro.[29] It included an interconnected organic view of nature, which declined in the Western world in the seventeenth century with the onset of the Scientific Revolution. Before the Enlightenment took hold of Europe in the seventeenth and eighteenth centuries, there was more overlap between European and Chinese worldviews and styles of thinking.[30]

Yet as we've seen, various experiments underline these differences between the East and West, and even scholars who caution about important historical nuances, such as the authors of *Individualism and Holism*, discern distinct East-West differences. One of the more extensive studies of the individual-collectivist dimension comes from Edward T. Hall in his book *Beyond Culture*.[31] Comparing various world cultures, Hall distinguished between "high-context" and "low-context" cultures. In high-context cultures people interact more deeply and intricately, and information is communicated widely across situations rather than being retained by people performing specialized jobs and distributed on a need-to-know basis. At a cremation ceremony in Bali in July 2012, I saw hundreds of people working fervently yet cheerfully together to make offerings and build temporary structures, including cremation towers and funerary bulls to hold the body. The work was not as hierarchical and divided among individuals as it might be in the United States. Rather, information and tasks were shared widely, and many complex decisions seemed to spontaneously emerge from the crowd by consensus. American Indians and other traditional cultures, and East Asians, such as Japanese, are examples of high-context cultures.

In low-context cultures, people have fewer direct and immediate interactions. Messages shared among them are not as highly enriched by contextual meaning—social, historical, and natural details of the time and place of contact—so they contain less real information. When messages are communicated via print or electronic media and not face-to-face, nuances of facial expression and body language are also lost. Low-context cultures circulate a lot more impersonal, generic information—messages tuned for larger audiences, not for a specific community or person. Indeed, many audiences today are global. Western modern cultures are mostly low-context cultures.[32] Low-context cultures dissociate valuable contextual knowledge more than high-context cultures do—knowledge, for instance, about the natural environment.

Hall identifies various components of context that become dissociated in low-context cultures: linguistic, kinesic (body language), proxemic (spatial), temporal, social, material, and personality. Cognition in high-context cultures explicitly includes these; low-context cultures omit them, instead isolating things from their lived-world contexts. He calls low-context cultures "highly individualized, somewhat alienated, fragmented cultures" in which humanity has become preoccupied with its own "extensions"—tools such as language and objects such as automobiles by which we participate in our own evolution.[33] Dissociating psychic processes arise in and support low-context cultures, Hall discovered, presaging later research in cultural psychology:

> We in the West are alienated from ourselves and from nature. . . . We live fragmented, compartmentalized lives in which contradictions are carefully sealed off from each other. We've been taught to think linearly rather than comprehensively, and we do this not through conscious design . . . but because of the way in which deep cultural under-currents structure life in subtle but highly consistent ways that are not consciously formulated. . . . Given our linear, step-by-step, compartmentalized way of thinking . . . it is impossible for our leaders to consider events comprehensively or to weigh priorities according to a system of common good.[34]

Western logic is dissociating, Hall observed, because it enables people to examine ideas, concepts, and mental processes according to low-context paradigms—separate from relevant information.[35] He also found that low-context cultures enable responsibility to be alienated from actors and diffused, thus facilitating destructive acts. In high-context systems, people in places of authority are personally responsible for the actions of subordinates, down to the lowest person in the hierarchy. Low-context systems diffuse responsibility throughout the system and make it hard to pin down. In these systems, Hall writes, often the lowest-ranking plausible scapegoat is chosen, such as the lieutenant who received the blame for the My Lai massacre.[36] The dissociating structures of low-context systems allow responsibility to float freely and dissipate.

Western styles of thought and cognition, as we've seen, are markedly more dissociating than others. And they provide the mental foundation that supports modern phenomenal dissociations and their destruction. But perhaps you're a European American whose mental habits more closely resemble the typically East Asian ones described above. Or perhaps you don't want to be pigeonholed into one of these categories. I entirely agree and want to emphasize again that these characterizations don't describe any particular person. They only pertain to average or typical qualities of large groups, entire societies. But the characteristics of groups—their cultures, shared assumptions of the world, mental habits—matter because they guide the direction of a society and the choices made in it. And the characteristics of thought typical in Western-style modern culture are particularly relevant given its deep and spreading influence throughout the globe.

INDIVIDUATION: CULTURE AND THE DISSOCIATED SELF

Western fragmented ways of seeing and thinking about the world extend to our basic understandings and experiences of personhood. Am I a kind of animal, or am I something quite different, the product of a unique and special divine intervention? Does my self belong only to this particular body? Our experiences of ourselves as persons, even what we think a person is, depend on beliefs, customs, practices, and norms unique to our culture. What it is to be a person varies dramatically across cultures in ways that can boggle the mind. The typical Western formation of personhood separates the self from society and nature.

The psychological and cultural processes by which a person emerges as an individual in the world—the production of personhood—is called individuation. Newborns don't perceive themselves as persons, or even as separate objects, according to understandings of early childhood and "psychogenesis." They apparently perceive no difference in the world at all. The psychiatrist Margaret Mahler (1897–1985), a pioneer in childhood development studies, established the separation-individuation theory of child development in the middle of the twentieth century. She outlined three phases of the "psychological birth of the human infant" beginning with biological birth: (1) the normal autistic phase (the first few weeks following birth), in which the infant has little awareness of anything beyond her or his own body and performs little or no object cathexis (the process of charging objects with mental and emotional energy); (2) the normal symbiotic phase (from about the second month to the fourth or fifth), in which the infant's mental image is still fused with the mother in a "dual unity," but the infant is vaguely aware of needs and satisfactions originating from the outside; and (3) the separation-individuation phase (about the fourth or fifth month until about the thirtieth or thirty-sixth month), in which the child gradually becomes aware of his or her separateness, first of the body, then of the whole identity.[37] These processes are assumed to be about the same across cultures, but the development of individuals diverges from there.

Recent research confirms that the emergence of the sense of self in each individual is culturally dependent.[38] Two psychologists at the forefront of psychological research about culture and the self, Hazel Rose Markus and Shinobu Kitayama, suggest that not only do conceptions of the self and personhood vary across cultures, but they also may be a key ingredient of cultures that makes them unique.[39] Markus and Kitayama think there are two overarching types of self, independent and interdependent, the first more prevalent in the West, the latter in Asia. Independent selves possess a sense of being autonomous and distinct from all others. Interdependent selves emphasize the inter-relatedness of the self to others. For interdependent selves, self-identity is partly diffused out into society, across other people important to the person.[40] Cross-cultural ethnographies of personhood put flesh on these ideas.

What Is a Person?

Would you believe that in some cultures a person might see herself as another version of her great-grandmother or a reincarnation of someone who lived ages ago? Or that she isn't a person with a self-realized personality but rather a personage with a fluid role that originates in the community? That she takes on characteristics from the living nature around her and never really dies but rather remains in an ongoing altered state? Or that her soul actually migrates beyond the confines of her body while she's alive? In the modern West these experiences and notions of personhood are alien, but in some other cultures they're not. We're fortunate to have a detailed and in-depth accounting of one such non-modern sense of self.

Beginning in 1903, the French Protestant missionary and ethnologist Maurice Leenhardt (1878–1954) lived for about a quarter century with the indigenous people of New Caledonia, a French colony established in the mid-nineteenth century on a Melanesian archipelago east of Australia. Leenhardt came to know the *Canaque* extremely well and carried out groundbreaking ethnography on them with startlingly new observations of what it means to be a person (I follow Leenhardt in using the French term *Canaque*, though today it's more commonly spelled Kanak). Leenhardt grew to develop a deep and abiding respect for the Canaque, as his sensitive ethnographies show. His research on their society, culture, and mentalities helped dispel simplistic notions of the "primitive" minds of indigenous peoples. By being so starkly different from most Western people, the Canaque tell us much about ourselves.[41] The following descriptions convey how Leenhardt recorded them a century ago. They certainly have been more influenced by modern ideas and lifestyles today.

The Person versus the Personage

To understand how Western senses of self differ from non-Western and non-modern ones, it's helpful to review the history of personhood in the West. Tracing the evolution of the person, in moral, legal, and philosophical terms, the French sociologist Marcel Mauss (1872–1950) writes, "Those [nations] who have made of the human person a complete entity, independent of all others save God, are rare."[42] But it happened in Europe beginning in ancient times. According to Mauss, the West's highly independent conception of personhood originated in the assumption of a separate, self-reliant persona and identity by sons in ancient Rome, "even while the father was still alive."[43] Previously, full personhood—being a complete and autonomous self—had depended on the death of a man's father. In both cases this status was reserved for men.

This moral sense of personhood eventually transitioned to a juridical (legal) one: a conscious being, autonomous, free, and responsible. The characteristically Western independent notion of the self became a basic fact of law in ancient Rome.[44] Religious developments then extended this sense of autonomy in the spiritual realm. In the seventeenth and eighteenth centuries, laypeople obtained the right to communicate directly with God, doing away with the cleric as intermediary.[45] This Christian metaphysical entity of the moral person is the one pervasive in Western culture today: an atomic model of the person—rational, indivisible, independent, and individual.[46]

Mauss's history of Western personhood depends on the notion of personage, which differs from person. He writes that in a "whole immense group of societies," particularly ancient Western and non-Western ones, the independent and self-directed type of person that we know today was uncommon. Instead, personal beliefs and actions were organized around something more like a role played by individuals in all spheres of life—from sacred rituals to family life. This role, the personage, comprises a full set of characteristics defined at least partly by social and natural forces outside of oneself. Although the personage is common in non-Western cultures, it's not universal, Mauss found. Some non-Western cultures also developed a notion and experience akin to the Western notion of the person.[47] Yet the contextually defined personage of most non-modern and non-Western cultures probably better describes people's experiences through much of human history. It stands apart from the self-defined Western person, independent of society and nature.

Indeed, in many cultures persons as we moderners conceive them may not exist. Leenhardt, a student of Mauss, found that the Canaque are decidedly personages, not persons as we understand the term. Similar to other societies based on personage, the Canaque are cast as actors, each "acting out the prefigured totality of the life of the clan."[48] Each Canaque personage is thought to manifest the sum of all Canaque—a holistic construction of the self. Shaped by socially defined roles and cultural features, the personage is deeply embedded in his or her society. In the West a sense of personage more akin to that of the Canaque probably prevailed before being replaced by independent personhood in ancient Greece and Rome, as narrated by Mauss.[49]

The Canaque's sense of connectedness goes beyond the social world. In Canaque ontology (basic assumptions about what exists in the world), the human body and other things in the physical world such as rocks, animals, and plants originate in similar structures. The Canaque "fails to delimit his body and ... to separate it from the world."[50] All elements of the material world are blended together in an identity of substance, a "flux of life." The

Canaque perceive self and other as distinct yet bonded: "The object adheres to the subject, and the subject adheres to the object."[51] The world of the Canaque is truly a participatory, engaged world.[52] As odd as this experience and idea of selfhood seems to us, it likely describes how most people experienced themselves through the vast majority of human history: integrated with and continuous with the entire surrounding world of matter and spirit.

In the 1930s, not long after Leenhardt lived with the Canaque, the anthropologists Margaret Mead and Gregory Bateson spent a couple of years studying the highly formalized social relations of Bali. They discovered that the Balinese think of themselves as playing roles, similar to the personages of the Canaque. These roles in Bali are crucially important to culture and social life.[53] Interpersonal behavior is always circumscribed by situational factors, particularly social status. The Balinese tend to act in close accord with their caste, economic class, age, and other aspects of their personal social milieu using language levels, posture, tone, and formalized patterns of behavior. Various conformity pressures regulate daily life.[54] Mead writes that the restrictions surrounding personal relations are so heavily codified that they have a quality similar to those surrounding sexual activity.[55] Balinese frequently use the term *desa-kala-patra*—place, time, and role—to remind each other of the sociocultural context that should guide their behavior.

During the several years I've spent in Bali, I've witnessed many examples of personage and role playing. It's evident in the detached, ego-free descriptions that people sometimes offer of themselves, for example. Westerners are surprised to hear a Balinese youth say casually, with a smile and no hint of shame, "I'm not very bright," as if describing the character he's playing in a drama. He accepts his role and understands that it's part of the fuller picture of the society to which he belongs. This Balinese comfort with social roles and class and caste positions contrasts with the social mobility that drives Westerners, though of course many Balinese today strive for social advancement or wealth.[56]

Although living with a role sounds restricting, perhaps even oppressive, to many of us growing up in Western cultures—and Balinese probably also feel that way occasionally—it's important to remember that it can also be liberating. Within a role, there's great freedom. In Bali I see a certain playfulness that's often missing in the West, where we may be too concerned with defining ourselves by our actions to be truly free. It's also been fascinating over the years to see some Westerners thrive on the more highly structured social conventions they find in Bali.

A Self Bound Outwardly

How is a personage situated in the world? Other generations of people, clans, and nonhuman nature bind the personage in its full physical and social context. Social life for the Canaque is a "continuous, relational, essentially reciprocal modality," writes Leenhardt.[57] These linkages are reflected in various practices such as naming. Names, terms for relationships, and intergenerational contracts anchor people in their society. In both Bali and New Caledonia, names given to people tie them to their clan and to other generations of their family. For the Canaque, great-grandparents and great-grandchildren are "brothers."[58] The same multigenerational tie is echoed in Bali, where one word, *kompiang*, means both great-grandparent and great-grandchild. Such terms express and reinforce kinship bonds. The Canaque also name children after ancestors.[59] Westerners sometimes do too, but for the Canaque the names do more than honor the ancestor— they convey identity and create generational cycles.

Various cultures use personal names to express one's position in the social network and relationships to other generations. In Australia, the Arunta and Loritja indigenous groups create an intergenerational cycle at the third and fifth generations: the terms for *grandfather* and *great-great-grandson* are homonyms.[60] The Zuñi (a North American Pueblo culture) identify older or younger sibling relationships through names.[61] Balinese names fix people socially in many ways: they indicate a person's caste, clan, and birth order. If your name is Nyoman, you're the third child of your parents—or the seventh because the names cycle after four. If your name begins with Gusti, you're a member of the third caste, *Wesya*. A person with Pande in his name hails from the traditional clan of metalworkers. Names also link people to events or conditions surrounding their birth in Bali, as elsewhere. A child born while the full moon is shining brightly might be named *Purnama*, full moon.[62] Balinese names record how people are triangulated into their social and natural worlds.

Names elsewhere also link people with nature. Native American names such as Sitting Bull, Tall Bear, and Red Cloud refer directly to elements of the natural world in which people are immersed.[63] The Zuñi use names that change with the season, recognizing the flux that occurs not only in the environment but also in people closely tied to the environment.[64] Names that anchor people to animals, the sky, or the seasons create an identity that goes beyond the self. They keep individuation in check by limiting the extent to which a person perceives and experiences himself or is perceived by others, as a separate, independent, self-defining subject.

Names are just one aspect of the intimacy that many old cultures experience with their lands and nature more broadly. The very identity of persons and their societies is often linked inexorably to the landscape. The *Kumulipo* chant, an origin story of the ancient Hawaiians, expresses a kind of evolutionary view of nature that makes humans kin to all other natural beings. Kinship within the human world is also paramount: some indigenous Hawaiians could name ancestors going back many generations.[65] Another Polynesian group, the Maori of New Zealand, traditionally conceived an intricate relation of self with tribe and of tribe with landscape. Land was seen as an ancestor and was a source of both tribal and individual identity—tribal identity merged with nature.[66] Harming nature becomes equated with harming the tribe. By contrast, in the West the landscape is divided into property to be owned by individuals. Individuals have limited life spans, so it's "rational" for them to exploit their resources for short-term gain and quick profits, turning the resources into cash. In old cultures people managing tribal and community landscapes must think about younger people and future generations, and pay respect to their ancestors who've nurtured the land. Strong tribal and identity-based connections to the land help explain why so many indigenous cultures meet their demise when expelled from their traditional landscapes.

Extending the Human Community and the Body

When a culture feels a deep connection with nature, it may experience its boundary with nature to be quite fluid and permeable. The Canaque term *kamo*, "the living one," means conscious, sentient spirit. Usually it refers to humans, but the boundary between humans and nonhumans, like that between the self and other people, is porous. Kamo can refer to animals, plants, and mythic beings. The Canaque are inspired to call others in the world kamo when they recognize in them human characteristics of posture or expression. A proud dog might be taken to be a chief, or a moonfish might be thrown back into the water because it revealed a "human expression." Leenhardt writes, "What is human transcends all physical representations of man."[67] Observing human qualities in other animals and giving them human status, even momentarily, blurs the boundaries between humanity and the rest of nature and thwarts any tendency to divide the two.

Transgressing boundaries, in this case between people, also occurs by way of physical interaction. In some societies, you must expect to occasionally lose some control over your own body and be willing to allow it to flow with others'. In extremely crowded situations, as in Bali at mass rituals, I've occasionally felt trapped and couldn't control my own movement. One time, I was pressed so hard against other people I could almost lift my feet off the ground and still be upright in the same location. This phenomenon sometimes happens in Western crowds, too, and like most Westerners, I find it unnerving. But Balinese people seem to enjoy it, laughing gaily even while pushing each other. They're indeed well known to love crowdedness and busyness, aesthetics that carry over into their arts. Painting contains little empty space, and music little silence. In such situations, you relinquish to the group some of your personal agency at its most basic level: bodily self-control.

Bodily compliance is probably learned by the Balinese from an early age. Mead describes the docility with which infants ride in slings in Bali, where the child "has learned to accommodate itself passively to the carrier's movements, to sleep, with head swaying groggily from side to side."[68] People carry babies in snug carriers on their fronts or backs in the West, but they usually don't work in the fields with them. A similar physical compliance takes place in teaching situations in Bali. In dance lessons, pupils are extremely plastic in the hands of the teacher, who guides their bodies, sometimes with sharp jabs.[69] Another way Balinese children seem to willingly give up their physical autonomy is by a practice of trying to merge with a parent or other bigger person. Bateson photographed Balinese children aligning in orientation and pressing up against the father's body, apparently trying to become one with him.[70] I've seen the same thing many times. The feeling of wanting to combine, to give up your separateness and flow with others, shifts the ego away from the center of your body to the group or to the cosmos. It's a visceral way of experiencing yourself as part of the realm beyond your skin.

The Personage through Time and Space

In the West we tend to conceive of history and transitions in linear terms and death as a divided, binary (dead or alive) state, so it's hard to understand the Canaque view of the personage extending out into space and back and forth in time.[71] The Canaque entertain "no opposition between the dead and the living."[72] Canaque death seemingly never finds a final form. A person who has died (in the Western sense) may be described in his last living state perpetually ("he's ill"), and Canaque language has no exact translation for the verb *to die*. Death isn't an absence of life but rather another form of existence. Leenhardt goes as far as saying that "the idea of death . . . does not exist."[73]

Like their notions of life and death, Canaque stories and myths also defy clear divisions and separations.[74] When a Canaque tells a story, she places herself in its landscape and situation so certainly and clearly that she can't reproduce the tale without recalling the landscape and scene in detail.[75] The Canaque mirror many indigenous cultures in enmeshing a story into its setting, for instance, the indigenous Pintubi people of Australia, whom the American poet Gary Snyder watched in amazement as they told stories at high speed while traveling in a pickup truck—the stories are usually recited or sung while walking through the landscape.[76] For the Canaque, the "background" scene is an integral part of the narrative and can't be neglected. To tell the story of an animal captured in hunt, a Canaque must describe the time of day, time of year, quality of light, landscape, the movements and responses of the prey, people present, and other features of the context. A description of the actions and characteristics of the hunter and prey alone would not make sense.[77] Story and myth, like human life itself, is inescapably immersed in cultural and natural surroundings.

Social Engulfment

In the modern world many people live intensely social lives, interacting with coworkers all day and with family and friends the rest of the time. But social isolation and alienation are a hallmark of modern living, and individualist cultures tend to involve spending more time alone. Sometimes this is intentional and desirable; other times, it's just an outcome of the fragmentation pervading modern society. By contrast people in less modern societies spend more time and energy in social interactions, and they seem to have more capacity for continuous social contact. In such cultures it's hard to forget that you're part of a social whole. You experience it almost continually.

Mead writes about the Balinese preference for anything that's noisily, crowdedly festive.[78] In my experience, most Balinese like to spend virtually all their time in the presence of others and prefer little solitary time.[79] Ceremonial work for a temple, family, village, or for any other purpose, is done in groups, often very large ones.[80] Mead noted that Balinese people in groups participate in a continuous chatter akin to verbal "doodling."[81] Women and girls can often be found in a kind of communal physical contact for delousing, where the hands become a social instrument. Balinese architecture exhibits an emphasis on group activities and meeting places. There are large, open halls in every neighborhood and many smaller ones along the road where casual, impromptu socializing takes place daily.[82]

Balinese performances are frequent opportunities for intense social interaction. Traditional music and dance are always performed at temple ceremonies, where food stalls are set up, people are entering the temple to pray, and various performing arts groups are playing for the gods or for the community. Audience members socialize festively while they watch. They comment on the performers, and since the audience usually spills around the performance space, they can see each other's expressions and reactions. A kind of communal response to the performance quickly emerges in these highly engaged situations. Often musical dramas explore social mores, with actors improvising jokes about current news events and community happenings, and audience reactions play a vital role in reinforcing or rejecting the values and politics of the play.

At one such festive ceremony, I came across a friend who had recently gone through the great indignity of losing his position in a gamelan group (musical ensemble) his father had helped found: a couple of upstarts joined the group, a conflict evolved, and he had to go. He'd been a major performer in the group, and now the entire community witnessed that he'd been cast out. At the ceremony, the group was performing for the first time without him. He was so upset he entered into a light trance. It was touching to see that while he was in this altered state of consciousness a group of his friends literally supported him, almost carrying him as they wandered through the temple compound. The support of his community enabled him to mentally remove himself and be at the event at the same time.

Balinese people learn music and dance in intensely social conditions, too. Balinese arts and culture are taught by rote—visually, aurally, and kines-thetically—using the eyes, ears, and full body.[83] Musicians learn long com-positions entirely by listening and mimesis (imitation). A musical leader performs a phrase, and then the musicians repeat it until it's grasped by all. Most parts are derived from the other parts in a holistic kind of organi-zation that allows some musicians to figure out their parts by listening to other parts. The process takes hours, and the musicians and dancers spend a great deal of time together socializing and learning, with many breaks. In the Western classical chorus I'm in, breaks are short and members are expected to learn most of the music on their own from notation. Members of the Balinese gamelan ensembles that I play with in the San Francisco Bay Area, mostly Americans, find the long, slow rehearsals with socializing a welcome diversion from the breakneck pace and more individualized tasks of our modern lives.

The Hyper-Individuated West as Death

With their freer, less codified social interactions and opportunities for anonymity, modern societies can seem like a refuge compared with the intense social immersion of non-modern ones. Anonymity and release from social constraints can be attractive for people who've met with unbearable circumstances in a more traditional society. Leenhardt observed that some Canaque contemplating suicide—for them an acceptable transition to another realm—instead chose to seek refuge in the West, implicitly drawing a comparison between the two. The modern world was one way out for them.[84] Either method would sever their social and natural connections.

Something resembling this step toward death happened for Canaque society as a whole when the French colonized New Caledonia and slowly displaced indigenous customs and cultures. Skillful artists and surgeons no longer knew the "Word of the clan" and therefore lost their talents and abilities.[85] Cultures worldwide have marched toward dissolution with the homogenizing forces of modernization, even while they retain some of their traditional elements. For people of more integrated, less dissociated cultures, loss of connections to nature and society undermines identity and purpose. The Canaque "cuts the moorings which connect him with tribal life" when he enters the modern world. One Canaque chief refused to leave his land, to dissociate himself from it, when Europeans came to evacuate and acquire the area. He chose suicide over separation from his landscape and people.[86]

Many Native Americans have chosen the same fate rather than being alienated from their land and culture. In the 1870s the California Modoc leader "Captain Jack" fought with his tiny band against the vastly larger US army. He ultimately jumped off a cliff rather than surrender for removal to a reservation.[87] In 1906 in Badung, Bali, and in 1908 in Klungkung, as the Dutch sought to extend their control from North Bali to the south, hundreds of Balinese royal family members and attendants committed ritual suicide en masse outside their palaces as the Royal Netherlands East Indies Army arrived. The Balinese royals chose death over losing connections to their people and lands, and the attendant humiliation. Provoked Dutch soldiers fired into the bloody crowd stabbing themselves and each other with daggers.

DISSOCIATED WESTERN PSYCHES

The personages of less modern cultures strain the imagination for many of us Westerners. Can we really believe that for a Canaque the self traverses from the body out into nature and society, across generations? Does she really not see death as an irreversible state, a clear-cut, one-way transition? It takes some effort to conjure such images. But even though our connections to the world beyond our flesh seem more attenuated than those of the Canaque, we aren't completely modern, after all. Many modern people believe in a transcendent soul that survives bodily death, for instance. Doesn't that idea echo our ancient ancestors' animist beliefs about their souls?

On further reflection, less modern conceptions of the self may not seem so radical. Don't you sometimes see human expressions in animals, especially pets? Have you ever trusted people for basic physical support, as did my upset friend at the Balinese temple ceremony? Don't we understand ourselves as having commitments and responsibility to family, community, and (to a lesser degree) nature in spite of our supposed self-interested motivations? Although our mental habits and concepts of self may differ markedly from those of less modern people, our basic brain structures arise from the millions of years our ancestors lived as gatherers and, more recently, farmers, when mutual support within a group meant life or death. As individualist as we are in the modern West, at some level most of us appreciate the importance of community.

Modern cultures and modern selves comprise a large spectrum of humanity. Personhood isn't experienced exactly the same way in any two cultures, and neither are we all the same within a single culture. Yet by comparison, the Canaque, the Balinese, and many other cultures outside the modern mainstream reveal just how far moderners have become individuated, separated in thought and in practice, from our larger human and natural communities. Perhaps these ideas of non-modern personhood are influenced by romantic notions of the virtuous native or noble savage who's one with nature and one with community. But Mead, Bateson, Leenhardt, and Mauss were all aware of these idealized images of less modern cultures and strove to ground their conclusions solidly in observations. Moreover, there's evidence that Leenhardt didn't necessarily even consider lesser individuation good for the Canaque. Greater individuation, he wrote, provides "the possibility of the person recognizing in himself a plenitude."[88]

But that plenitude comes at the price of a fragmented view of the self and the world that ruptures important relationships. As we moderns perceive our physical environments, we dissociate things out of the contexts in which they're actually embedded. As we try to understand why things happen, we ignore context, relations, and the larger picture and emphasize actors, objects, and their inherent qualities. So we neglect important spatial, social, and ecological relationships. Truth, we've come to assume, is a cleanly divisible, discrete realm, not the complex, difficult-to-unravel continuum premised by Eastern dialectical thought.

Hyper-individuated Western personhood—independent, autonomous, and full of rights and inherent, context-independent qualities—may be the deepest, most personal expression of dissociation in many of our modern lives. We strip relations from our very conceptions of ourselves. We downplay ties between generations as if we're born *ex nihilo* (from nothing) and see our bodies as independent of other bodies and separate from the rest of nature, from the flows of space and time, from myth and landscape. On the whole, Western cultures strive for the ideal of the fully individuated person and get closer to attaining it than most others. Dissociation inhabits Western psyches, starting from our most basic experiences and conceptions of the self, which in turn support our destructive modern divides from nature and consequences. In the next chapter I explore the history of ancient ideas that arose alongside this peculiarly fragmented experience and practice of the self.

CHAPTER 5

ANCIENT TRACES OF DISSOCIATION

In [the] task of education, the emphasis must be on the analysis of modernity: yet that analysis is best carried out in full recognition of where our current models originated, and of earlier notions, in a less complex world, of how humans should live together.

—G. E. R. Lloyd,
Ancient Worlds, Modern Reflections, 2004*

However different we think we are from the people of the past, our knowledge and experiences of the world emerge out of thousands of years of human history. To better understand the forces shaping us and our environments, we must seek out the origins of our most fundamental ideas about the world and the human place in it. In the modern West that means going back at least to the ancient Greeks. Philosophers and historians agree that intellectual developments in ancient Greece are deeply embedded in Western thought. The Greeks' philosophy and approach to life created the ground on which much of the modern West stands: Western-style either-or logic, a predilection for categorizing things, winner-take-all debate as an important way of discovering the truth, and, not least, individualism and the nuclear family. These Western heirlooms are united in their divisiveness.

The environmental philosopher J. Baird Callicott says that discontinuity and discreteness have typified Western thinking about nature for the last two and a half millennia.[1] Dissociating ideas and practices have permeated Greece from the time of Homer (perhaps the seventh or eighth century BCE) and colored the thinking of Socrates (c. 470–399 BCE), Plato (428/7–348/7 BCE), and Aristotle (384–322 BCE), three pillars of

Western philosophy. In the 1940s the British historian and philosopher Robin George Collingwood (1889–1943) identified the period during which these three lived—the classical—as the first of three broad eras of pronounced "constructive cosmological thinking" that shaped Western ideas of nature and the cosmos with major intellectual and scientific developments.[2]

Certain aspects of Greek culture and life remain vital and continue to mold current Western culture and thought. The ancient Greeks' turn from the mysterious and mythological toward the rational and mainly secular foretold a central characteristic of modernity. Our marked individualism echoes (though much attenuated) Greek life.[3] Aristotle's logical principles still serve as gatekeepers to the truth. The basic principles of the modern scientific method—a systematic interrogation of nature—are to be found in ancient Greece as well.[4]

How did the Greeks begin to fracture the world? They did so by conceiving of it and living in it as a profoundly disconnected place, by thinking of themselves as completely independent of one another and by behaving that way. They set the mind, envisioned as the human essence, apart from the body and nature, and imagined that free (male) citizens—and nobody else—possessed reason.[5] Although Greek practices of social alienation and ideas of disconnection were, of course, not universal, and various thinkers had divergent ideas of the world, Greek living and thought exhibited a peculiarly fractured essence.

Where did the Greeks get these ideas and practices from? The evidence is less clear for their origins than for their existence, but they seem to have come from the particular ecological and social conditions the Greeks experienced: the disconnectedness of the physical geography of the Hellenic peninsula and the islands of Greece, the intense social strife of Greek communities, and the wide-scale adoption of written language. Before we consider Greek concepts, let's first examine the milieu of Greek life leading up to the classical period.

DISSOCIATED LIVING IN ANCIENT GREECE

> The Greeks came into being and the world, as we
> know it, began.
> —Edith Hamilton, *The Greek Way*, 1930†

Being a male citizen in ancient Greece from the time of Homer through the classical period of the fifth and fourth centuries BCE meant competing to build your prestige and that of your household (*oikos*, pronounced "ecos") over all others in your community. One method was by winner-take-all debates in which you'd try to demolish your opponent. But the eminence of your *oikos* was based largely on your personal property, much of which you gained as war booty. So you'd war with other *oikoi* to try to gain their possessions. The material wealth of your oikos would determine your social class.[6] The intense strife and tumult of ancient Greek society was captured in a Corinthian's statement about Athens, the epitome of Greek culture: "One might truly say that they were born into the world to take no rest themselves and to give none to others."[7]

As you and other Greek men sought to establish your identities as victorious warriors, conflict between communities and households would become a major part of your experience. You'd deem your autonomy and that of your oikos as paramount,[8] and you'd seek to ensure both by establishing the supremacy of your own *polis* (a city-state such as Athens or Corinth) over all others and by disparaging cooperation with other communities. Because the emergence of urban life in the polises had led to a breakdown in the older tribal orders, kinship ties weren't as important. Instead, there was a focus on individual identity and identification with the polis.[9]

A zest for competitive struggle between communities pervaded Greek culture and found many expressions, particularly in the numerous civic and Panhellenic festivals, the most famous of the latter being the Olympics.[10] Not only would you, as an ancient Greek man, see war and fervent competition as normal, but you'd also prize the valor earned by participating in them.[11] I say "man" because these contests concerned only men and, moreover, only Greek male citizens. Foreigners, women, and slaves made up most of the population, but they had few rights and privileges and weren't eligible to be citizens.[12] Because they were deemed inferior from the start, it would be senseless to allow them to compete.

Greek individualism was forged in this intensive contest system. Men's overriding concern with their personal fame and honor led them to seek not only victory but total and conclusive wins against opponents in artistic contests, debates, inter-oikos competitions, and wars. Competition was ruthless. Greek contests tended to be "zero-sum games," winner-take-all events, not activities that seek to advance the welfare of all participants. So the debate form prevailed over the deliberative process, in which truths are admitted from various sides and middle-ground solutions are possible. Zero-sum games are anathema to cooperation. In the social context they manifest a belief in the individual person as the primary fact of existence, more real and relevant than anything else, such as family, community, and nature.[13]

The Greeks' heavy investment in competitively established self-esteem discouraged stable and enduring attachments to others. Your friend could quickly become your enemy, so social connections were shallow. Narcissism and concerns about personal status dominated, and most Greek citizens talked about themselves and issues of personhood often. They exhibited a persistent and nagging preoccupation with the concealment of, search for, and revelation of identity. The ardent individualism of the Greeks set the precedent for modern Western individualism.[14]

The shift in focus from community to the individual as the source of value and the ground of human existence played out in other kinds of relationships. The ancient Greeks had loose connections with various kinds of things—people, animals, nature. Indeed, the Greek style of relating can be thought of as "low object attachment" because it devalued relationships in general.[15] Studying the Greeks, the American sociologist Alvin W. Gouldner wrote, "Diffuse, low object attachment means there are fewer objects in the world to which individuals are irretrievably committed or from which they can extricate themselves only with great effort; conversely, it means that there are many more objects in the world that can be used, treated primarily as resources, without being restricted by sentiment in the uses to which they may be put."[16] For the Greeks, the self and its attributes rose to the top, above relationships.

This tenuous connection to things meshed well with the distinctly Greek brand of rationality, which Gouldner called "total-commitment rationality": all resources are put into a single rationally determined choice without reservation.[17] Rationality must supersede the emotions when making choices, and logical choices must be given precedence over emotionally motivated ones. Low object attachment and total-commitment rationality both dissolve connection and downplay context. Emotions might bind us with other people, animals, and the natural world, limiting our choices, so they must be overcome. The Greeks remained committed to choices built on cold reason even when such decisions broke the bonds of family and city-state.

In 480 BCE, during the second Greco-Persian war, the Greeks activated the powerful combination of low object attachment and total-commitment rationality. Thinking they would be overwhelmed by the Persian army, the Athenians vacated their city as a war tactic, deserting their *polis*, farmlands, gravesites, and the seats of their ancestral gods. They chose to divert the battle to the sea, to leverage the strength of their powerful navy. The tactic worked. With three hundred ships against a thousand, Athens defeated their foes. This brazen act was driven by zero-sum-game thinking: the belief that all is lost unless you win.[18]

Low object attachment and total-commitment rationality had their dark sides, however: they created great instability in society as relationships among people remained in flux and choices were frequently countermanded very quickly by new rationally deduced ones while traditions and norms took a back seat. As I discuss further on, the ideas of two of Greece's greatest thinkers can be seen as a response to all the instability created by these peculiarly Greek approaches. Divisive ancient Greek experiences and thought styles set the stage for the Western view of individual people as separate from and independent of nature and other people.

Dissociation by Written Word

Various forms of writing preexisted classical Greece, but the Greeks made writing more user-friendly. Adding vowels and other innovations to the Semitic alphabet, they invented phonetic writing, perhaps the most important technological innovation of ancient Greece. It made both reading and writing easier and more popular. In the transition from Plato's to Aristotle's generation in the fourth century BCE, books came to be read more and more by a growing general audience. Writers including Herodotus ("the Father of History," c. 484–425 BCE), Plato, Aristotle, and other influential Greek thinkers used it to record history and to create and share new ideas and knowledge; phonetic writing enabled their millennia-long influence. But there was a downside to writing, one that Plato alludes to: the increasing use of written language in Greece divorced knowledge from lived experience and removed the senses from nature.[19]

Before phonetic writing came on the scene, the stories that shaped European societies, such as the epics of the great poets including Homer, were propagated orally. Storytellers memorized long tales in poetry or verse form and recited them back to audiences. It seems almost inconceivable that anyone memorized the *Odyssey*, the *Iliad*, and other long epics, but they did so, with much practice—the stories were told frequently. Telling stories differs drastically from writing them down to be read later privately. In the telling, a story picks up the contexts in which it's "performed": the time and place of the storyteller and the audience. Each village or town has its own style and concerns, so different parts of a tale might be emphasized, left out, or changed to refer to recent events. Orally communicated stories have lives. They participate in the natural and social contexts in which they're told and heard.

Written texts reflect the more contingent (context-independent and open to free will) realities of their writers because they're not produced, reproduced, and re-imagined before a live audience. Writing remains static and immutable as it travels. Oral texts absorb power and truth from their telling—they take in the world. Words on a page become hardened and refuse the living world that continues past their creation. The written word is an abstraction of the spoken word, two steps removed from the phenomenal world. Phonetic writing is yet another level removed because it depicts the *sound* of the spoken word. By contrast, many non-phonetic writing systems consist of pictograms, characters that visually resemble the thing to which they refer (the word for *house* might look like a house), possibly mixed with alphabetic characters. With its invention phonetic writing thus removed the reader multiple times from the subject and allowed for easier expression of abstractions. It's hard to imagine Aristotle's metaphysics propagated in a strictly oral tradition.[20]

The media theorist Robert K. Logan researched the effects of alphabetic writing on human history in collaboration with the renowned philosopher of communications Marshall McLuhan, who coined the expressions "The medium is the message" and "global village." In *The Alphabet Effect: The Impact of the Phonetic Alphabet on the Development of Western Civilization*, Logan argues that the abstractions of phonetic writing profoundly influenced Greek culture beyond displacing oral culture.[21] They served as a model for other kinds of abstractions. From about 2000 BCE to 500 BCE, in the narrow geographic region between the Tigris-Euphrates river system and the Aegean Sea, many crucially important developments foundational to Western thought arose alongside the phonetic alphabet in Greece: codified law, monotheism, abstract theoretical science, formal logic, and individualism.[22] In Greece, abstract ideas such as beauty, justice, and reason took on new meanings and became the subject of a fresh type of discourse, as Plato's and Aristotle's writings show. Timeless analytical statements that depended on alphabetic literacy emerged, expressing universal truths—truths independent of the contexts in which the writings were written or read.[23] A new systematic approach to the natural world arose, one in which phenomenal experiences in nature could become not only secondary but suspect.

Abstraction tramples the particularities of lived experience by codifying them into universal laws—those that transcend any particular place and time. Moreover, the experience of reading differs markedly from that of learning in an oral tradition. The shift in learning from the spoken to the written word heralded a move from a more outward and socially attuned focus to a more inward one. In oral traditions, we engage directly with parents, poets, and other teachers. Interaction with written text makes learning—propagating myths, stories, and knowledge—a more solitary act. Lived social relations play a smaller role.

Writing a thought commits it to a particular expression, one that will live indefinitely in that form. The widespread adoption of writing may have contributed to the entrenchment of argumentation and conflict in ancient Greece by giving the impression that ideas and positions must not change. Even Plato, one of the West's first great writers, seemed uncomfortable with the increasing popularity of writing during his lifetime: "Written words . . . seem to talk to you as though they were intelligent, but if you ask them anything about what they say . . . they go on telling you just the same thing forever. And once a thing is put in writing . . . [it] drifts all over the place, getting into the hands not only of those who understand it, but equally of those who have no business with it; it doesn't know how to address the right people, and not address the wrong."[24] Plato laments that written words go on to have a life of their own, disconnected from the intent, intelligence, and experiences of the writer. Perhaps we should read his writings—and all writing—in that light.

Phonetic writing also transforms our experiences of the phenomenal, tangible world. David Abram observes that spoken language, with its base in breath and voice, embodies a sensuality that connects people to the world around them. Speech often imitates the sounds of the physical world or enacts a gesture, a bodily movement. The fully phonetic writing of classical Greece enabled the replication of words without actually speaking them. Previously, non-phonetic or partially phonetic alphabets required that written text be spoken to be understood. The Phoenician alphabet, for example, in use in the Middle East from around a thousand years BCE, included only consonants and never evolved vowels or spaces between words. Lacking these enhancements, Phoenician texts, like old Hebrew documents, had to be spoken aloud to be understood. By adding a full complement of vowels and spacing to the Semitic alphabet they borrowed from the Phoenicians, the Greeks could represent the sound of speech on the page well enough that it didn't have to be spoken. Although it was initially common for people to speak text while reading it, this last vestige of physicality in writing eventually disappeared, removing language another level from the physical world.[25]

Dissociating Greek Ideas

In the Greek context of intense individualism, with an overriding concern for the self and a turn toward the solitary acts of writing and reading, Greek thought and life established for the West a new disconnected, fragmented style of relating to and understanding the world. Theory became ascendant over empiricism (investigations of things founded on observations of real-world phenomena). Plato even sought to sever the present from the past by disparaging tradition and seeking to build a society starting from ideas rather than experience.

In their theories about the world, moreover, the Greeks began to invent models of the cosmos consisting of independent, inert, autonomous parts. A hallmark of the new Greek worldview, the atomic theory introduced by Leucippus and Democritus in the fifth century BCE, saw the entire physical world and even the soul as consisting of passive corpuscles—atoms—devoid of sensory attributes, interactions, and context.[26] The atomic theory sees nature as inactive, mechanical, and inherently quantitative. It can be reduced to its component parts because they're only externally related to each other (there's no essential quality of each part that relates it to any other).[27] This focus on separate, autonomous, and independent parts in ancient Greek thought both reflected and informed the Greeks' fractured perception of the world.

Ancient Greek natural philosophy and practices come into focus when compared with the very different style of thought and customs prevalent in ancient China. Ancient Chinese philosophy, science, and culture tended to invoke qualities of similarity, interdependence, and complementarity among things—correlations rather than oppositions. Whereas an ancient Greek might see himself as a self-made, autonomous, and independent person, for instance, an ancient Chinese person tended to think of herself foremost as a member of a family and a society. The oppositional and dissociative qualities of Greek thought and life would likely have appeared strange in ancient China, where the notion of yin and yang held great importance. The dual unity of yin and yang expresses relations of mutual interdependence and reciprocity. While Greeks sought to define ideas and physical objects in opposition—according to mutually exclusive features—Chinese people focused on features of interdependence and complementarity: both parts, yin and yang, depend on the other for their very existence.[28] They're *internally* related.

While the Greeks divided things, including classes of people—superior from inferior, citizen from non-citizen—the Chinese held that a sense of seamlessness must be respected, even while political power was held by an elite stratum of society. The Chinese valued compromise, avoidance of confrontation, respect for authority, and a desire to work within established norms of knowledge. They built on what could be taken as common ground. Although exceptions certainly existed, on the whole the Greeks valued theory, competition, ruthlessness, and raising the self above others at any cost.

Chinese values of connection, interdependence, and complementarity didn't slow their advancements in science and technology. On the contrary, in empirical science and technological innovation the Chinese were much more sophisticated than the Greeks, who remained more concerned with abstraction and universals (entities that are the same everywhere and always, independent of particular times and places) than with empiricism. The Greeks seemed to think that taking real-world observations seriously would taint their theories, which would need to be repeatedly reconciled with observations in the messy, flowing phenomenal world.[29] Chinese acceptance of interrelatedness and the importance of the unfolding of events in the material world allowed their grounded science to bear fruit more quickly than did the Greek commitment to abstractions disconnected from lived, physical, observed reality.

ENTER PLATO

> Plato's presentation of the philosophy of Socrates and his development of it through fusion with Pythagorean doctrines is the first great coherent treatise on man within the Western tradition.
> —Alvin W. Gouldner, *Enter Plato*, 1965‡

> The safest general characterization of the European philosophical tradition is that it consists of a series of footnotes to Plato.
> —Alfred North Whitehead,
> *Process and Reality*, 1929§

Although Plato is long dead, his thought echoes in our Western minds. Beginning with Plato, together with Socrates (whom we know mainly through Plato's writings), topics of intellectual concern in the West shifted from the study of nature to the study of human life. Out of this shift arose a sharper distinction between nature and culture, the realm of the human.[30] Without adopting the simplistic and reductionist assumption that Plato's thought is entirely a product of the civilization in which he lived, his philosophy must be seen as rooted in the political dilemmas and social tensions of Greek society and in his own political concerns.

In 1965, Alvin W. Gouldner published a rich, though neglected accounting of the social circumstances of Plato's thought, *Enter Plato*—the name conjuring Plato as a headlining actor entering the stage of the great drama of Western intellectual history. Plato came of age at a time "when the established order and accepted standards seemed on the verge of dissolution" from continuous political challenges and theoretical criticism.[31] As a young aristocrat, Plato, son of Ariston, was oriented by birth to the responsibilities and services of the state.[32] He grew up in a society with many ills, including intense competitiveness and the corresponding vulnerability of close relationships, a lack of accord on the best political order, the flux of a "quick turnover" society, frequent strife and inter-polis war, and class struggle. The injustice of his teacher Socrates's trial and execution also made a deep impression on Plato. His utopian designs in *The Republic* and his philosophies of transcendence and death (discussed further below) can be seen as attempts to produce a rational substitute for politics and a stable replacement for the tumult of Athenian society.[33]

Plato's thought departed dramatically from the nature philosophy of the Ionians (founded by Thales in the seventh century BCE, two centuries before Plato) that had been prevalent in Greece. Plato's reasoning, in which a person seeks to the know pure, eternal truths of the mind, and to know oneself carefully through introspection, contrasted sharply with the Ionians' outward focus on the substance of nature as it's known primarily through the senses. The Ionians sought to submit reasoning to the test of observations, an approach more like that of the ancient Chinese. The senses are necessary if not sufficient tools for knowledge, they believed.

For Socrates and particularly for Plato, in contrast, the mind is supreme. It infuses all nature and provides the logic that shapes and moves the world. The senses are suspect. They corrupt the mind and imperil logic: "Observation by means of the eyes and ears and all the other senses is entirely deceptive."[34] Reason must be on guard not only against the self and the senses but also against tradition and beliefs received from the social environment.[35] It's a force to be used to overcome tradition, to cleave the present from the past.

The new form of reason promoted by Plato reflected values of mastery and control over nature and society, a rising above and beyond both. The mind and its products, the Ideas, when divorced from the body and physicality, became sources of certainty, stability, and continuity. The emotions—prized in the Greek Dionysian tradition (Dionysus was the god of such things as wine, ecstasy, ritual madness, and intoxication)—were too close to the body, so Plato denigrated them. He followed closer to the Apollonian tradition, which elevated the intellect and reason. Plato sought detachment from the culturally prevailing use and meaning of real, physical-world objects. That is, he sought to dissociate himself from Greek traditions. In doing so, he also separated from the people who related to objects in conventional ways—people following customs and living by accepted norms.[36]

Theories of an Elite Citizen

Plato traveled in the upper echelons of Greek society. He was a free, powerful, slave-owning, elite Athenian citizen, and his theories reflect his position of privilege and his detachment from the lower ranks of society. The freeman's prejudice toward manual labor oriented his views. He performed little or none of it and probably no craft besides writing. Physical work was done by others, including the women, servants, and slaves in his household. His perspective was that of a spectator, a consumer rather than a producer. Plato's lack of participation in growing food or making everyday objects such as bowls and sandals probably led to his seeing things only in terms of their external performance or finished form. Whereas an ancient Greek potter might have looked on a bowl as a combination of the raw materials and work that went into it—and as something that would one day become shards—Plato imagined things in a kind of permanently complete state, which became represented in his philosophy as the Forms, the idealized, final, eternal versions of objects and concepts.

Plato's view contrasts sharply with the Ionian concept of nature. The latter imagined humans as technicians trying to exercise limited control over a material world that's infinitely varied and ingenious yet inexorable in following its own laws. By contrast, Plato saw humans (really, white men) as masters of nature. The classicist Benjamin Farrington writes, "The new [Plato's] conception of Nature, as a power with ends in view, which enforces its will on a subordinate but refractory matter, is the conception of a master who governs slaves."[37] The desire evident in Plato's philosophy for control over both lower-class people and nature reflected and fortified his privileged social position. Another classicist, Geoffrey Ernest Richard Lloyd, writes, "What the elite activity of pure [theoretical] science presupposed in terms of inegalitarian hierarchical and authoritarian political and social institutions was not lost on them and is not lost on us, and the splendors of the life of pure theory, find, hardly surprisingly, their most eloquent spokesmen among those who were the beneficiaries of those inequalities."[38] The domination of lower classes went hand-in-hand with the domination of nature. It required active separation from both.

Plato famously advocated "philosopher kings" in *The Republic*, in which only a highly trained elite could rule society; proposed using eugenics to create a specialized, more capable ruling class; accepted slavery (except for enslavement of a free Greek, as opposed to a "Barbarian"); and drew extreme and exaggerated distinctions between masters and slaves.[39] He clearly and actively endorsed a stratified society that gave select people control over others. The authoritarian ideal is a key component of Plato's thought and runs through the dialogue. He even instills within reason itself an authoritarianism that holds sway over proper thought, creating a double authoritarianism in the process.

Plato's conception of reason is authoritarian because it denies that any of the other faculties have any legitimacy. Intuition, imagination, and insight are excluded from the start. This authoritarian reason matches the authoritarian social structure that Plato also favors. He projected the master-slave relational archetype onto other spheres of society to formulate a hierarchical order of control. The rational part of the psyche is master to the impulses. Likewise, master rules slave, man rules woman, man rules nature, and reason rules emotion.[40] The master separates from and rises above all.

At the root of this double authoritarianism of reason and the male citizen lies the fact that reason isn't supposed to be used by and for all people. It's the province of the citizens of the polis, who have full legal rights. Plato and his followers closely associated women, aliens, and slaves, the majority of people living in Athens, and saw them as base and lacking in certain capacities.[41] The reason inherited in the West premises the system of slavery and divisions between producers and consumers and between people and the sensuous world.[42] It denigrates the emotions, women, the senses, and relation.

Plato's Problem with Change

Plato developed his philosophy, in part, to contend with the tumultuous change happening in his society, both in material life and in the world of ideas. The Ionian philosophy popular when Plato was born saw nature as incessant change knowable through the senses. True to the Ionian tradition, the pre-Socratic philosopher Heraclitus famously said, "No man ever steps in the same river twice." The river is always flowing and thus is continuously becoming something other than what it had been.[43]

Heraclitus expressed the dynamism inherent in the nature of the river through a doctrine of universal flux, which saw change as an essential and pervasive aspect of the natural world. Plato's teacher Cratylus, a student of Heraclitus, extended this concept to an extreme yet logical conclusion. Cratylus said that if everything is constantly changing, no statement about anything could be true—a position of extreme skepticism. Heraclitus was wrong to think that one could step into a river even once because the river never has a stable identity—it's always already becoming something else.[44] The world had melted into a confusion of flowing, pulsating senses.

The notion that nothing is really knowable (an epistemological assumption) because nature is constantly changing (an ontological assumption) dismayed Plato. He thought if you allowed yourself to become obsessed by the constant flux in the perceptible world, the result would be confusion and ignorance. His response to this bad news was rather extreme and characteristically decisive. He concluded that the whole phenomenal world should be distrusted, held in suspicion. In his writings he resoundingly denounces the perceptible world, the whole physical world, as an unintelligible, turbulent realm that allows no knowledge.[45]

Severing himself from his teacher's disconcerting legacy, Plato aligned himself instead with Pythagoras, who taught his influential philosophy several generations before Plato. Pythagoras thought the stable elegance of mathematics was the truth underlying all of nature. He and his followers held little esteem for the sensible, material world: "A joyless place where murder and vengeance dwell, and swarms of other fates—wasting diseases, putrefactions, and fluxes—roam in darkness over the meadow of doom."[46] The Pythagoreans sought to transcend earthbound existence, and since the body is part of the physical sphere, they sought liberation from it, too. Plato added to this totalizing disparagement of the material world by denigrating the senses, the part of humans closest to, most connected with, loathed phenomenal existence. To complete his break with the whole of physical nature, Plato also rejected Socrates's immanence theory of form, in which the Form is manifest in each phenomenal, material instance of it—that is, every chair expresses and contains within it the abstract Form of a chair. Plato wanted to rid the Forms of any remaining vestige of lowly physicality.

In part to escape from sense data—sights, sounds, tastes—which bind us to nature, as a basis of knowledge, Plato posited a new transcendence theory of Forms, which he called Ideas.[47] Plato's Forms or Ideas are models that particulars (actual real-world things, which exist only in a particular time and place) resemble.[48] The Forms descend from the abstract mathematical, geometric view of Pythagoras.[49] They're knowable only through disciplined, dialectical practice and are eternal, static, unchangeable, and perfect.[50] These qualities would bring stability to the tumultuous world of classical Greece. Plato took this conception to the extreme—he imagined the Forms as more real than the world accessible by the senses. The phenomenal world is mere illusion, he taught. The idea of a chair is more real than a chair could possibly be. Change, the central attribute of things according to the Ionians, had become for Plato the attribute by which things refused to be real. Plato emphasizes these points throughout his writings, particularly in the *Symposium*. But how exactly did he excise reality from the phenomenal world, turning the latter into an illusion and renouncing any concern for it?

In the *Timaeus* Plato writes that the visible world is a copy, an image of what is eternal and true: the Forms.[51] The timeless and enduring character of the Forms provides a foil against which to judge the relentlessly shifting perceptible world. Physical things don't manifest the Forms—they don't make the Forms real. On the contrary, physical things are merely poor imitations of the Forms, which are the *only* real things. With Plato sensation and observation no longer held the key to knowledge about the world. The Forms are more perceptible and intelligible than things in the sensible world because they're unhidden, unconcealed, and undeceptive.[52] One consequence of this worldview is that only rational human thought can be used to discover the true world (the world of the Forms) and only things that are unchangeable are knowable.

The dialectical quest for knowledge of the Forms, in which reasoned argument resolves differing assertions, entails a flight from experience and a reliance on purely mental or symbolic manipulations. Studying and experimenting with real-world things through observation becomes unimportant.[53] The primacy of the Forms suggested to Plato the need for a great task: classifying the entire sensible world into categories represented in the Forms. He relishes classification, traversing hierarchies of categories to arrive at the true definition of a thing.[54] His only concern with physical-world objects seems to have been fitting them into categories, considering foremost their morally relevant characteristics: Is this object beautiful? Is it good? In this way, Plato brought together the moral and existential orders. Incorporated into Ideas, values gained a purity that denied the subjectivity and contentiousness of lived experience.[55]

Just as Plato's social class let him remain detached from subservient people, his denigration of sensory input and the physical cosmos marked a radical dissociation from nature and all its flux, complexity, and messiness. Although Plato's disregard for empirical observation set him in opposition with the Ionians, who considered observation a path to knowledge, it was consistent with much classical Greek thinking in physics and metaphysics, which tended toward abstraction and the creation of new knowledge through ungrounded thought. Plato believed the truth was available internally, without special experience, and could be reached through careful intellectual work.[56] His philosophy marked a retreat from nature.

Freed from the epistemic (knowledge-generation) limitations of the chaotic sensible world and turbulent human history, Plato and other Greek thinkers could assume their quest for foundations, certainty, and axioms by developing systems of thought with universality and stasis as core traits. Plato's Forms or Ideas are so universal they take precedence over and transcend even the gods. In the *Euthyphro* Plato states that the good isn't good because the gods approve it, but rather the gods approve it because it's good.[57] In his social theory Plato strives for universality by leaving out history, development, evolution, and laws of change—or any change at all, really, as change is mere appearance.[58] The problems of humanity could not be decisively solved on Earth because of flux and uncertainty, so Plato sought to transcend them by separating the soul from the body in life and in death.[59]

Plato's thinking on human identity, death, and the relation of the soul to the body have remained immensely influential in the West. Val Plumwood, the environmental philosopher, sees Plato's "philosophy of death" as a source of modernity's troubled environmental relationships.[60] Apparently following Socrates, Plato claims that the soul exists independently of and with a superior status to the body. The soul lives separately from the body before, during, and after bodily life.[61] This was a revolutionary claim, judging by the frequent reactions of surprise and disbelief that Socrates encounters in the dialogues when uttering it.[62] Note that this conception of the soul differs from that of reincarnation or of the transmigration of the soul because it exists indefinitely (and quite contentedly) without any body at all.

For Plato, it wasn't enough to make the soul independent of the body—the latter also had to be deemed inferior. In the *Phaedo*, Plato offers a wealth of disparaging commentary on the body, as well as on physical nature. The soul, he writes, only gets a clear view of the facts when it's free of the body and the senses: "When it [the soul] tries to investigate anything with the help of the body, it is obviously led astray. . . . Then here too—in despising the body and avoiding it, and endeavoring to become independent—the philosopher's soul is ahead of all of the rest. . . . In fact the philosopher's occupation consists precisely in the freeing and separation of soul from body . . . true philosophers make dying their profession."[63] Plato has Socrates uttering these words shortly before Socrates's death, which the elder welcomes. The body, and by extension the whole physical world, is merely an obstacle. Plato's worldview provides a way around the impediment even during life: the soul need not continue to associate itself with the burdensome body. The active disparagement and refusal of the body even becomes a requirement for entrance into heaven:

> If at its release the soul is pure and carries with it no contamination of the body, because it has never willingly associated with it in life, but has shunned it and kept itself separate as its regular practice . . . then it

departs to that place which is, like itself, invisible, divine, immortal, and wise, where on its arrival, happiness awaits it, and release from uncertainty and folly, from fears and uncontrolled desires, and all other human evils, and where, as they say of the initiates in the Mysteries, it really spends the rest of time with God.[64]

In these chilling passages, Plato leaves no doubt about his contempt for embodiment and physicality—and their correlate, material nature. Although he believes it's not only possible but also preferable to actively dissociate from your body during life, doing so can be but an interim step. Only death can achieve the most desirable, greatest distance from the nature within and the nature without.[65] Life, as even the best live it, is a crippled thing. Death will release and fully heal it: "For is not philosophy the study of death?"[66]

In the *Phaedo*, we can hear Plato working hard to convince us of the persistence of the soul after the body's passing.[67] In one of the most disturbing images from the dialogues, Socrates remains calmly dispassionate in the face of death. He refuses to put off drinking the hemlock and says he will gain nothing by spending more time with his friends in this world.[68] In accepting his death gladly, Socrates shows great appreciation for and trust in both the universal nature of the law that has condemned him and the universal, transcendent nature of the soul.[69] These universals are ironic, though, because they eschew the immanent particularity of a person who stood out in his time and in history as well. This cheerful acceptance of a death sentence may be Plato's (and perhaps his teacher's) most vivid expression of disregard for life in the phenomenal world, something clearly not worthy of care.

Socrates's dispassionate approach to his own death also shows how these philosophers relegated emotions to the trash heap. Plato not only radically separates the mind from the body and gives the mind independence, but the whole messy affair of severing the two results in the expulsion and demotion of the parts of the mind most closely associated with the body: the emotions and the senses.[70] They're not to be trusted. Plato even blames the body for the existence of emotions and the senses.[71] He struggles to devise a particular kind of mind that can be liberated from the body, a mind wielding a reason so pure that it can approach the Forms and find the beauty in them. It's not a mind meant for observing, enjoying, or caring for the physical world.

Plato allows a few emotions to remain with the mind, but he transforms them into rationalistic versions of their more complete manifestations. For instance, he strips love of its Dionysian qualities and makes it a thing of reason rather than something of the deepest emotions or of the whole embodied person.[72] Fully dissociating the rational mind from body and nature meant expelling the elements that linked them and cleaving other connections. The senses physically connect us with immediate knowledge and awareness of everything and everyone outside ourselves. The emotions inform and shape how we interact with human and nonhuman others; they tell us much about the world around us. By casting the senses and the emotions in a lesser role that only interferes with the higher activities of the self, Plato also diminished connection with others and the environment.

In Socrates's death scene, Plato leaves no doubt about his poor view of the emotions (and perhaps that of his teacher). Socrates literally expels the emotions from his death chamber. As his friends arrive for their last visit, Socrates's wife Xanthippe sits beside him with a little boy, possibly their son, on her knee: "She broke out into the sort of remark you would expect from a woman: Oh, Socrates, this is the last time that you and your friends will be able to talk together!" Hearing her emotional outburst, Socrates instructs that she be escorted away, and "Some of Crito's servants led her away crying hysterically."[73]

Shortly before Socrates drinks the hemlock, his children and the women of his household arrive. He instructs them in carrying out his wishes and then tells them to go away.[74] After Socrates drinks the poison, his friends break down crying, to which the great philosopher responds, "Really, my friends, what a way to behave! Why, that was my main reason for sending away the women, to prevent this sort of disturbance. . . . Calm yourselves and try to be brave."[75] In this scene Plato (and perhaps Socrates, if Plato is a faithful reporter) identifies the emotions with women. Both must be controlled, even at the momentous occasion of death. Reason (personified in the story by Socrates), the province of the male citizen, must keep women, children, and the emotions in check. They must not be allowed to bind people and things together.

But Plato and Socrates don't merely drive out the emotions. They assign them to a lower realm of existence where they also put the feminine, slaves, the body, and nature. The rational mind, by contrast, exists in a higher sphere, along with reason, the master, the masculine, the freeman, and culture. Later Western thinkers built on these dualistic constructions, creating a "set of interrelated and mutually reinforcing dualisms" that runs like "a fault line" through the West's entire conceptual system, as Plumwood instructs.[76] Male ascends over female, master over slave, reason over emotions, and culture over nature. Dualisms dissociate. They fracture our thought through a set of techniques that Plumwood calls "the logical structure of dualisms."[77] They emphasize the differences between classes of things—such as humans and nature—and the similarities among members within a class—such as Plato and Socrates. Dualistic thinking trivializes the dependence of more powerful people, such as George Washington, for example, on less powerful ones, such as his wife and slaves.[78]

All the various separations and divisions in Plato's thinking, taken together, point to dissociation as a key organizing principle. Divisive constructions permeate Plato's thought: the transcendent Ideas constituting the one true reality, separate from lived experience; an elite, eugenically produced, imperious ruler and ruling class above and separate from the general populace; a doubly authoritarian reason—one that only male citizens may properly use and that emphatically excludes all other forms of knowing; separation of the emotions, the body, sensation, women, and nature from the true ground of existence, the rational mind; disparagement of particulars and observational data; and an erotics of death as a liberation of the true reality, the mind and its productions, from the illusory reality of the phenomenal world. In Plato, the technique of dissociation—the dividing of the world into hyper-separated, isolated, and alienated parts—found its earliest, most complete, and most significant expression in Western culture.

Plato on Nature

Plato places little value on living nature. He replaces bonds of affection and respect toward concrete and particular natural things with an abstract and universal love of the imagined Forms, and he repudiates sensory experience and observation of the natural world as a way of knowing. Nature is properly understood only by contemplating the Forms. Moreover, philosophizing and contemplating the Forms is the most important activity a person can undertake. Information "from below," from Earth, must be suppressed in the process. In the *Phaedrus*, the eponymous character criticizes Socrates for "never leaving town to cross the frontier nor even . . . so much as setting foot outside the walls" of Athens. Plato has Socrates respond, "You must forgive me, dear friend, I'm a lover of learning, and trees and open country won't teach me anything, whereas men in the town do."[79] Physical nature can neither educate nor be a valid field of learning.

His elite position allowed Plato to disparage direct knowledge of the soil and natural processes and devalue the lives of people who got their hands dirty. He did little manual labor and spent little time, we can surmise, inhaling the rich aroma of humus, tending livestock, getting soil under his fingernails, or molding clay pots. Plato's immediate environment consisted mainly of the truly glorious cut marble and brick temples, agoras, and houses of Athens, just as our phenomenal worlds are filled with artifice, though of a different type. His detached relationship with nature may reverberate in the lives of us moderners for whom nature is "outside the walls." Plato had women, slaves, and wage seekers to work the land for him, and he thought of them as a lower class of existence, along with the nature they worked.[80] Today in the United States another subordinated class, "illegal" immigrants, do much of agricultural labor. They remain invisible to most of us, along with the nature they work.

Plato's view that the philosopher should welcome death as a release from the inferior realm of the body and senses—his "philosophy of death"—emphatically rejects physical, sensuous nature. No longer part of the cycle of life, dying becomes a departure from the prison of the body—an escape.[81] When he banishes the emotions of mourning and loss surrounding death, Plato also removes much of its human and animal meaning.[82] It becomes a permanent release from unwelcome embodiment, not a cyclical return to the physical world in another form, as it is in so many animist cultures.

If all of physical nature is an inferior realm from which we must distance ourselves while we're alive and from which we must seek to escape by dying, then why care for it? Why nurture this place we're trying to leave permanently? To put it mildly, Plato's nature-despising philosophy doesn't exactly promote an ethic of environmental benevolence. And to the degree that his views of the human place in nature shape our modern sensibilities, we're left to wonder how an environmental ethic could fit into modernity at all. But there's reason for hope if we choose to guard against not only Plato's view of the human-nature relationship but, even more so, his affinity for dissociating thought—and that of his star pupil.

ENTER ARISTOTLE

> Aristotle's physical doctrine was accepted as dogma for sixty generations. No other person in the history of science, and very few in the whole course of human culture, had so deep and long-lasting an influence on subsequent thought. . . . The chief interpreters of the three great religions finally blended the main principles of Aristotle's philosophy with their religious conception of the universe, thus turning the whole of that philosophy, including its physical and cosmic aspects, into unquestionable dogma.
>
> —Samuel Sambursky,
> *The Physical World of the Greeks*, 1956**

Aristotle's thinking diverged significantly from that of his teacher, Plato. He rejected Plato's theory of Ideas, which are transcendent—existing above, beyond, and independent of physical nature. Aristotle criticized the way Plato arrived at his theory of transcendence. Plato had learned through Cratylus and the Heraclitean doctrines that all sensible (observable by the senses) things are always in a state of flux, so no science of them is possible. Plato responded by taking what Aristotle implies was as a path of convenience. Only non-sensible things are valid objects of inquiry: "He called things of this other sort Ideas and believed that sensible things exist apart from Ideas and are named according to Ideas. For the many sensibles which have the same name exist by participating in the corresponding Forms."[83] Aristotle is explaining that Plato thought the Forms and Ideas are the only real things, and everything in the material world, the sensibles, are mere imitations of them.

More of a scientist than Plato, Aristotle was concerned with the material and human reality around him and with the sensory functions that make such things known. Criticizing Plato, Aristotle says that confusing and illogical results ensue from Plato's theory of transcendent Ideas: "Non-substances will be prior to substances," and "both the elements and the principles" will be "prior to that of which they are the principles and elements."[84] Aristotle ends *Book XIII* of the *Metaphysics* with a detailed argument against the Ideas or Forms and other universal principles.[85]

He rejects transcendence and instead adheres closely to a doctrine of immanence: it's the actual, phenomenal world that is real, and God or the Ideas are in it, not separate from it. Mental images originate in, or are derived from, material reality. Aristotle writes in *Book IX* of the *Metaphysics* that actuality (the sensible world) is prior to potentiality (ideas) in several ways: in formula, in substance, and in time.[86] The upshot of this complex statement is that for Aristotle, the physical, sensible, material world of nature comes first, before any mental constructions of it. After Plato's denial of the reality of the physical world, Aristotle's refreshing cosmology seems to offer the possibility of caring about and connecting with nature.

Knowledge for Aristotle is grounded in lived, observable, phenomenal reality. But unfortunately his science and philosophy tend to create unnecessary and unrealistic conceptual divisions by de-emphasizing or ignoring relationships. Like Plato, Aristotle was an elite Athenian citizen and a member of the Academy, producing volumes of conclusions and teachings from comparatively little empirical data from a secluded sanctuary north of Athens. Contrary to the popular image of Athenian academicians pursuing knowledge for its own sake, these scholars worked not just for personal edification but also with an eye toward power, to influence contemporary policy making. Alvin W. Gouldner called the Academy the RAND Corporation of antiquity.[87] The students of the Academy, like their teachers, came from the privileged class of citizens, and Aristotle's later pupils would include the Macedonian elite, most notably Alexander the Great, who deployed his knowledge for extreme political power in building an empire.

Unlike Plato, Aristotle promoted the acquisition of knowledge through observation and use of the senses.[88] But like Plato, Aristotle valorized and reveled in abstraction and theorization more than in empirical study. He does report the particular features of a variety of animals and other aspects of the cosmos in several of his works.[89] Yet he departs from studies of sensible things to engage in a formal systematization of knowledge that becomes so extensive that it seems to leave lived experience far behind. His heart seems to be in the abstract realm of thought.

Aristotle was endlessly concerned with universalizing and abstracting rather than with knowing particular animals, plants, and other real objects. A's and B's abound in his writing, as do statements such as "X is a man." He writes about "every man" and "every kind of pleasure."[90] Particular things, beyond particular thinkers, don't seem to attract his interest, as he's concerned with producing a systematized and thus necessarily abstract accounting of the phenomenal world. He often seems obsessed with classifying things—he only seems to want to know enough about a thing to place it into one of his categories. Once a thing is categorized, its qualities, especially those not used to assign it to a category, need no further consideration. As with Plato's Forms, Aristotle's abstraction and systematization seem motivated by a desire to wipe one's hands clean of the chaotic flux and complexity of the natural world.

Aristotle's approach to classification reduces the living, pulsing landscape to an "automobile-parts warehouse."[91] His taxonomy depends more on logical relationships among things—whether they all have fins or long tails—rather than on functional ones—whether they all graze on clover and therefore might be in competition for resources. His focus on abstraction and defining logical categories in a way aligned him with other ancient Greek thinkers such as the Pythagoreans, who not only sought to describe the world through mathematics but also considered the cosmos to be fundamentally mathematical in structure—making nature itself an abstraction. Not all ancient Greek writers neglected phenomenal observation. A century before Aristotle, the historian Herodotus described in vivid detail such things as the conflict between Greece and Persia and the places and people he learned about in his travels.

The scientist-historian Samuel Sambursky argues that Aristotle's dogmatic inclination for classifying at any cost petrified science.[92] His tendency to hyper-distinguish objects—to emphasize differences and ignore similarities—as he worked to arrange the whole world into categories probably influenced early modern projects to set humans apart from all other animals.[93] In truth, we resemble and overlap with other animals, particularly other great apes, far more than we differ from them. Only in the past few decades has modern science rediscovered the conclusion intuited by most indigenous, animist cultures: no single capability such as consciousness, reason, or upright walking separates people from the rest of nature. Gorillas, for instance, can learn sign language and then invent their own sentences, words, and signs, using them to converse with each other as well as with their trainers. Crows and other animals can make simple tools and solve problems. Examples of overlap between humankind and other animals are virtually endless.

It's easy to admire Aristotle's monumental contribution to the clarity and organization of thought in various fields from logic to physics to ethics. He's remained influential for over two millennia and has informed our modern approach to the world. But it's also important to recognize that Aristotle's methods may have led us to prefer abstractions over sensory perceptions of real things as they change and grow. His science seems oriented toward eventually obviating the need for such observations. And not only Aristotle's methods in science but also some of his contributions to logic urge us to disregard important relationships and contexts.

Dissociating Logic

Aristotle's formal logic continues to influence reasoning in the West and beyond. But his three logical principles—the law of identity, the law of noncontradiction, and the law of the excluded middle—create distorting divisions in the field of truth, as I discuss in chapter 4. And they represent only one way among many of introducing new knowledge: reasoning styles throughout much of the non-Western world contrast sharply with Aristotle's precepts. The law of identity, which says the universe is a natural order, advises us to conceive of things frozen in time, dissociated from the flux of the material world. The law of noncontradiction says a statement can't be both true and false and thus drives us to make clear-cut distinctions where none may exist. The law of the excluded middle, which says that every statement must be either true or false, denies the gray areas of truth in the messy realm of flux. It urges us to ignore processes by which things come into being, change, and decay. Our sensuous experiences reveal a full spectrum of truths. The sky is blue, but it's also painted a whole continuous palette of colors by some sunsets. By refusing the flux and continuity of the lived material world, Aristotle's logic divorces us from it.

Aristotle goes so far as to insist that the laws of noncontradiction and of excluded middle are axioms presupposed by all intelligible communication. As axioms, he tells us, they can't be proven, but anyone inclined to deny them may be refuted *ad hominem*—by attacking the person's character.[94] Yet Aristotle's de-contextualizing logical principles are challenged not only by non-Western logics, but also by more recent developments in formal logic by Western philosophers, including intuitionist logics, relevance logics, sociative logics, and "other systems that deny bivalence or the principle of non-contradiction or both."[95] Sociative logics demand, for instance, that there be some actual connection between the antecedent (the fact being assumed) and the consequent (the conclusion) other than the abstract rules of logic that bind them. That is, context and real-world situations must always be considered. These non-Aristotelian logics don't fragment our understandings of the world quite the way Aristotle's laws of logic do.

Oppositional Aristotle

Aristotle's thinking exhibits another tendency to divide: a strong emphasis on opposition. Opposition is a fundamental way of thinking about the world for Aristotle, who believed, "in a sense all thinkers posit contraries as principles, and with good reason."[96] In the *Physics*, Aristotle arranges various qualities of nature into pairs, such as Solid-Void, Rare-Dense, and Hot-Cold. The Pythagoreans also considered opposites essential in their natural philosophy, and in the *Metaphysics*, *Book I*, Aristotle builds on the Pythagorean table of opposites to arrive at his own set: Finite-Infinite, Odd-Even, One-Many, Right-Left, Male-Female, Resting-Moving, Straight-Curved, Light-Darkness, Good-Bad, Square-Rectangular.[97] The two elements of each pair possess mutually exclusive qualities, he writes, and all the elements on the same side share qualities, so Male, One, Light, and Good are closely associated.

Aristotle strains to place things into opposition. Actually, compared to the similarities between them, the differences between males and females are certainly few. Complementary would be a more apt understanding of their relation. He endows his worldview so heavily in oppositions that two pairs of opposites, hot-cold and dry-moist, in all their possible combinations, are the basis of the four fundamental elements that constitute the physical world.[98] Like Plato's dualisms, Aristotle's opposites dissociate because each side is fully contrary to and independent of the other, rather than admitting both differences and similarities, as well as complementarities.

Aristotle also minimizes relationships in his physics, preferring to define the qualities and movements of objects without reference to others. So an object is heavy in an absolute sense, not just compared with others. Aristotle's difficulty with relations appears clearly in his theory of motion. Things move due to their inherent qualities rather than to interactions. An object floats because it has the property of lightness. Why, you might ask, do large ships float? Today, science understands floating as an interaction of an object and a medium with different densities. But for Aristotle, things move because of their own "nature," such as fire with its upward locomotion.[99] In dissociating worldviews, relationships become subordinated to the innate attributes of objects, on which the focus remains.

With Aristotle relations take on a limited and almost unreal status. He writes, "We deny that a genus of relations exists by itself."[100] Physical outcomes in the world such as motion and position must therefore be explained merely by the attributes of objects. Strange explanations ensue. Consider how Aristotle sees a building: "What is heavy travels down by its nature and what is light travels up by its nature, and so the stones and the foundation are down, then earth right above because it is lighter, and finally wood at the very top since it is the lightest."[101] In another passage, you can tell that Aristotle's working hard to explain how inherent qualities of things, rather than relationships, cause their movements: "And air is actually light and will at once realize its proper activity as such unless something prevents it. The activity of lightness consists in the light thing being in a certain situation, namely high up ... the motion of light things and heavy things to their proper situations. ... The reason for it is that they have a natural tendency respectively towards a certain position and this constitutes the essence of lightness and heaviness, the former being determined by an upward, the latter by a downward tendency."[102] Rejecting that relations are true and real means Aristotle must awkwardly and forcefully insert the causes of motion into the objects themselves: "A motion is in a movable [object], for it is of the movable that it [motion] is the actuality, and it [motion] is caused by that which can move [the movable]."[103] Aristotle goes to great lengths to omit relationships, context, and the surrounding environment of events from his science.

Aristotle was not alone in overdrawing distinctions among things, focusing on oppositions, and placing all the forces that make objects move and change into the objects themselves. His Greek contemporaries felt compelled to create conceptual divisions and oppositions. The Greek contest system and other antagonistic, divisive aspects of Greek life seemed to carry over into various aspects of Greek worldviews. Separation and conflict were the order of the day. Lloyd writes, "In one instance after another, the converse of the Greek recognition of the potential hostility between pairs of opposites was a desire to separate them, even when their very opposition connects or joins them."[104]

Forcefully separating the two members of a pair lays the groundwork for a moral schema that elevates one over the other. Power was never far from the motivations driving Aristotle, Plato, and their colleagues. Aristotle aligns Male, Odd, Right, Light, and Straight with the Good and Female, Even, Left, Darkness, and Curved with the Bad.[105] Even when he recognizes that male and female must collaborate in procreation, he defines male as a capacity and female not by a complementary capacity but by an incapacity: "The male provides the moving cause and the form in generation . . . while the female provides merely the matter—and that is a mark of the greater 'divinity' of the male."[106] Women had few rights compared with citizens, who were men. Philosophies that divide, demote, and disempower reinforced the dominion of Athenian citizens.

We can credit Aristotle with acknowledging the reality of the sensible, phenomenal world around him that includes people in their bodies as well as meadows, ants, deer, sand, and rivers. That's more than Plato gave us. But his taxonomies and systematizations of knowledge about the natural world lead us to an outcome similar to the one we encounter with Plato's Forms—a turning away from lived experience toward the workings of the human mind. Products of mental labor may seem like the most real or relevant thing to someone who works little with soil, animals, and cloth. Aristotle tells us expressly that intellectual activity and contemplation (not examination, observation, experience, or relation, not caring and nurturing in the material world) are divine activities that provide the best path to happiness: "The activity of a god, then, which surpasses all other activities in blessedness, would be contemplative. Consequently, of human activities, too, that which is closest in kind to this would be the happiest."[107]

FROM THE ANCIENT WEST INTO MODERNITY

> Man's separateness from and superiority over nature
> suggest that he has a right to radically rearrange and
> to violently transform the natural environment and that
> he is disassociated from the harmful consequences of
> doing so, if any there be.
> —J. Baird Callicott, *Earth's Insights*, 1994††

Out of the divisive ancient Greek milieu of extreme competitiveness, social stratification, and intellectual tumult arose Plato and Aristotle, two men of powerful intellect who used their minds and the newly popularized technology of phonetic writing to try to create order in their world. Greek society and culture had refashioned persons as highly autonomous individuals endeavoring foremost to build their wealth and esteem at the expense of others, in contrast to most human cultures that have ever existed. The elite citizens of Athens denied their dependence on nature and those who worked with it: women, slaves, and other laborers. Their denial took perhaps its ultimate form in Plato's rejection of the rational mind's dependence on nature, the body, and all things that attend the body, including the emotions.

Plato achieved his peace by envisioning the rational mind, the province of the male citizen—and its productions, the Forms or Ideas—as the true reality. That's how he escaped the chaos and messiness of the ever-changing and perhaps ultimately unknowable phenomenal world. The stable and transcendent Forms also helped him avoid the instability of thought created by the ascendance of reason, which questioned everything, over tradition. Embracing the products of the human mind, Plato spurned nature and the "lower" orders of humanity who worked with it, justifying neglect or abuse of both.

Although his ideas were more grounded in the phenomenal world, divisions also permeate Aristotle's thinking. As his intellectual heirs, we may find it difficult to see how his science dissociates us from nature. But compared with the incessant observation and experimentation of the ancient Chinese, or against the depths of observation, respect, and engagement practiced by many indigenous people, Aristotle's science seems far

less concerned with actual experience. He sought intellectual stability and control through steadfast categorization and systemization that's no match for the dynamic chaos of the material world captured through the senses—the constant flowing and ebbing of a river or the fluid and intricate flight path of a sparrow.

The ideas of Plato, Aristotle, and other ancient Greeks live on today. Plato might be gratified to see that in industrialized societies, many workers—perhaps most—spend their days working not with crops or potter's clay but with the immaterial productions of the human mind: spreadsheets, sales numbers, accounts, software designs, mechanical engineering drawings, and management plans. Some do manual labor, but they work mainly with artifice shaped by human minds—cars, passenger jets, computers, office buildings, or lightbulbs—building, cleaning, or repairing these things. Few work with wild and productive nature. Plato also seems to have presaged the information age and the disembodiment of human relationships now lived through telephones, email, Facebook®, and Twitter®. Children text "no oatmeal" from their bedrooms to parents making breakfast in the kitchen.

Yet our liberation (or alienation) from our bodies and nature doesn't seem to have progressed enough for some people. Many technologists and futurists predict, and some such as the scientist-inventor Ray Kurzweil even anticipate, the day when people will be able to live without bodies (Plato's prisons). They welcome the Singularity, the moment when machine intelligence reaches and then quickly surpasses all human intelligence, reproducing and growing independently of us. According to Kurzweil, it's right around the corner: the year 2045.[108] Supposedly, soon after, humans will be able to upload their minds into computer hardware, where they can live indefinitely. Greg Egan's novel *Diaspora* can be read as a celebration of this union of humanity and machines. Techno-optimists seem untroubled by alternate scenarios like the dystopian world of *The Matrix*, in which the machines get the upper hand. Plato might approve of the merger.

MODERN TRACES OF DISSOCIATION

A fter a couple of centuries the classical Greek civilization of the Golden Age that Socrates, Plato, Aristotle, and their brethren knew had fallen. The first of the three periods of flourishing thought about nature in the Western world was over, though its ideas would not be forgotten. We now turn to the other two periods: the early modern period of the Renaissance and the Scientific Revolution and the later modern period of ecological theory, a new philosophy of organism, and twentieth-century physics. During these periods, as in Greece in the classical period, philosophers and scientists subjected the idea of nature to "intense and protracted reflection," virtually reinventing it.[1] The new thinking about nature during the European Scientific Revolution was particularly transformative.

The sixteenth and seventeenth centuries in Europe saw remarkable intellectual and scientific achievements. In the previous century, a period of intense artistic, engineering, and cultural activity amounted to a rebirth of European culture and society, the Renaissance, driven in part by a resurrection and celebration of ancient Greek ideas. Many thinkers made major contributions to science and philosophy that culminated in a new, modern worldview, including the Italian philosopher Bernardino Telesio (1509–1588); the Prussian astronomer Nicolaus Copernicus (1473–1543); the Italian philosopher, mathematician, and astronomer Giordano Bruno (1548–1600); the English philosopher, statesman, and scientist Sir Francis Bacon (1561–1626); the English philosopher Thomas Hobbes (1588–1679); the English physicist, mathematician, and natural philosopher Sir Isaac Newton (1642–1727); the French philosopher and mathematician René Descartes (1596–1650); the German astronomer Johannes Kepler (1571–1630); the Italian physicist, mathematician, and astronomer Galileo Galilei (1564–1642); and the English philosopher

John Locke (1632–1704). Their ideas dramatically reshaped human-environment relationships.

Rapid and profound scientific developments, including the transition to a heliocentric (Sun-centered) cosmological view and a machine-like conception of nature, took hold in the Western imagination and resulted in the expression of vast human power over the natural world. Nature went from a living organism that partnered with humans in their activities to a dead and inert realm awaiting our manipulation. These protagonists and the changes they effected gave birth to the current historical era—the modern—a time of unprecedented development of human technologies, population growth, ideas, and global interconnection—and environmental change many times more intense and rapid than anything seen before. The widespread adoption of dissociating thought and practices also distinguishes our era.

EARLY MODERN DISSOCIATED THOUGHT

Early modern thinkers reacted strongly against the teleological (goal-oriented) view of nature formulated by Aristotle and advanced by medieval Christian scholars. In the teleological perspective, nature is permeated by a tendency to create or realize certain forms. These forms don't actually exist in the physical world—they're more like ideals that nature strives toward. Aristotle named this view the theory of final causes. It explains movement as a "natural tendency" toward a particular effect. Water has a natural tendency to run downhill, for example, and the sun has a natural tendency to be round. Francis Bacon and the other first modern scientists rebelled against this doctrine. They wanted to discover the details and characteristics of the phenomenon in question rather than just say that things simply tend to achieve some predestined form. They turned their attention to the physical conditions already existing in the world, such as two billiard balls heading toward each other, that could lead to a particular result. These existing conditions are called "efficient causes."

The newly emerging theory of nature in the sixteenth and seventeenth centuries insisted on explanations based on efficient causes. It assumed that all change and process could be explained by the action of material things

existing at the onset of the change. You may recognize the theory of efficient causes from high school or college physics laboratories, where you
record precisely the starting conditions of an experiment, for instance, the
height of a ball and the angle of the inclined plane at the top of which it
initially rests, then the final conditions, for instance, the distance the ball
travels before it stops. You might then check your results against those predicted by Newton's formulas of mechanics. The small difference, you were
probably reminded, resulted from the fact that the real world has friction,
whereas Newton's laws of motion pertain to the ideal-world situation of
objects isolated in a frictionless void—objects dissociated.

Rejecting Aristotle's teleology, the seventeenth-century philosopher-
scientists aligned themselves with Plato even though Aristotle's philosophy dealt more extensively and explicitly with natural science. They took
up Plato's Pythagorean cosmology, which suggests that mathematical
structures, like Newton's equations of mechanics, represent the true nature
of things. For the early modern philosophers, change was the outcome of
these structures rather than simply the expression of a tendency.[2] The idea
that numbers and formulas underlie nature and cause all change in the
physical world prepared the way for nature to be thought of as a machine.

A Machine World

> The order physics has been using is the order of
> separation.
> —David Bohm, "Postmodern Science and a
> Postmodern World," 1988*

In Europe, prior to the emergence of modern science, people generally
considered nature a living organism whose immanent (actual and present,
dwelling with) energies and forces were vital and psychical in character.
Nature was endowed with agentic qualities—it was an aware and self-
directing force possessing reason and sense and experiencing love and hate,
pleasure and pain. These faculties and passions were the causes of natural
processes. People of the Middle Ages called this alive, active, self-producing
conception of nature *natura naturans* (nature naturing, or nature creating
itself). *Natura naturans* was the immanent force that animates and directs

the "complex of natural changes and processes"—the actual, perceivable manifestation of nature in the physical world. The latter was called *natura naturata* (nature natured, or nature created).[3] In the modern period the passive form of nature, *natura naturata*, eventually replaced the active form, *natura naturans*, which had infused Greek thought and held sway through to the Renaissance.

The passive and mathematical views of nature gradually submerged the organism model and, together with the rise of machines in people's experiences, inspired a new, mechanical metaphor to replace it. Early modern scientists began to think of nature as a machine made of interchangeable parts, a view that eventually dominated modern cosmology. Gone was the living, vibrant organism that demanded respect and care. Conceiving nature as a machine widened the gap between it and humanity—it's harder to relate to a device than an organism, to respect it, for instance. Carolyn Merchant, the environmental historian, called this shift to a mechanical model "the death of nature."[4]

The Catholic Church also played a role in distancing early modern Europeans from nature. Early in the modern era, God was the divine creator and ruler of nature and, as a watchmaker to a watch, came to exist outside of his own creation—he was not immanent in nature. This early modern view of God as clockmaker and the cosmos as clock built on the Christian idea of a creative and omnipotent God. It also reflected the increased experience of early modern Europeans in designing and constructing machines. By the fifteenth and sixteenth centuries, machines came to inhabit more widely the everyday lives of common people.[5] God the clockmaker and, later, nature as a machine were the products of the changing experiences of European clergy and intellectuals. The removal of God from nature and the new relationship of God to his mechanical creation further alienated people from nature. Samuel Sambursky writes, "By divorcing man and his vital interests from natural phenomena the Church helped to create the feeling that the cosmos was something alien and remote from man. It was this feeling that prepared men's minds for the next stage in which the investigator faced nature as its dissector and conqueror and thereby ushered in our own scientific era which still, after four centuries, retains its vigour undimmed."[6]

Before nature became a machine, people conceived of it in terms of astrology and magic. They approached the natural world as a soul engaging another soul. In reconceiving nature as a machine, Europeans claimed mastery over it. The human mind became the master of lifeless, mechanical nature.[7] Invoking exhaustive accounts in texts and artwork from the period in *The Death of Nature*, Merchant reveals the will to power and control that motivated philosophers and scientists such as Bacon, Descartes, and Hobbes to replace the organic worldview of the cosmos with a mechanistic one. Nature became passive material awaiting and even inviting manipulation by "man."[8]

Gender played an important role in the re-visioning of nature. Bacon and others cast nature in female terms and used imagery of sexual conquest, perhaps forceful, to encourage the unconstrained application of science for human ends, including pleasure. They didn't seek to "exert a gentle guidance over nature's course" but rather to "conquer and subdue her." Bacon urges us to "bind her to your service and make her your slave." Nature, he says, "betrays her secrets more fully when in the grip and under the pressure of art [technology] than when in enjoyment of her natural liberty."[9] Early modern philosophers drove an antagonistic wedge between people and nature. Nature envisioned as dead, passive object allows exploitation; nature imagined as female invites domination.

No Center, No Body

Several developments in cosmology and natural philosophy paved the way for the mechanistic worldview. Copernicus's study of the solar system, *De Revolutionibus Orbium Coelestium* (*On the Revolutions of the Celestial Spheres*), published in 1543, removed Earth from the center of the universe, replacing the geocentric (Earth-centered) model with a heliocentric (Sun-centered) one. This shift diminished the importance of Earth and its residents, who were no longer physically—and thus morally—at the center of everything. But it also sent a subtler message: the universe has no center at all. The Sun became the center of the solar system, around which the planets revolved, but not necessarily the center of the universe. Now any point could be regarded as a center.[10]

The newly decentered cosmos could no longer be an organism, as Robin George Collingwood notes. The idea that the entire world is an organism implies differentiated organs, so one point in space must be inherently different from all others. With the new decentered cosmos, the whole world was made of the same kind of mundane matter. Gravitation applied everywhere, and the stars became homogenous with Earth rather than possessing a divine substance of their own. Although removing Earth and humanity from the center of the universe humbled humans, in one sense it also made them more powerful. Now universally applicable laws and a science to discover them became possible. Thus, Newton could relate the gravitational force holding the moon in orbit to the one that draws a terrestrial body, say, an apple, to the ground.[11]

The Italian friar Giordano Bruno extended from the solar system to all the stars this notion that there's no difference between terrestrial and celestial substances. Arguing that the world isn't divine but rather mechanical, he completed the machine-like view of nature by extending it throughout the cosmos. The Sun is merely a star. There's no transcendent God who designed and constructed the world, Bruno thought, and God isn't immanent in all creation. This dispiriting of nature paralleled the decline of the organic worldview. "The identification of nature with God," writes Collingwood, "breaks down exactly when the organic view of nature disappears."[12] Both God and life were removed from nature, leaving behind mere substance, to which humans related very differently. Ejecting the divine from nature cost Bruno his life: he was burned at the stake by the Roman Inquisition.

The English philosopher Thomas Hobbes also extended the mechanical metaphor, this time to human society. Just as the desire to bring order to chaotic nature had inspired the mechanistic view of nature, the desire to maintain order in potentially chaotic human society motivated Hobbes. He imagined people as interchangeable, replaceable parts, each fulfilling a specialized role subservient to the overall function of the machine. Today's highly specialized workforce matches Hobbes's ideal far more than that of seventeenth-century England. The mechanistic model makes it easier to divide society into classes according to economic function—loggers, farmworkers, butchers on the kill floor, supervisors, agribusiness executives. That, in turn, facilitates subordinating entire classes of people who work more closely with nature.

Mere Appearances

Johannes Kepler, the German astronomer and mathematician, also contributed to the new mechanistic worldview. The organic view of nature had rested on the notion of an internal vital energy that animates nature and drives change in the natural world. Kepler rejected this ancient idea. He said that quantitative mechanical energy—not qualitative vital energy—drives nature, causing measurable, quantitative change. Galileo, the "father of modern science," conclusively established the new conception by restating the Pythagorean-Platonic standpoint of transcendence. The universe, he famously declared, is a book written by God in the language of mathematics. It's open before our eyes, but it can't be read until we've learned the language of mathematics in which it's written, with letters of triangles, circles, and other geometrical figures, without which it's impossible for humans to comprehend.[13]

Galileo followed Plato's theory of transcendence by believing that the only real and comprehensible aspects of nature are quantities: the radius of a circle or the mass of an apple. Qualities and qualitative distinctions such as those between colors or sounds—things that actually involve human bodies in acts of perception—are simply the effects of objects in the physical world acting on our sense organs. They're secondary and don't actually exist in the world—they're mere appearances. The real, quantitative world acquires its qualitative appearances only by the intrusion into it of living and sensible beings. According to this line of reasoning, both God and humans transcend nature. The human mind stands apart from the material world, conferring on it its qualitative aspects.[14]

How bizarre it is to contemplate that colors and other qualities that we access through our senses are unreal and are merely side effects of what truly exists: quantities, numbers, and geometrical figures. It's stranger still to envision the entire natural world as a vast book written in the language of mathematics. Imagine yourself in a place of wild, pulsating, vibrant nature. Perhaps you're sitting near a gurgling stream in a forest. Insects skim across the water, tree branches sway with the breeze, sunlight filters through the leaves, small plumes of algae flow with the current, the scent of pine needles wafts in the air, and clouds form and re-form, drifting across the sky. It's all unreal, a deceptive illusion, according to Plato and these early modern philosopher-scientists. Mathematical structures underlying it all are the true reality, producing all the detailed nuance and flux that you perceive.

In the excitement of their stunning cosmological discoveries, these early modern thinkers, brilliant as they were, seem to have confused the tool they used to study nature and document change in the natural world—mathematics—with nature itself. Edward T. Hall called this phenomenon "extension transference": people confusing human creations or mental models with what they're created to represent.[15] The extension in this case is mathematics, and it's being transferred onto nature as a whole. In this extension transference, Galileo and those following suit subordinated the human body and its senses and dissociated them from nature by inserting the tools of science between the two. Today, their program moves ever closer to completion. Seldom are human senses used directly to know nature in ways vital to science and the economy. Microscopes, electronic sensors, and other devices instead tell us what's out there.

Knowing the Machine

The new assumptions about the basic makeup of the world (ontology) introduced with mechanism were matched by new assumptions about how we can know anything about it (epistemology). The new epistemology further divided humanity from nature. Merchant summarizes the mechanistic assumptions of the period thusly:

1. Matter is composed of particles.
2. The universe is a natural order.
3. Knowledge and information can be abstracted from the natural world.
4. Problems can be analyzed into parts that can be manipulated by mathematics.
5. Sense data are discrete.[16]

The first two are ontological assumptions. The first says that the universe is atomic—all matter is composed of particles. Leucippus and Democritus introduced the atomic theory in the fifth century BCE. Newton and other modern scientists institutionalized it in the seventeenth century CE and later. It dissociates by eradicating interconnection from ontology. The second assumption implies the law of identity, which dissociates as I describe in the previous chapter.

Number three is the assumption of context independence: valid and proper knowledge can be produced by considering events and things out of the contexts in which they occur. It says we can study how things behave in a universal sense, by considering their innate qualities, not the situations they're found in, not the relations they're embedded within.

Assumption four says all problems can be broken down into parts that can then be analyzed separately. Assembling the results from the separate parts produces a complete and valid conclusion. Parts of a problem can thus be considered without concern for the rest of the problem. Taking assumption four seriously can lead to strange conclusions. We might be able to ignore, for instance, the notion of an environmental crisis or the idea that there's some type of problem with the human-nature relationship. We just have separate environmental problems to solve.

The fifth assumption implies that even at the fundamental level of perception and sensation, nature comes to us in discrete quantities. There's no continuum of colors in the sky, for instance, only a series of discrete colors. It almost seems as though early modern scientists dreamt of computers and digital technology, where data are indeed all discrete. Perhaps we invented computers to fulfill their dream.

These dissociating assumptions have been shaping how Westerners create new knowledge about the natural world for centuries. It's the norm for scientists to claim that new substances or technologies are safe based entirely on observations in the laboratory, a place and situation dissociated from context by definition. Scientists in agricultural biotechnology, for instance, claim that their genetically modified crops are safe after testing under controlled laboratory conditions. But no lab can reproduce the overwhelming complexity, chaos, and unpredictability of actual ecological and social environments. In the early 2000s, scandal erupted in the field of agricultural biotechnology over reports that transgenes (genes transplanted from one species to another using genetic engineering techniques in the laboratory) from genetically modified maize ("corn" in the United States) made their way into local maize "landraces" (wild ancestors of maize) in the center of maize diversity near Oaxaca, Mexico. Breeders use landraces as a kind of genetic stockpile, so their possible contamination by transgenes was alarming and certainly unintended.[17] Laboratory experiments, dissociated as they are from ecological and social contexts, could not have stopped the contamination or predicted its course.

No Human, No Mind

Descartes goes further than Plato in dividing mind from body. He says that not just rationality but consciousness itself sets the human mind apart from all else. There's no consciousness anywhere else in nature—not in dogs, not in lions. Everything besides God and the human mind is machine, even the human body: "So, also, the human body may be considered a machine, so built and composed of bones, nerves, muscles, veins, blood, and skin that even if there were no mind in it, it would not cease to move."[18] The mind is separate from the body, which is "only a composite," produced by "accidents." In contrast, the human soul isn't dependent on any accidents but is "pure substance." Mind is immortal.[19]

Severing the mind from the body opened up a new possibility for understanding human identity as independent of the body and nature. Descartes wrote, "I have a distinct idea of body in so far as it is only an extended being which does not think . . . this 'I' is entirely distinct from my body and . . . it can exist without it."[20] The body is almost an artifact for Descartes: "Absolutely nothing else belongs necessarily to my nature or essence except that I am a thinking being."[21] The thinking mind becomes the basis for existence itself, hence Descartes's famous dictum "*Cogito ergo sum*" (I think, therefore I am). Val Plumwood calls this the "Cartesian birth of alienated identity."[22] The body becomes irrelevant to identity and to knowing the world. The mind alone, not the whole being composed of mind and body, determines truths.[23] Descartes thus deepened the divorce of humans from nature.

The extreme division between mind and body created a problem in Descartes's theory: what to do with the senses, which seem to operate somewhere between mind and body, connecting the two. Descartes solved this difficulty by rendering the senses into two separate categories, one associated with the body, the other with the mind. This division mimics the one that Galileo, following Plato, constructed between the quantitative, measurable aspects of the world, which are real, and the qualitative distinctions, which are mere appearances. Descartes concluded that the bodily aspect of sensation is the impression that objects create on a sense organ such as the eye. The mental aspect was the mind's awareness or contemplation of this sensation, which Descartes, true to the mind-over-body formula, called "sensation proper."[24] This delicate work of division between the mind and the body reveals the pains that Descartes took to cleanly excise mind from body so that body could be conceived as a machine, along with the rest of nature.

According to the late-twentieth-century philosopher and ecological psychologist Edward S. Reed (1954–1997), the Cartesian two-stage theory of perception outlined here led to "the complete undermining of primary experience" in the Western world.[25] Primary experiences give us direct sensory information about our artificial, natural, and hybrid environments. That's what ecological psychologists call *ecological* information. When Descartes elevated the mind over the body and its senses, he demoted ecological information. The senses could not produce certainty, Descartes believed, and he seemed to dread uncertainty above all else.

The Cartesian subordination of experience can be seen in our everyday modern lives. Secondhand experience, the acquisition of processed information about the world through telecommunications, printed matter, and machines, is displacing primary experience. Today, we engage mainly with secondhand information.[26] Processed information isn't intrinsically bad, but to be well grounded its meaning must be based on primary experiences. Ecological information is particularly important in personal interactions because we're highly perceptive observers of other people, capturing subtle changes in facial expression and so on, and these cues matter.[27] Similarly, ecological information about the natural world allows us to understand more fully what nature can tell us—and empathize.

The long-held Western disdain for primary experience has consequences Descartes probably never imagined. As our lives increasingly fill with secondhand experience, we gradually become more like automata, writes Reed. As we commute to work by car, other people become "metal-shelled" obstacles to our progress.[28] The 1992 "non-narrative" film by Ron Fricke, *Baraka*, strikingly compares the bustle of people on subways, escalators, and city streets with the workings of industrial machines—including some that process baby chicks by the millions, tossing them down chutes as if they were unconscious parts of Descartes's machine. The mechanization of our everyday lives diminishes encounters with others. The result, says Reed, is "widespread conditions of confinement, meaninglessness, and empty regularity in our lives."[29] More primary experience—more direct, sense-based engagement—with nature and people can make our lives more meaningful.

On the other hand, the loss of primary experience, combined with the understanding of nature as a machine and the assumption that only the human mind possesses consciousness, makes communication and empathy between humans and nature impossible. That's why Descartes's followers could ignore the screams of the animals they were dissecting alive. They're soulless machines incapable of feelings, according to Descartes. Fortunately, we haven't completely expelled the non-modern intuition that the rest of nature is sentient and alive, so most of us can still sympathize with animals.

Yet the Cartesian idea that animals are not conscious perpetuates the abject misery of factory-farmed animals in the United States and elsewhere. Treated like soulless machines, they're corralled into huge lots without grass, fed a diet that makes them sick or grow so fast that their legs collapse under the weight, de-beaked or castrated without anesthesia, pumped with antibiotics, and slaughtered on immense, blazing-fast *dis*-assembly lines. Weak or sick animals are sometimes culled with cruel methods. Few people ever see these actions carried out on their behalf.[30]

The political scientist Timothy Pachirat did, though. He studied an American industrial slaughterhouse firsthand by working there. Out of the eight hundred workers in the plant, most of whom work dirty, hard, dangerous jobs, only eight see the cattle alive. Zones in industrialized killing centers "segregate the work of killing not only from the ordinary members of society" but also from most employees, who are excluded from "the most explicitly violent site of all: the kill floor."[31] As each animal enters, an operator on the kill floor aims an air-powered gun to its forehead and shoots a metal rod into the animal's brain. It doesn't always work on the first try, so the animals sometimes thrash about while the worker tries again. The moment of transition from life to death isn't always clear on these gruesome lines.[32] But what does that matter if the animals have no consciousness?

Cleaving the World

With their dissociating ideas the early modern philosophers extended the dualisms inherited from the ancient Greeks. It's important to understand how these constructs work because they're the intellectual basis of the destructive separations we experience in everyday life. Dualisms differ from simple distinctions or dichotomies: they exaggerate the differences between the members of a pair and understate the similarities. A dualism is a dichotomy gone awry and imbued with power. One member of the pair eclipses the other. In her book *Feminism and the Mastery of Nature*, Plumwood systematically exposes their structure and their role in important relationships.[33]

Plumwood found the archetypal dualism in Plato's Allegory of the Cave. The Cave resembles a womb as Plato describes it, engulfing and mysterious.[34] The "great task" for "man" is to emerge out of the Cave into the blinding light of Reason and *logos*—the sublime, eternal, and incorruptible world of the Forms. The Cave is full of darkness and illusions; it's a world of mere Appearances (sensory knowledge). When you abandon it, you cast off the "nature within" by rejecting and separating from "the lower order, which includes the mother, primal matter, the earth, and all that is conceived as belonging to it."[35] Only then—when you actively disparage and disavow this inferior realm of material nature—do you enter true selfhood, in a kind of intellectual birth.

The allegory provides the foundation for Plato's mind-body and reason-body dualisms, which have influenced Western philosophy for more than two millennia. The two pairs germinated into a whole "set of interrelated and mutually reinforcing dualisms which permeate Western culture and form a fault line which runs through its entire conceptual system," says Plumwood. Some key dualisms are

culture/nature
reason/nature
male/female
mind/body
mind/nature
master/slave

reason/emotion
reason/matter
rationality/animality
universal/particular
production/reproduction
human/nature
subject/object
self/other[36]

In each of these pairs Western mainstream philosophy elevates the member on the left and diminishes the one on the right. It also aligns all the members on the left with one another and all the members on the right with one another. Thus, reason and culture, thought Plato and other Western philosophers, belong to the male. Nature and reproduction align with the female, all parts of the lower sphere of existence. Plato said male reason should concern itself with the Forms, which are universal. Women and slaves in his society did much of the work with nature, providing for others and allowing men to engage in more mental labor.

Plumwood finds that the most potent streams of Western thought, the "Platonic, Aristotelian, Christian-rationalist, and Cartesian-rationalist traditions," conceive a profound "hyper-separation" of humans from nature. The mind-body and reason-nature dualisms lie at the center of the divide. She writes,

> The natural world is *homogenized* and *defined negatively and in relation* to humans as "the environment." . . . The body and the passions belong in Plato's account to a sharply distinct lower realm, homogenized and defined by exclusion, to be dominated and controlled by superior reason, and to be used in its service. . . . What is virtuous in the human is taken to be what maximizes human differentia, and hence what minimizes links to nature and the animal. . . . Discontinuity is obtained via an account of human identity and virtue which eliminates overlap with the "animal within," or polarizes this as not truly part of the self or as belonging to a lower, baser "animal" part of self. The human species is thus defined out of nature, and nature is conceived as so alien to humans that they can "establish no moral communion" with it.[37]

In other words, qualities imagined to distinguish humans from the rest of nature, such as reason, gain the spotlight and are deemed the greatest and most relevant ones. Features we may have in common with the rest of nature are ignored or downplayed. People thus become separate from nature *by definition*. Whereas Plato emphasized mainly reason-body, reason-emotion, and universal-particular distinctions, Descartes favored divisions of mind-body, subject-object, human-nature, and human-animal.[38]

These dualisms are infused with power. Plumwood points to the master-slave pair as the central organizing principle of them all. It arose, perhaps, from Plato's experience as the master of slaves (and of women and his household). The very structure of dualisms is founded on an *instrumental* power over the subordinated, disempowered others on the right-hand side of each pairing. That is, the master and those aligned with him on the left-hand side of the fault line use those on the right-hand side as instruments of their will—tools to satisfy their needs and desires. Slaves and rivers are merely devices for economic gain. Western humans divorced themselves from nature with domination in mind.

The quest for power propels overwrought distinctions between the self and others. Separated from the whole material world, the human mind becomes free to manipulate, control, and abuse it because all moral value shifts to the human side of the divide. "Nature in the West," writes Plumwood, "is *instrumentalised* as a mere means to human ends via the application of a moral dualism that treats humans as the only proper objects of moral consideration and defines 'the rest' as part of the sphere of expediency."[39] The conceptual divide opens the space for a moral schism that permits us to treat nature as passive tool rather than active partner. Plumwood writes of the "Dream of Power" that motivated Descartes and other Enlightenment thinkers: "In the Cartesian dream of power, the subject is set over against the object it knows, in a relation of alleged neutrality in practice modeled as power and control. We are yet to awaken from this dream, which has formed modern conceptions of scientific knowledge and rationality, human, as well as nonhuman identities."[40] Awakening from the dream means seeing nature as a realm worthy of respect and care.

Power-laden dualisms also legitimated and paved the way for colonization, the mainstay of modern history. From the fifteenth through the twentieth centuries, European nations colonized humanity and nature throughout much of the globe, taking precious metals, crops, indigenous knowledge, workers, and slaves to expand their economies. To justify their often violent actions, they imagined non-European people as distinctly and inherently different from themselves, creating colonizer-colonized and master-slave dualisms that presupposed separate, discontinuous orders of existence.[41] So Thomas Jefferson, a slave-owning founding father of the United States, could write that blacks are "much inferior" to whites in reason and other faculties, suggesting even that blacks are inferior overall in "the endowments of both body and mind."[42] Relegated to a lower order, non-Europeans and their nature became available for virtually limitless exploitation. Nineteenth-century residents of India died by the millions of starvation and related diseases while huge grain exports made their way to England.[43] Besides skin color, colonial apologists used traits associated with apes to distinguish themselves from colonial subjects and to align the latter with the lower realm of nature. In the nineteenth century the English even used such techniques against the white residents of their oldest and first colony, Ireland, reimagining them as apelike and barbaric.[44]

A Violent Structure

When does a distinction turn into a dualism? It's helpful to understand how dualisms are made so we can avoid these destructive constructions. Plumwood developed a framework that reveals their "logical structure," which consists of five ways of mentally splitting things apart. The master-slave dualism is the archetype. In *Backgrounding*, or denial, the master denies his dependency on subordinated others in various ways, such as refuting the importance of the other's contribution and conceiving the other as expendable or insignificant. Plato, for instance, doesn't acknowledge his slaves' contribution to his life or to his ability to spend time philosophizing. *Radical exclusion*, or hyper-separation, means emphasizing differences and downplaying similarities. Colonists even considered the dark-skinned people they colonized as subhuman: wild, savage, uncivilized, and still part of nature. Plato similarly set male citizens apart from others in his society with his concept of reason.

Incorporation, or relational definition, defines the subordinate member of the pair with respect to the dominant member but missing some important qualities. Plato believed that women were originally created out of men, minus certain capacities, such as bravery and rationality.[45] *Instrumentalism*, or objectification, conceives a slave or other object in relation to the needs of the master or self. A slave has no needs of her own and is little more than the one who does the work. Ancient Athenians even considered slaves to be nonpersons or simply tools, which made them look even more irrational in the master's eyes when their actions didn't accord well with his.[46] The master uses *Homogenization*, or stereotyping, to minimize or eliminate differences among subordinated others so they don't have to be treated as whole persons with particular personalities, needs, and desires. The logic of homogenization makes them appear uniformly inferior and thus interchangeable, like parts of a machine.[47]

These dualism-creating techniques appear repeatedly in the thought of famous Western thinkers. Bacon instrumentalized nature, saying it must be brought into the service of "man," made a slave. Descartes radically excluded consciousness from the entire rest of nature besides humankind. Both denied human dependence on nature, assuming men could master her. Bacon homogenized nonhuman animals also by conceiving them all, from ant to chimpanzee, as one sort: machines that lack consciousness. When Descartes proposed the *über*-dualism of the human mind, the one true substance, against the whole cosmos, he deployed several of these techniques. He radically excluded mind from nature, denied that the human mind depends on nature for its existence (thinking, as did Plato, that it can live without a body), and homogenized nature as a mechanical realm made of a single substance.

The logic of dualism diminishes the lived aspects of relationships, and the dualisms running through Western thought continue to shape our thinking and action. Radically excluded and instrumentalized, nature is no longer a partner we wish to relate to in a rich, subject-to-subject way. Following Descartes's lead, we somehow hover above nature, which is made into our slave. So animals live like machines in Concentrated Animal Feeding Operations (CAFOs) in the United States, awaiting their turn through the *dis*-assembly line. But the tendency to dissociate important things has been neither absolute nor uniform in the modern world. Important countercurrents have arisen.

REASSOCIATING MODERN THOUGHT

In the nineteenth and early twentieth centuries, significant new ideas emerged in modern cosmology and natural science that began to counteract some of the dissociating aspects of Western natural philosophy. The German philosopher Georg W. F. Hegel and the British philosopher Alfred North Whitehead (1861–1947), among others, began to reintegrate the cosmos into a seamless whole that subsumed humanity and all of nature into a flux. Hegel proposed an organic, anti-mechanistic *Naturphilosophie* that regarded the flux of nature as real. It probably inspired Whitehead's "philosophy of organism," which is permeated with interconnection.[48]

Whitehead's cosmology centers on a concept he calls the *occasion*, a confluence of space and time in a unique entity or moment bound inseparably together with all others past, present, and future. "Actual occasions," also called "actual entities," are the final facts of the universe; they're "drops of experience, complex and interdependent."[49] All occasions together form a continuously extensive world that appears as an organism due to its interconnection and coherence.[50] Each occasion individually connects to all other occasions in the universe and through history, creating a fully interwoven fabric of time and space. Reality is flux and process. The cosmic organism can't be equated with the sum of its parts, and its parts can't be conceived as matter in isolation. Rather, occasions comprise a unity of continuous time and continuous space. In Whitehead's epistemology, an occasion can't be alienated from its space-time continuum to study it in isolation, as one would do using the analytical approach. Interconnection remains an essential aspect of each occasion.

It's difficult to imagine a more integrated—less dissociated— cosmology. Whitehead's reconnection of space and time into an inseparable whole creates a new level of connection absent in prior modern Western worldviews—though common in non-Western ones. Nature is an unstoppable, indivisible dynamic flux.[51] The concept of a unique, instantaneous state of nature, as in a snapshot, doesn't exist. By envisioning a universe infused with connection and change, Whitehead's cosmology runs counter to Plato's static and eternal Forms. It rejoins mind and nature out of their Platonic and Cartesian alienation.[52]

The new science of biology in the nineteenth century also bucked against dissociating, mechanistic modern science. It elevated living things to a higher plane of morality by creating a separate science for them. Descartes, thinking the world mechanical, had tried to explain biological facts in terms of physics (synonymous with mechanics at the time).[53] Charles Darwin (1809–1882), Alfred Russel Wallace (1823–1913), and other biologists set aside mechanistic physics to study the dynamics of life directly. The French philosopher Henri Bergson (1859–1941) even hoped to invert Cartesianism by proposing a new cosmology that attempted to explain the whole physical cosmos in terms of life. Bergson saw matter as a by-product of life but failed to explain how life could come before matter.[54] The life sciences of the late nineteenth and early twentieth centuries— particularly ecology, a science fundamentally about interconnection— provided the larger intellectual context in which Whitehead could reassociate the cosmos. These threads nourished one another in asserting a new connectedness in the world.

The new late-modern cosmologies came with some familiar, unwanted baggage, however. Both Hegel and Whitehead allowed for some type of eternal forms in their theories, immaterial essences resembling Plato's Forms. They seem to be split-off relics of the human mind that could not be fully accounted for elsewhere in the new cosmologies. But from the perspective of the West's intellectual past, it's less surprising that Whitehead retained this semblance of a dualism between mind and nature, if that's what it is. Western philosophers have mostly been loath to conceive mind as completely subsumed within and pervading nature.

From Plato's dissociation of reason from body, through Descartes's mind-body dualism and Christian theologies that conceive mind as the product of divine intervention, and Whitehead's placement of forms outside of nature, directing its process, Western cosmologies have consistently refused to allow nature to incorporate the human mind. The possibility that forms are artificial mental abstractions—highly reduced, imperfect, incomplete, power-laden, and simplified renditions of reality created through human experiences that differ across cultures in a phenomenal world that is far too complex to be grasped directly—this possibility, which humbles the human mind, seems to continually evade Western cosmological thought.

Although these late-modern cosmologies retain dissociating characteristics, they would truly advance Western thinking toward a more associating, connecting worldview if they were ever broadly adopted. Alas, except for biology and ecology, they remain outside the mainstream of science and popular understandings of nature. Whitehead's cosmology has not shifted school curricula much. The conceptions of nature espoused by scientists and philosophers of the Scientific Revolution remain the predominant organizing principles of our thinking on nature today.[55] Popular understandings of nature follow more closely the notion of a passive realm ready for exploitation by humans than do Whitehead's idea of a total, interconnected organism. Whitehead might approve of the *Gaia* hypothesis introduced by the scientists James Lovelock (b. 1919) and Lynn Margulis (1938–2011), in which the whole Earth resembles a self-replicating, complex organism.[56] But *Gaia* has yet to make it into the mainstream of ecological, biological, and cosmological thought.

Whitehead's conception of an interconnected, organic reality may make intuitive sense, but modern institutions continue to operate according to the mechanical cosmology, which is attuned to practical utility—to exercising power. They treat nature as a collection of machine parts, as when a logging company clear-cuts a forest, breaching its interconnections; or an agribusiness fumigates a field with methyl bromide, killing the communities of microorganisms that make the soil fertile and productive. As Collingwood writes, the "tendency of all modern science of nature is to resolve substance into function"—we care about what works more than what is.[57] Nature usually takes on a quality of wholeness only when we experience it aesthetically, when we approach it in our leisure time. But then its productive workings remain hidden and unknown, and the wholeness we sense remains superficial and illusory. I may visit a wild area and see it as a seamless, vibrant flow of organisms and materials all interconnected and interdependent, but then I return home and touch the natural world through economic and other institutions that so far refuse to see nature that way.

Modern science and technology continue to be guided by the instrumental drive of early modern cosmology. One of the newest sciences, biotechnology, is founded on mechanistic molecular biology, which treats genes like machine parts. Genetic engineers developing new crops move genes between unrelated organisms in the friction-free biological oasis of the laboratory. Genetically modified organisms (GMOs) developed in labs operated by or funded by Monsanto and other agribusinesses, such as the transgenic maize that contaminated maize landraces in Mexico, are planted over vast acreages across the American heartland and elsewhere with little prior testing for human and environmental health. Agribusinesses gloss over the differences between the controlled environments of laboratories and the uncontrollable, complex ecological relationships of the real world with sales pitches about "feeding the world" and other promises. They engineer popular GMO varieties to be resistant to herbicides (that they also sell), which farmers then apply more liberally. The resulting overuse of pesticides has already created resistant "superweeds" on many farms in the United States. Applying biocides heavily in an ecosystem causes weedy species to evolve new varieties that resist them, much as overuse of antibiotics causes microorganisms to evolve resistant strains. Some ecologists are also concerned that by transferring herbicidal resistance to weeds, "genetic drift" from transgenic organisms to other species in the field may create new weeds that are extremely difficult to control.[58] These are the problems that result from ignoring connections and context.

The reductionist and mechanistic model of molecular biology echoes early modern science in another way. Advocates for the human genome project popularly referred to the genome as the "book of life." The project worked to decode the book for the good of humanity—and many life-saving discoveries have already been made. But the idea is eerily similar to Galileo's assertion that the universe is a book written in the language of mathematics. In the case of the universe, the language is mathematical symbols including numbers and geometrical figures. In the case of the human genome, the language consists of strings of the letters A, C, G, and T representing the four nucleotide bases of a strand of DNA. In both cases mathematical or alphabetical symbols and equations are taken to adequately represent nature. But real-world nature involves far more complex relationships among substances within cells and among organisms.[59] The idea that messy, complex biology can be reduced to sequences of numbers readily stored, managed, controlled, computed, and used for prediction has seduced biology into treating nature—even human nature—as an instrument of human will. To the mechanistic modern sciences, which serve as our proxies for relating to nature, nature has become not a living presence but rather an assemblage of stuff.

THE FANTASY OF PURE SUBJECT AND PURE OBJECT

> Philosophy stopped dead when it came face to face with the "subject" and the "object" and their relationship.
> —Henri Lefebvre, *The Production of Space*, 1991†

The dissociations that run through the history of modern Western thought shape our assumptions about the human place in nature and how we can know the world around us. Nature has fallen from its status as an active subject that requires respect, consideration, and close attention to become an instrument of human minds that lie beyond it. As the passive object of human needs and desires, mechanized nature is stripped of its own intelligence and telos (goals), so it can no longer be a communicative partner. We don't converse *with* machines, as nature is now conceived; rather, we converse *through* them. The magic with which people related to organic nature has no place in a relationship with a machine; the shaman has no role.

Everything removed from nature when it became a machine, the categories of mind and qualities, had to be placed elsewhere in the new modern worldview. Mind achieved its own special status outside of nature, in accord with the Judeo-Christian religious tradition in which God creates "man" separate from the rest of nature. Following Plato, Descartes invented a version of human identity built solely on the unnatural mind, separate from the natural body, which remained associated with the rest of machine-like nature. Early modern philosopher-scientists, the would-be masters of nature, leveraged the dissociating rationality of the ancient Greeks to complete the divorce of the human mind—true humanity—from all else.

The fault line of dualisms running through Western thought widens and lengthens with each new dissociating concept. Separated from nature, the human mind can assume the role of master of the material world, its dissector and conqueror. The Copernican decentering of the cosmos, along with the Galilean conception of the universe as a book written in mathematics, removed additional moral constraints to human action. These ideas that separate humans from nature and make nature a dead, mechanical realm open for exploitation justify—and therefore reinforce and perpetuate—our phenomenal dissociations from nature.

These various conceptual dissociations in the history of Western thought seem motivated by a particular fantasy, which I call the fantasy of Pure Subject and Pure Object. Pure Subject is the idealized, omnipotent, transcendent self, complete, independent, whole, and self-contained, static and eternal—a godlike figure free from dependency and constraint in the material world. Pure Object also exists separately and independently of all other existence, but it's inherently passive and lifeless. Pure Subject can use and manipulate Pure Object freely because doing so affects nothing else in the universe—it's so thoroughly disconnected. Because it has no agency, will, or direction of its own, Pure Object bends completely to the will of Pure Subject. Its lack of will almost invites manipulation by Pure Subject. Pure Subject is master; Pure Object is slave.

Pure Object exists in isolation from its surroundings, its environment, because all that's important about it comes from its own inherent attributes, not the context within which it exists at any given moment. Moreover, there's nothing special about an individual instance of Pure Object because there's always another like it—it's replaceable, replicable, repeatable, and unchanging, devoid of any defining relationships. And it places no demands on Pure Subject. The ideal of Pure Object was tacitly at work when slave traders yanked an African out of the social and economic contexts of his life and sold him as an alienable commodity in the North American colonies. From the traders' perspective, only the man's inherent qualities as a worker, and thus his market value, were relevant. Today, the ideal of Pure Object undergirds global commodities markets, where forests become abstract, context-free, interchangeable board feet and pigs become mere pork futures.

Dissociation is the structure and principle of the Pure Subject–Pure Object fantasy. Absolute dissociation is a form of purity: fully dissociated, Pure Subject may wash his hands of the needs of Pure Object. Thus I can live my life without really considering nature. Qualities that might connect Pure Subject and Pure Object, such as sense and feeling, are extirpated— ignored, de-emphasized, and expelled to an inferior realm—like the emotions Plato sought to banish. Feelings for nature are left out in decisions about whether to drill for more natural gas or cut a forest. As a fiction, Pure Object indeed belongs to the realm of Platonic ideals, which are also fictions. The dissociated worldview underlying this fantasy, which releases us from the messy and constrained material world, may be a secular version of heaven, a place free of material constraints and needs. But it's an illusion because our damages are quite real and do affect us, other people, animals, plants, and the whole of nature.

The Pure Subject–Pure Object schema suggests there's no need for an ethics of nature. But we're beginning to feel more intensely the limits of this model of interaction. The dissociating fantasy continues to shape much of the modern planet, creating human and ecological casualties. More people are noticing and finding these outcomes intolerable. In the final chapter I examine ideas and practices that heal the fractured modern relationship with nature. But first I discuss one more fundamental way that dissociations shape our modern world: through reconfigurations of space.

CHAPTER 7
MODERN SPACES

S trolling around the old hill town of Perugia, Italy, with its meandering, narrow streets, stone-paved pathways, and Medieval and Renaissance stone and brick buildings, feels altogether different, it hardly needs saying, from walking in Manhattan or Los Angeles. Much of Manhattan envelops pedestrians in skyscraper-lined canyons following straight, wide streets filled with cars and buses. Los Angeles, a sprawling automobile city with no real center and few places that invite ambling, contrasts even more with Perugia. These towns exhibit starkly different histories and logics of space. Perugia rose slowly over thousands of years as an aggregate of myriad decisions made by residents building incrementally on the past. Streets curve according to the slope of the hill or the easiest paths to the market or church. Built mostly by hand, stone by stone, the town feels organic and intimate.

Built entirely in the modern era, New York and Los Angeles express Western ideals of rationality and mechanization. Street layouts following Cartesian coordinate systems order space into equal, quantifiable units. Buildings align perfectly for miles. Urban street grids are not new—they've been around since ancient times—and they don't cover these cities entirely, but they haven't been used on this scale before. Underscoring the rationality of Manhattan's layout, most of its streets and avenues are named with numbers, and there's even a third dimension to the grid: the floor numbers of buildings. In San Francisco, another modern city, the street grid refuses to yield to the hills; it goes straight across them as if they weren't there, creating the steep slopes famous for movie car chases. Los Angeles is a vast patchwork of rectilinear blocks that stretch for miles, forming wide expanses of homogeneity. By contrast, no two buildings in Perugia look alike, and on a satellite image it's hard to find any oriented in the same direction. Instead of a coordinate system, a plate of penne pasta seems like a more apt metaphor. Yet architectural sensibilities and materials used there create a distinct visual and tactile harmony. In the more modern American cities, the mechanical order prevails. Machines built them, and machines—subways, taxis, buses, elevators—power movement through them. They're monuments to regularity and bureaucratic efficiency.

New York, Los Angeles, and San Francisco are great cities that have contributed major developments in literature, finance, entertainment, technology, and many other aspects of modern society. And a resident of Manhattan, living in a compact apartment with only one external wall to heat and using public transit regularly, can easily consume fewer resources and create less environmental damage than a typical resident of suburbia. The point isn't to disparage these cities but rather to understand how the logics of dissociation and modern mechanical rationality shape them. It's particularly important because their influence is so enormous. Cites around the world from Shanghai to Mumbai to São Paulo are becoming more modern every day, gradually resembling one another with their glass-sheathed high rises; imposing scale; zones of wealth and poverty; and mechanized flows of people, materials, and products.

Societies organize their spaces not only according to local physical conditions such as weather and readily available materials, but also according to their histories, assumptions, and ideals. Accordingly, the modern Western tendency toward dissociating concepts is manifest not only in our separations from nature but also in the way we organize our spaces. The story of modernity could even perhaps be written almost entirely from the perspective of space, beginning with European maritime explorers breaking down geographic limits by connecting their nations with remote and disparate places; through the expulsion of commoners from communally managed spaces in Europe; to the division of vast tracts into rectilinear monoculture farms meant to exclude all but one species of life, such as corn in much of Iowa.

Modern transformations of space trample long-existing relationships, such as the connection between the Ahwahneechee Indians and their ancestral lands in what is now Yosemite National Park or the ecological association of eucalyptus trees with their native Australia when brought to California. But they don't just disconnect—they also reconnect in problematic new ways. Eucalyptus trees diminish biological diversity in California and other places they've been introduced by, for instance, changing the soil chemistry so it's less habitable for local species. As its empire grew in the nineteenth century, Britain increasingly appropriated food produced in its colonies to the extent that by 1900 it was producing only enough grain to feed its population for eight weeks of the year.[1] Mike Davis makes the case that alienating food resources from the colonies and global trading partners, through ruthless colonial administration and by sweeping up agricultural production into the market, led to the great famines of the late nineteenth century in India and other places—the "Late Victorian Holocausts."[2]

Modern reconfigurations of space compartmentalize, dividing areas into single-use, single-ownership, specialized zones—often rectangular ones such as American plains-states farms and housing plots on street grids. They homogenize, too, making spaces in one location resemble those in another that had historically been quite different. In the process objects and forms, such as architectural styles and agricultural practices, become diffused across the globe. Modern spaces also consciously simulate existing ones in new locations, so an African landscape is given life in North America, as I describe further on. These transformations have intertwined consequences. Local expanses give up some of the originality and uniqueness that express local culture, and location loses relevance. Space is transcended; it becomes less meaningful. So we in industrialized, wealthy nations can have virtually any product in the world delivered, even entire places from elsewhere.

Moderners are not the first people to make major changes to our landscapes and seascapes. Ancient Mesopotamians deforested much of the Middle East, leaving desert-like conditions behind. By the close of the Middle Ages, most of Europe was converted to pasture, farmland, and village. Throughout the same period, Polynesians transformed the ecologies of the Pacific islands they colonized, for instance by bringing dogs and rats with them, which devoured ground-dwelling species, and by cutting down the last palm trees on Rapa Nui (Easter Island).[3]

But exploiting new sources of energy in the last two centuries, particularly fossil fuels, has enabled the modern economy to remold the planet to an astounding extent. Vast farms, ranches, feedlots, sprawling cities, suburbs, big-box retail stores, parking lots, and airports sprout in place of forests and grasslands. Great dams hold back expansive reservoirs while water diversions for irrigation shrink water bodies, such as the Aral Sea in Central Asia, to a fraction of their former size. Agricultural runoff and sewage flow into seas, creating large algae blooms that consume most of the available oxygen in the water, making it uninhabitable for many aquatic species. You can easily see the resulting dead zones, like many of these modern changes to Earth's surface, in satellite imagery. Modernity has transformed the surface of the planet on a massive scale.

For this tour through the spaces of modernity, I take as my guide the French sociologist-philosopher Henri Lefebvre (1901–1991), who thought deeply about how spatial practices both express and shape culture.[4] Modern spaces mirror our political and philosophical assumptions and hold great sway over how we engage with the landscape and with each other.

BROKEN SPACE

Modern regimes of space sever and reconstitute relationships at scales from the molecule to the landscape. They remove genes from one organism and put them in another, and they re-create landmarks far from the lands they originally marked. In the nineteenth century the US government violated the relationships between many Native American tribes and their lands, forcing some to move hundreds of miles to reservations. In the 1990s Disney built a New England–style town in Florida. The modern worldview holds spatial connections in contempt and seeks to nullify them, to alienate objects, styles, ideas, and people from the places that gave them birth. Lefebvre writes, "Alienation . . . is so true that it is completely uncontested."[5]

Over the centuries of the modern era, colonists and markets have repeatedly disrupted people's traditional and ancestral relationships with their lands. Karl Marx, who developed the first comprehensive understanding of the functioning of capitalism, wrote of "primitive accumulation," the process by which land enters the market system, becoming a commodity. Production for markets then replaces subsistence practices, and without land to subsist from, people must sell more of their labor on the market. Alienation from productive land thus supports alienation from the products of one's own labor. Throughout most of the two hundred thousand years or so of human history before the modern era, attachments

of people to the lands supporting them were complex, spiritual, and heredi-
tary. Of course, they usually had to be defended and were often stolen
through war. But the widespread conversion of working landscapes, along
with wild landscapes, from subsistence use into capital for market production and exchange is one of the hallmarks of the modern period.

Outright theft by colonization, capitalist incursion, and European enclosures of the commons are some of the means of primitive accumulation.
Environmental historians write about this process, such as when European
colonists came to New England and erected fences and palisades to keep
their livestock in and wild game out. The colonists convinced themselves that
the lands weren't being used and were thus not owned by anyone. Actually,
Native Americans were using the countryside throughout New England, but
traces of their use were too subtle for the Europeans to notice—or inconspic-
uous enough for them to disregard.[6] In accord with John Locke's definition
of property, colonial subsistence farmers and mercantile producers mixed
their labor with the soil, creating property *ex nihilo*.

When subsistence producers are alienated from lands they and their
ancestors may have nurtured for centuries, deterioration often ensues.[7]
Their accumulated knowledge of the particular ecological conditions of the
locale is lost, replaced by generic agricultural practices that may not be
closely attuned to the ecological characteristics of the local landscape and
social characteristics of local people. For instance, in China from 1949 to
1976, Chairman Mao's "Great Leap Forward" compelled millions of Han
Chinese to settle ethnic minority areas in a nationalist project of ethnic
dilution. Many brought agricultural methods, such as deep tilling, that
were inappropriate to the settled lands and caused soil erosion and other
problems. With a sense of "utopian urgency" to build a powerful socialist
economy, Mao also ordered millions of acres of forests cleared for conversion to agriculture and other uses, denuding much of the country's forest
heritage. Thus, not only capitalist but also socialist encroachment can cause
environmental damage by alienating traditional productive connections
with nature. Tens of millions of Chinese died of famine as long-standing
agricultural practices were disrupted.[8]

When people are alienated from the places where their ancestors have
lived and their cultures have grown, the myths and tales that bind them to
the land are also lost. Lefebvre writes, "Every social space has a history, one

invariably grounded in nature."[9] Through most of human history, mythopoeic landscapes overlay physical ones with stories that bound generations to place. Modern places don't have as many stories attached to them, at least not ones that connect to current habitants and their ancestors. And modern place names exhibit these disconnections. A Japanese philosopher once said, "Your streets, squares, buildings, have ridiculous names which have nothing to do with them." Fourth Street, Tenth Avenue, AT&T Park, and other names unrelated to local events and people form an abstract layer over the landscape, displacing and erasing natural and social history.[10]

The landscape loses meaning when it's converted into commodity. All its wealth of soil, plants, animals, and history gets reduced to a one-dimensional quantity, a price. Other natural things—crude oil, cattle, cocoa—likewise appear as quantities in commodity markets, following Plato's and Galileo's lead, to make them more amenable to the rationalist processes of modern instruments and institutions. Observing how science works, the French historian of science Bruno Latour writes of "inventions . . . that transform numbers, images, and texts from all over the world into the same binary code inside computers . . . the handling, the combination, the mobility, the conservation, and the display of the traces will all be fantastically facilitated."[11]

Evidence abounds of nature and "resources" being converted into commodified, computed, and rationalized forms. The settlement of the US plains states, prompted by the Homestead Act of 1862, imposed a lattice of easily registered, quantified, and administered rectilinear units onto natural forms that were anything but rectangular. You need only fly over the central plains or view satellite imagery to observe the immensity of it all: the midsection of the United States is a continent-sized Cartesian grid, with half-mile-diameter circles drawn by giant center-pivot irrigation systems filling many of its rectangular farm plots.[12] The landscape looks more like a machine than a living countryside. And that's how it functions—as a crop-producing engine.

At the opposite, microscopic end of the scale, bioinformatics and molecular biology apply a mechanistic approach to living organisms that in a sense removes them from space. When a living thing is reduced to its genome, a code, it exists outside nature. That seems particularly strange from the evolutionary biology perspective that views organisms as a product of the historical, ecological relationships within which they evolve. A bird may have a long, curved bill because that shape fits better into flowers common in its ancestors' environment, enabling them to get nectar more easily. The bill and flower incorporate a spatial relationship.

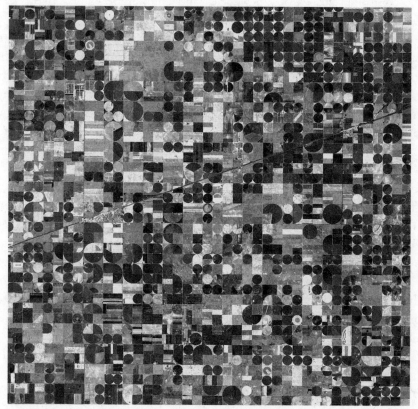

Figure 7.1. The mechanized landscape of Kansas from space. Haskell County, Kansas, June 2001. (Image from NASA/GSFC/METI/ERSDAC/ JAROS and the U.S./Japan ASTER Science Team.)

But markets don't care about biological space, which is more of a nuisance. Genetics and bioinformatics make biology more malleable, controllable, manipulable, and ownable—open to the techniques Latour describes. Corporations rush to patent organisms such as genetically modified canola created by small alterations to the genomes of plants domesticated over centuries or millennia by farmers using traditional plant breeding and husbandry. Even while genetic engineers and companies remove the embodied, spatial aspects of plants as they reduce them to genetic codes, those same aspects nevertheless remain relevant. Maybe uncontrolled gene flow and other ecological problems can't take place in the controlled spaces of laboratories and computers where GMOs are developed, but they can in the real spaces of nature.

Our minds and bodies are also born of spatial relationships. The innate human ability to perform an enormous range of tasks probably arises from millennia of experience in a diverse mosaic of grasslands, savannas, woodlands, and dense forests where our ancestor species lived in Africa. They encountered hundreds of varieties of plants and animals, including many large, life-threatening mammals.[13] Being able to stand up and look far into the distance, and to invent various tools with nimble minds and use them with agile hands, gave an advantage to those from whom we descended. Spaces molded us.

Bodies and cultures together link people to natural places. Australian Aboriginals recite "songlines" while walking through various landscapes. For each particular path through the terrain, there's a specific song that must be recited; the path, the terrain, and the songline are one. The songs give direction and guidance to Aboriginal travelers, telling them where they can find water, for instance. The practice attests to the importance of bodies in nature, together creating and living story and myth; it bonds culture, space, and the body together as one.

For modern and indigenous people alike, walking through culturally meaningful spaces creates personal identities.[14] But many of the modern zones we traverse are commercial ones, such as shopping malls, which calls into question what kind of identities we're creating for ourselves. The removal of our minds and bodies from landscapes also leads to strange consequences. Space becomes dissociated and abstracted. The cultural historian Rebecca Solnit writes that with treadmills, for instance, space is "measured entirely by time, bodily exertion, and mechanical motion. Space—as landscape, terrain, spectacle, experience—has vanished."[15] But space doesn't merely vanish in modernity; it becomes reconstructed in particularly modern ways.

MODERN CONSTRUCTIONS OF SPACE

Removing traditional spatial arrangements makes spaces available to be reconfigured in the image of the dissociating modern worldview. Modern spatial practices dissociate by compartmentalizing, creating artificial boundaries that control the flows of products and ideas. They homogenize by reproducing common forms across the globe, reducing diversity and reinforcing uniformity in the process. Songs and food products travel as if they belong at the same time everywhere and nowhere; the Great Sphinx of Giza can be copied in Las Vegas. Modern spaces differ dramatically from traditional, non-modern ones; the latter come together while the former break apart.

Compartmentalization

As Europeans settled the New World, they entertained various ways of thinking about the land that justified reshaping it according to their needs and desires. In the first half of the seventeenth century, the English Puritan and governor of the Massachusetts Bay Colony John Winthrop (c. 1587–1649) built on the idea of stewardship as set forth in Genesis 1:28 of the Bible—"Increase and multiply, replenish the earth and subdue it"— to legitimate transforming nature in North America. Winthrop asked the English, Why work so hard to earn a living from crowded lands in England and "in the mean tyme suffer a whole Continent as fruitfull & convenient for the use of man to lie [empty and unimproved]"?[16] Dispelling from their minds the vibrant, productive nature and indigenous inhabitants already living there, colonists considered the New World a blank slate—a place almost compelling them to take it and use it.

The blank-slate conception alienated meaning from the land and transgressed social and natural connections.[17] European settlers put up fences throughout the landscape to keep their livestock bounded and to exclude game; the fences frustrated Native Americans because they made hunting harder by blocking the free run of animals. Europeans also divided land in their villages with stockades around yards and garden plots. These divisions were foreign to the American countryside and initiated the long process of divorcing indigenous people from the lands of their ancestors. The rectilinear shapes foreshadowed the vast grid that would grow to cover most of the half-million square miles (1.3 million square kilometers) of the Great Plains three centuries later.

Before settlement by the descendants of Europeans, the American Great Plains was an undulating continuum of rich ecologies, with enormous herds of buffalo and other creatures roaming far and wide. Today, they're fenced off allotments of sections (1 square mile, or 640 acres, or 260 hectares) and quarter sections, each producing a particular crop or serving a particular rancher, mechanistically putting out food as interchangeable elements in a giant, matrix-like agricultural machine. Divided into Cartesian blocks, modern spaces resemble machine parts, each created for a separate purpose. Assigning single, specific functions to zones mirrors the specialization of labor into jobs such as teacher, firefighter, accountant, and doctor. Cutting up space into a grid makes it easier for rationalized market and bureaucratic systems to treat it like commodities or administrative units.[18]

On the plains artificial boundaries created by roads, fences, and earth-turning machines have replaced natural borders such as the edges of forests and rivers. Early in the nineteenth century cowboys drove cattle on open ranges across the Great Plains, but the new limits inscribed on the landscape eventually brought a different kind of order to the movements of cattle and humans alike. In the winter of 1886 a fierce blizzard and cold spell drove cattle southward for warmth and access to grasses (buffalo, long adapted to the plains, can brush aside snow with their enormous heads to eat the grasses lying beneath, but cattle can't). Immense herds of Texas longhorn cattle drifted up against a barbed-wire fence along the right-of-way of the east-west Union Pacific Railroad and could progress no farther south. According to a firsthand account, when the blizzard cleared, you could walk along the fence for 400 miles (about 650 kilometers) from Ellsworth, Kansas, to Denver, never touching ground, stepping only on cattle carcasses.[19]

Laws, courts, corporations, and other social institutions create an administrative grid that overlays the physical one. Institutions inscribe their ideologies of separation on the land, as the Black Act did in 1723 in Great Britain when the new private owners of once-public lands planted hedgerows to demarcate ownership and prevent others from passing through or using the property. The enclosures of the commons in Europe, from the sixteenth through the nineteenth century, marked a sea change in European spatial practices. Commoners became excluded by law, market pressures, and fences from access to historically communal lands. Enclosures raided this public inheritance for private wealth, privilege, and economic advantage. Commons management traditionally involved chance meetings and cooperative management and enriched the lands with symbolic and cultural meaning. It brought together communities for a shared cause. When commoners were excluded and management became a private affair, some of the glue binding the community was also lost.[20]

The drive to industrialize accelerated the alienation of the populace from common-pool resources. English industrialists believed the commons created masses of lazy and dissolute people. They complained about the sense of independence that being able to produce on communal lands gave commoners and argued that enclosing them would help secure "the subordination of the lower ranks which in the present times is so much wanted."[21] That is, besides wanting to make lands available for market production by commercial farmers, eighteenth-century capitalists sought to drive citizens into wage labor by taking away their access to the commons, where they could produce some of their own food and wool, and meet other needs. One bureaucrat bragged, "The labourers will work every day in the year, their children will be put out to labour early" when prevented access to the commons.[22] Today, most people in industrialized nations work for wages and have little or no means of subsistence.

Enclosing lands creates a self-perpetuating condition by removing people from both sustenance and community. When the landscape is broken up into private allotments, it becomes more difficult for the public to prepare for resistance or social change of any kind. Public spaces, sometimes large ones, are needed in order to organize. In the fall of 2011, the Occupy Wall Street movement began in Zuccotti Park, a "privately owned public space" in New York.[23] The park's owners eventually called on the police to remove the protesters' encampment, and occupy-movement supporters were gradually expelled from city parks across the United States, often at the behest of private businesses. The lack of open and communal spaces in the United States and much of the industrialized world hinders spontaneous public organization and cooperative movements.

In the modern world, we increasingly live our lives in private zones rather than public ones, and doing so inhibits spontaneous social interaction. Rebecca Solnit notes that as everyday life has become more privatized and compartmentalized in nuclear-family homes, suburbs, and cars, made possible by television, telephone, and the Internet, it becomes less necessary for people to go out into the world.[24] Cars pit drivers—unable to communicate with each other—against other drivers in deadly machines. Such artificial divisions disintegrate our social and natural environments.[25] Modern spaces are strategically compartmentalized to promote consumption. Shopping malls have replaced the village square as places where community members can freely congregate, but the malls are private zones with use restrictions and hours set by their owners. Social engagement in them remains subordinate to corporate interests, community peripheral to commerce.[26]

With all its divisions, sprawl, and car dependence, Los Angeles may be the paradigmatic space of modernity.[27] Mike Davis vividly describes the compartmentalization of this far-flung metropolis. High walls surround residential neighborhoods, and lawn signs threaten outsiders against trespassing. "Richer neighborhoods in the canyons and hillsides isolate themselves behind walls guarded by gun-toting private police and state-of-the-art electronic surveillance," Davis writes. Restrictive covenants enforce neighborhood exclusivity.[28] Homogenous, barricaded neighborhoods punctuate the city.

In downtown Los Angeles, Davis writes, a "monumental architectural glacis" (a phalanx of large, glass-sheathed buildings) maintains the functional purity of the corporate citadel by keeping out nonbusiness types, including the area's poor. Elsewhere, police barricade streets and seal off impoverished neighborhoods in the war on drugs. Architecture, urban planning, and city administration conspire to compartmentalize. The architect Frank Gehry's design for the "heavily fortified" Goldwyn Regional Branch Library, "the most menacing library ever built," shows the irony of public architecture that manages exclusivity.[29] Public spaces recede, and people are alienated from the landscape. Spatial dissociations both reflect and reinforce social division.

Modern alienations of space shift power from the public realm to the private and enable domination and oppression, writes Lefebvre. Modern urban life physically disperses community and makes people into something more like pure workers and consumers, not citizens. It thereby deflects class struggle (workers seeking to empower themselves) and the possibility for revolutionary change that might place community back at the center of life.[30] Lefebvre decries the consequences of enclosure and compartmentalization: "The basis and foundation of the 'whole' is dissociation and separation ... such dissociation and separation are inevitable in that they are the outcome of a history, of the history of accumulation, but they are fatal as soon as they are maintained in this way, because they keep the moments and elements of social practice away from one another. A spatial practice destroys social practice; social practice destroys itself by means of spatial practice."[31]

Solnit thinks the compartmentalization of modern urban, suburban, and rural landscapes makes walking over long distances a rebellious activity. People moving through the streets with others who share their beliefs isn't merely a leisure activity—it's how community is created and maintained.[32] Resistance to the spatial divisions of modern life can take many forms. The British organization "Reclaim the Streets" aims to recapture the link between open, communal spaces and community. Parades, marches, festivals, and other events that continue the long tradition of communal gathering in public spaces resist compartmentalization.[33]

In the early 1990s bicycle riders in San Francisco launched Critical Mass, a communal bike ride that temporarily reclaims the streets from the life-threatening cars and trucks that usually fill them. Riders in groups ranging from ten to tens of thousands now regularly celebrate Critical Mass in hundreds of cities around the world, displaying the power of community and human-powered transportation. Their costumes, decorated bicycles, and portable audio systems create a festive atmosphere of solidarity that momentarily replaces the heartless and divisive feel of many urban thoroughfares. Smaller groups of riders often pass through my neighborhood with music playing, reminding residents that streets and other public spaces can be places for gathering and enjoyment, too.

Homogenization

When I visited Egypt in the late 1990s, I was startled to see the Great Sphinx of Giza, about 4,500 years old, gazing on a Pizza Hut®, the American chain restaurant. A decade earlier in Munich, German colleagues confused me by asking whether I wanted to eat at "Pizza Hoot" (the German pronunciation of Pizza Hut). In the mid-1990s, I first saw a Dunkin' Donuts® in Bali, alongside shops selling statues of protective demons and locally crafted handwoven cloth. In 2006, Balinese friends came home to the village with shopping bags from Ace Hardware®, an American chain store that had recently set up on the island. Today, Esprit, Guess®, Polo®, Prada®, and just about every other international chain is established there. Back in the United States local shops, department stores, and websites let me buy seemingly anything from anywhere. Besides food, most things I buy, it seems, are made in other countries. Retail companies, products, and services today are radically globalized. But they're not the only things that are. Organisms from every corner of the planet travel and establish themselves in new places, thanks to the great mixing machine of modern transportation.

The defining technologies of the modern era may be the space-collapsing ones: high-speed, long-range transportation and telecommunications. From the dawn of modernity centuries ago, it was European maritime technology, combined with weapons technology and scientific methods, that unleashed the Age of Discovery and then European colonization. Regional and even continental trade was common long before the onset of modernity, of course. Just think of the Silk Road, established during the Han Dynasty, in the third century BCE. But modern transportation and telecommunications, dramatically accelerated by the nineteenth- and twentieth-century exploitation of fossil fuels—and inventions such as the steam engine that could convert them into mechanical energy—amplified millions of times over the worldwide movements of raw materials, goods, and biota. Social institutions such as financial networks, international legal frameworks, and commodity markets support the mobilization of stuff on this unfathomable scale.

These movements and transmissions make space and location less relevant to people's lives. That is, modern technologies and cultures *transcend* locale, which becomes simply another parameter to be managed. So, for example, many people can do their jobs from anywhere; first, because their work is fully mental, involving no material productions (at least none that can't be shipped quickly); and second, because telecommunications such as the Internet allow them to interact with other people in real time and deliver their mental productions instantly. Various companies with no physical location, such as online magazines, have emerged in recent years. Staff members can be anywhere. I and many others benefit from the new trans-locality, and being able to work from home can have ecological benefits, as fewer people drive to work. But driving and telecommuting aren't the only options. Many people live within walking, biking, or public transit distance of their workplaces, where they can interact with others face to face.

The great modern mixing machine is homogenizing spaces worldwide. Landscapes are becoming more alike, as are ecosystems, home interiors, and cultures. When things become available everywhere, the uniqueness of any particular space diminishes. You might argue that the greater variety of goods available in most places allows people to construct ever-more unique spaces for themselves: furniture from IKEA®, art from Cuba, cuisine from Thailand, films from Bollywood, and music from Los Angeles. But this particular combination could be assembled in New York, Athens, Beijing, or Johannesburg.

Diversity can create a certain uniformity at the same time it makes our personal and public spaces less particular to the locations they're in. This same effect happens in ecology. Zebra mussels, native to southern Russia, have established themselves worldwide by riding in the ballast water of oceangoing ships. They've colonized the San Francisco Bay, the US Great Lakes, Lago di Garda in Italy, and harbors worldwide, making these places and their biota increasingly similar. As ecologies worldwide become more alike, more species become extinct and global biodiversity falls.

Not just living organisms but genetic materials have become detached from locale in the modern era. In the Age of Discovery, European explorers began to bring back dead and living specimens and seeds from exotic locations around the globe by the thousands. Their massive mobilization of plant genes had major consequences for ecology and the power of European colonial centers.[34] A plethora of domesticated plant species from the New World, for instance, have sustained and enriched Europe through the modern era. Europeans took home tomatoes, potatoes, squash, maize, various beans, tobacco, and many other crops developed over millennia by New World farmers. The potato was an enormous gift to poor and hungry people in Ireland because it grows unusually well in bad soils and can be left in the ground until needed. But the mobilization of plant genomes can spell disaster as well. The eucalyptus, brought from Australia to California in the nineteenth century and planted as a source of timber, has diminished ecological diversity over large areas by creating inhospitable habitat for local species.

Homogenization also impacts cultural diversity as modern landscapes become more alike. Since fifteenth-century Italian Renaissance architect and engineer Filippo Brunelleschi (1377–1446) designed important buildings for various cities, star architects have created standardized forms they move freely among communities.[35] Driving homogeneity in urban landscapes is the aforementioned proliferation of international commercial brands. When Prada, Benetton, KFC, and McDonald's move into a town, they remove a little bit of its uniqueness. In recent years my friends in Ubud, Bali, complain of transnational corporations making their village resemble the historically more commercial beach town of Kuta. Even when they put up locally inspired facades, these stores replicate basically the same product, ambiance, and architecture, with minor variations only as necessary. The stamp they imprint on global cityscapes mirrors the centralized organization of multinational investment in New York, London, Paris, Tokyo, and Milan.[36]

Quintessentially modern cityscapes blur the boundaries between profit making and personal life. Zones that "used to stand alone [as leisure and public spaces] . . . now mix social and commercial functions, sponsors, symbols."[37] Shopping malls, pedestrian zones lined with stores, and stadiums where fans cheer corporate sports teams while buying shockingly overpriced concessions have become the sites of social interaction. Although markets have long been places of chance meetings and casual get-togethers, there are fewer noncommercial spaces where unplanned friendly encounters take place besides schools and churches. Moreover, the markets of old sponsored by local communities are now replaced by strictly corporate spaces that infuse our lives with global capital.[38] Whether you live in Des Moines or Singapore, your teenager's familiars may include Abercrombie & Fitch, Burger King, and other international brand names and logos, along with shopping-mall friends.

Homogenization is also taking place within, not just between, cities.[39] Municipalities historically have had public centers and layouts that orient people as they move about—distinct neighborhoods, specialized market districts, perhaps a port, and nighttime entertainment areas. In the United States in particular, the city center has receded as the nexus of social life with the growth of suburbs and the flight from the urban core.[40] Lefebvre decries the ensuing destruction of street life and neighborhoods in American cities and the "disappearance of many acquired characteristics of city life . . . security, social contact, facility of child-rearing, diversity of relationships, and so on."[41] The treadmill discloses a stark reality about homogenized suburban and urban life, writes Solnit: "a device with which to go nowhere in places where there is now nowhere to go."[42]

Diffusion

Recall the discussion in chapter 1 about the great spatial diffusion of materials in high-tech electronics and the global spread of those products once they're sold and then again when they're recycled. Such goods are commonly assembled from parts originating around the globe, are transported over great distances once built, and later are discarded far and wide. The origins of their materials and components and the paths they take to merge into a single commodity are so long and complex that in many cases they can't be completely known. Consequently, the true costs of things, including their environmental costs, are also unknown.[43]

The same holds for nonmaterial aspects of culture such as literature, music, television shows, and movies. People from hundreds of different societies listen to Lady Gaga or Adele and watch Bollywood or Hollywood films. Because these products are so dispersed, their meanings are, too. They reflect attitudes and speak to people in many places and take on new meanings wherever they go. Perhaps it's a good thing ideas and designs travel so widely today. Many people hail the emergence of a "global culture" that unites the planet. But the more we tune in to a world culture, the less we're enmeshed in our particular places in the world. We're left to wonder, Does the diffusion of culture bring us gradually closer to a generic experience that's indifferent to place—and indifferent to any nature that might be there?

Intensive diffusion has been a key aspect of modern life since the time the earliest European transcontinental explorers shipped material and cultural objects around the globe and back to colonial centers. In the fifteenth century few people depended on products from other continents, but today many societies would quickly collapse if completely cut off from the rest of the world. The small desert nation of Dubai in the United Arab Emirates makes a superb example: it has few natural resources besides fossil fuels, so it's almost completely dependent on imports, including the materials to build its rapidly expanding skyline. Utter reliance on the global dispersion of goods gives power to the term *strike*, which originated in 1768 when English sailors collectively chose to strike (haul down) the sails of their ships to suspend commerce and the international accumulation of capital.[44] The ability of truckers, longshoremen, and other transportation workers to quickly shut down economies carries one of the most potent threats to the machinery of capital.

European improvements in technology—ships and navigation instruments and techniques—spawned the "commercial network of global scale," which arose in about the fifteenth century.[45] Since then, various technologies have sped up and expanded the network. For instance, in the nineteenth century meat became a national product in the United States and then a global one with the invention of a new transportation technology: refrigerated train cars. By allowing for great distances between livestock roaming the fields and the dinner table, refrigerated train cars made possible a centralized meat-processing center—the vast Union Stock Yards in Chicago—where cattle and profits could accumulate. Live animals arrived there from various origins throughout the American Midwest. The new meat corporations shipped slaughtered animals via refrigerator cars to distant cities such as New York.[46] These transportation technologies spatially dissociated animal bodies from the landscape. Today, fresh specialty meats are flown long distances, dispersing animal carcasses globally for consumption. They include wildlife and endangered animals (parts or whole, live animals), traded on the Internet and elsewhere.[47] Spatial diffusion helps make nature into a commodity and ruptures various relationships in the process.

At the center of modernity's myriad spatial diffusions lie plentiful fossil fuels—the coal, oil, and gas that power transportation. They make it extremely cheap to transport people and products, but only when environmental and health costs are excluded, as they normally are. The enormous total energy of available fossil fuels makes location almost entirely irrelevant and thus stimulates the modern transcendence of space. Digging up and burning these remnants of ancient organisms requires exploration and mining that leads to spills, leaks, and blowouts such as the 2011 Deepwater Horizon disaster in the Gulf of Mexico, which killed eleven people and released about five million barrels of crude oil into the sea. Transporting oil leads to devastating accidents such as the Exxon *Valdez* spill that badly polluted Alaska's Prince William Sound, an ecological treasure, in 1989. Burning fossil fuels releases into the atmosphere—from smokestacks and tailpipes—great quantities of carbon dioxide, the primary greenhouse gas driving global climate change, our most pressing environmental problem. We're using the atmosphere as a sewer.

But perhaps even more consequential for the environment than all these problems is that by breaking down spatial relationships and limits, fossil fuels distribute the economic demand of billions of people onto environments and natural resources everywhere on the planet. Forests disappear and fisheries become depleted as products from them flow cheaply and easily around the world. Consider the irony: we burn ancient forests (in the form of gasoline refined from petroleum oil) in chainsaws to cut down today's forests. Then we burn ancient forests in trucks and ships to move the wood around the globe to its destinations.

How did all this spatial diffusion come about? How is it embedded in our products and manufacturing processes? A stepwise historical look at a single type of product traces the progression of diffusions leading to the modern situation.

The Stages of Spatial Diffusion: The Lamp

The lamp holds great significance in human history. It subverted the natural limitations placed on humans by the nightly darkness. Greek legend has it that the Titan Prometheus stole fire from Zeus and gave it to humans, earning the wrath of Zeus and never-ending torment: he was chained to a rock where an eagle descended every day to peck out his liver, which would grow back in time for the eagle's return. Fire is a pivotal technology that unleashes most others and thus makes human civilization possible. It's necessary for metal smelting and forging and has been used over eons for a connected set of purposes: cooking, heating, and our current focus, lighting.

The historian Wolfgang Schivelbusch wrote a history of the lamp that reveals a lot about its spatial diffusion.[48] Artificial lighting first established itself with wood fires, he notes. The bulk and weight of wood, together with the absence of large-scale transportation technologies, meant that fires had to be made from locally collected materials. As resin and other materials were applied to the wood to make torches, lighting became separate from heating and cooking. The use of resin marked the first important step in the distancing of the source of the fuel (the resin) from where it's burned (on the wood). The candle and the oil lamp came next, making light more easily transported and versatile. The invention of the wick, for both the candle and the oil lamp, completed the removal of the site of combustion from the fuel source. The wick transports the fuel during use from the reservoir in the oil lamp (or the pool of molten wax at the top of the candle), to the tip of the wick, where it burns. With these separations, fire was gradually tamed. Schivelbusch observes, "In the candle flame, burning steadily and quietly, fire had become as pacified and regulated as the culture that it illuminated."[49]

In the early nineteenth century the invention of gas lighting further removed the fuel source from the lamp. Gas was initially produced locally (albeit using transported materials), usually in the same building as the lamp, with coal as the basic ingredient of the distillation process. The gas was conveyed to different parts of the building through pipes. The drive to industrialize gas production eventually led to more distant gas manufacture. In England, also in the early nineteenth century, Friedrich Albert Winsor spearheaded centralized gas production and delivery through gas mains, an idea adapted from the delivery of water. The network of pipes satisfied the growing demand for gas in factories and homes with the efficiency and control of industrial production.[50] One observer wrote, "The whole difference between the greater process of the gas-light operation and the miniature operation of a candle or a lamp, consists in having the distillatory apparatus at the gaslight manufactory, at a distance, instead of being in the wick of the candle or lamp . . . and . . . in transmitting the gas to any required distance."[51] Schivelbusch concludes that the technical qualities of nineteenth-century gas lighting can be summed up in a single word: distance.[52]

As homes connected to a centralized gas supply in the early nineteenth century in Europe and the United States, they lost some of their autonomy—gas production moved beyond the control of the head of household. Each house was now tied to a particular industrial energy producer and no longer made its own heat and light self-sufficiently.[53] A psychological reaction set in among the public: "To contemporaries it seemed that industries were expanding, sending out tentacles, octopus-like, into every house. Being connected to them as consumers made people uneasy."[54] But in truth they'd already been dependent on networks of coal companies and other commodity producers for their heating and lighting. The pipes just made their dependence more tangible. Today, our reliance on utility companies and other producers is so extensive and taken for granted that it rarely enters consciousness.

When gas mains were still new, households severed their connection to them nightly while they slept to restore safety and autonomy for a few hours.[55] Fears of the dangers of gas production rose along with its centralization and consumers' reduced control. Two centuries later, misgivings about utility networks reverberate in the idea of "getting off the grid"—permanently disconnecting—at least in parts of the United States. Yet the electrical and gas "grids" are just two of the many networks of sustenance to which modern homes connect, including all those that produce the food, clothing, housing, medical care, and other essentials of modern life. Like the industrial coal networks of the early nineteenth century, they're not less real just because they're not connected by a physical pipe.

Returning to the history of the lamp, arc and incandescent electric lighting followed gas in the mid to late nineteenth century. Arc lamps pass electricity through a gas, incandescent lamps through a filament. Electric lines could transmit energy great distances from the site of its production—coal, oil, or a river. But the object of illumination could now also be further separated from the lamp due to the greater luminescence of electric lights. The first electric incandescent lamps passed electricity through a bamboo fiber. Therein lies another spatial diffusion: the increasingly global sources of materials to construct the lamps themselves.[56] Lightbulbs continue to become more and more complex today. Steel, tungsten, and glass incandescent lamps are giving way to compact fluorescents and now light-emitting diodes (LEDs). As they advance, their materials and components come from ever-more diffused sources.

Over history, the light emitted from lamps gradually moved away from the fuel through a progression of technologies: resin, wick, gas pipeline, and electrical cable. Likewise, materials, components, and processes that create lamps came from increasingly distributed sources. Today, high-voltage electrical lines carry power hundreds, sometimes thousands, of miles to lamps (lightbulbs) manufactured on another continent from materials that may originate on yet another continent. These networks are too long and complex for consumers to trace and thus easily block consequences from our view.

Simulation

What's it like to ski downhill on real snow in the middle of a city in flat, dry, desert Arabia where temperatures are sweltering? You can find out on the indoor mountain at Ski Dubai, part of a large shopping mall in the United Arab Emirates. The weather outside may be over 104 degrees Fahrenheit (40 degrees centigrade), but inside it's more like the Alps in winter. Tobogganing and penguin encounters are also available. Just one example of dozens of indoor ski slopes around the world, Ski Dubai vividly illustrates *anatopism*, something out of its place (whereas *anachronism* means something out of its time). Some indoor slopes take advantage of actual inclined terrain and cooler weather, but they all proclaim the quintessentially modern divorce from—perhaps contempt for—places and their limitations. We're unlikely to care for places when we're so willing to reject and utterly transform them.

The anatopistic simulation of a space both rips it out of its original context and creates strange dissociated relationships in the new one: picture frozen water on a fake hillside in a sizzling city. These aberrant juxtapositions can require a lot of resources, such as the massive amount of concrete to make an artificial hill and the energy to keep it as much as 80 degrees Fahrenheit (27 degrees centigrade) colder than the outside. The act of reproducing entire spaces—sometimes complete with cultural symbolism—treats the place where the re-creation is done as if it's blank and empty. And it's a characteristically modern thing.

Simulation complements compartmentalization and homogenization as modern ways of reconfiguring space. Compartmentalization divides zones for individual ownership and industrial mechanization.

Homogenization and simulation both result from trying to make everything available everywhere. All three processes inscribe into the material world the extreme contingency assumed in the modern worldview: nothing is given; everything can be redefined and remixed at will. Constraints need not be accepted. Resources may be expended boundlessly to transcend the limits that space and distance impose.

Simulations abound in the modern world. In Las Vegas you can now visit parts of Egypt, Paris, and Venice simulated with large-scale monumental reconstructions or stylized buildings, including replicas of Venice's Piazza San Marco and Paris's Eiffel Tower. Also in Las Vegas is a giant pyramid-shaped hotel-casino with over four thousand rooms, the Luxor, which mimics the great pyramids at Giza. To underline the extravagance, skyward out of its apex shines the brightest light on Earth: the 315,000-watt "Luxor Sky Beam."[57] Out front stands a giant replica of the Great Sphinx—this one staring at a faux Egyptian obelisk and parking lot rather than at a Pizza Hut.[58]

Theme and amusement parks are especially good at simulating distant and exotic locations. Using the insensitive moniker "the Dark Continent" until the 1990s, Busch Gardens brings Africa to Tampa, Florida, through landscaping and the continent's famous wildlife. The iconic castle at Disneyland and Disneyworld consciously mimics the fairy-tale castle Neuschwanstein built by King Ludwig II of Bavaria on an Alpine mountaintop. But Neuschwanstein itself, built in the late nineteenth century, was intended to emulate an older medieval style long anachronistic in Ludwig's time.

In Miyazaki, on the island of Kyushu, Japan, there resided until recently the largest indoor beach in the world, only about a mile from the sea: Ocean Dome. Closed for renovation in 2007, the facility featured a warm "ocean" of chlorinated water and a 460-foot (140-meter) beach of artificial sand (ground marble) that doesn't stick to the skin. Patrons could sunbathe below an artificial sun in a fake sky warmed to a constant 86 degrees Fahrenheit (30 degrees centigrade) or bob in adjustable artificial waves viewing nearby palm trees and a tropical horizon painted on the far wall.[59] Although a real beach isn't far away, its setting isn't tropical, and it can't be used in comfort year-round. The limitations of local space and nature can be overcome by modern technology, but only if we're willing to buy into the illusions demanded by simulations.

Today, we simulate natural spaces because they're far away or somehow inconvenient. How long will it be before we replicate landscapes because they've been ruined by modern industry?

It's not only public and commercial zones that copy a different time and place—new private homes do, too. Housing developments often fill spaces stripped of history and tradition with elements trying to re-create them out of thin air. In New England I've seen many new houses ape colonial styles with architectural accents such as fake columns that amount to little more than kitsch decorating oversized boxes with enormous heated garages. These simulations express a detachment from the actual physical and social landscape, which in New England would demand smaller homes that can be heated more efficiently in the frigid climate and unique styles developed by local architects who don't simply tack on the symbols of a bygone era.

Computer technologies have opened up whole new possibilities for transcending space. Many people seem to live inside their Facebook® or MySpace® pages, where they can build new identities. But online environments such as Second Life® create entire physical worlds for people to realize fantasy lives where they can interact with virtual others. Real-life technology is gradually approaching the most fantastical science-fiction technologies for creating realistic simulated worlds. The prime example of the latter is the "holodeck" of Star Trek, an empty stage that can be programmed to reproduce virtually any scene, complete with characters so realistic that users can be injured by them. The holodeck's accuracy and completeness sometimes defies even the ability of its users to distinguish it from reality. But even now in the twenty-first century, we're beginning to lose track of what's genuine—and thus what we should care for and about.

The French philosopher Jean Baudrillard (1929–2007) writes that simulation has become so pervasive in our era that objects and spaces being imitated are already themselves simulacra. We're left with a "precession of simulacra." In building Neuschwanstein, Ludwig wanted to emulate the old-fashioned fairy-tale castle of his fantasies and sought to create spaces true to his friend Richard Wagner's operas—a clear case of life imitating art. In some cases it's no longer possible to even know what's being reproduced. Baudrillard invokes Jorge Luis Borges's fable, in which "the cartographers of the Empire draw up a map so detailed that it ends up covering the territory exactly." Deemed normal and acceptable, simulations become confused with the truth.[60] They block out our knowledge of the "the real," argues Baudrillard. Like Borges's map and the amusement-park version of Africa in Florida, spatial simulations cover and obscure more authentic places full of social or natural history.

One of the most remarkable projects of imitation comes from the experts of make-believe, the Disney Corporation. In the early 1990s, Disney created an entire freestanding master-planned municipal entity, Celebration, Florida, on about five thousand acres in the middle of the state. The town is created, owned, and managed by the company's subsidiary, the Celebration Company. In its marketing and planning, it evokes visions of "authentic" community life, which is odd considering how much simulation is involved. Prospective residents can purchase homes in one of six preset models that emulate architectural styles "reminiscent of classic American towns": "Classical," "Victorian," "Colonial Revival," "Coastal," "Mediterranean," and "French." Designs and planning throughout appeal to nostalgia for small-town America with quaint, pre-1940s New England–style white picket fences and broad porches: "Overall, the look and feel of a warm and friendly hometown." Small-town imagery entices buyers: "Morning coffee on your front porch. An afternoon stroll to Market Street. Family evenings in the neighborhood park."[61]

Disney is homing in on (and perhaps honestly trying to supply) something often missing in modern life: community and authentic connection to place. Advertisements say residents are "connecting in ways that build vibrant, caring, and enduring traditions." But is that something a corporation can sell? The website features images of well-groomed, financially comfortable, happy white people. The pretense is never far away, however. Every night during the holidays, for instance, the town center, Market Street (which is corporate-owned and has its own logo), sees a Celebration "tradition" unfold: the "Nightly Snowfall," which runs hourly. In fall, concealed fans blow weathered leaves gently through the streets. Neither snow nor autumn leaves are true to the subtropical climate of central Florida— local nature is spurned. Hidden speakers pipe in easy-listening music to public areas. A large tome of bylaws produced by the corporation, not by citizens, delimits the daily round—local community is spurned, too, in a sense.[62] The invented character of the town, built from the landscape up over a few years, makes a charade of the company's attempts to market its authenticity and historicity. How could all the kitsch and commercialism of the place not dissociate people from the history and nature of central Florida—even while it furnishes aspects of community people hunger for?

LESS MODERN SPACE

Taking a look at how less modern spaces are organized sheds light on all the disassembly, compartmentalization, homogenization, and simulation going on in modern ones. Japan is a good starting point because, though it's the decidedly modern society that would build Ocean Dome, it also retains some distinctly non-modern ways of organizing space. Indeed, Japanese culture carefully cordons off, protects, and honors traditional spaces, preserving them as an element of everyday life. Like those in many traditional cultures, they better connect people to history, culture, society, and nature.

Visitors to Japan notice immediately that Japanese architecture tends to mix modern elements with styles from the country's distant past. I've had the good fortune to spend time in several homes in Japan, including weeks spent with the family of a friend near Nagoya. Some of the rooms in their home resemble those of a typical American residence, with the addition of Japanese decorations. Nobody followed ancient tradition by sitting on the floor in these spaces, including the living room and some of the bedrooms, because they're furnished with Western-style couches, easy chairs, and beds. But they're among the least-used rooms in the house.

Two other rooms look like they would have two centuries ago. The parents' bedroom has a floor of *tatami* (straw) mats and a simple futon on the floor as a bed, with closets built into the walls. A tatami-mat floor and *shoji* (rice-paper) walls also define the most formal room of the house, a traditional living and dining area. Here, we ate the *osechi*, the elaborate New Year's Day meal. In the kitchen the traditional and the modern come together. The working area of the kitchen looks much like an American or European one (with the addition of a large rice dispenser). But dining (and most of the casual socializing) takes place on a raised area of tatami mats enclosed by shoji screens. The dining table, a *kotatsu*, is about a foot off the floor with drapes on the sides holding in the warmth from the heater underneath, which warms your legs while you eat—even many upper-class homes have no central heating despite winters that go well below freezing.

Like many other families in Japan, my friend's doesn't retain such traditional spaces as a kind of memento of the past. They continue to create and use them much as their ancestors did, letting in modern practices of space, along with contemporary objects such as sofas and flat-panel TVs, in ways that don't completely supplant tradition. Lefebvre writes, "In Japan . . . traditional living-quarters, daily life, and representational spaces survive intact . . . not as relics, not as stage management for tourists, not as consumption of the cultural past, but indeed as immediate practical 'reality.'"[63] In much the same way that Japanese spaces insist on retaining traditional elements, they also preserve bits of nature. Rather than expelling nature, they incorporate it. One Japanese philosopher writes, "There is no house in Japan without a garden, no matter how tiny, as a place for contemplation and for contact with nature; even a handful of pebbles is nature for us—not just a detached symbol of it."[64]

Traveling in Japan, I've been struck that traditional spaces always seem to be nearby, no matter how modern or industrial an area is. Japanese culture limits and confines modern change so that detachments and reconstitutions, which do occur, are kept perpetually incomplete. The country certainly has embraced industrialism. Factories, smokestacks, bullet trains, electronics, pollution, and high-tech toilet seats are abundant. Yet the Japanese manage to conserve traditional spaces that connect, and not just as museum pieces.

In Bali, space also retains particular characteristics that resist modernization, and Balinese experiences of space are distinctly non-modern. As they move about, the Balinese remain constantly oriented to the cardinal directions of the island: *kaja*, *kelod*, *kangin*, and *kau* (mountain-ward, seaward, east, and west). In much the same way, they're always aware of their current social location, or context, and use various language levels to speak to people of different social status. Spatial directions depend on current location. When in the north, *kaja* is south because the mountains are central; when in the south, *kaja* is north. When moving something together, even indoors, a Balinese person might shout, "to *kaja!*" Space in Bali isn't the absolute, uniform, rationalized space of the industrialized world. It's always relative and contextual. The Balinese remain perpetually in tune with the terrain of their island.

Other aspects of Balinese spaces make them starkly different from modern ones. Most Balinese architecture exhibits an emphasis on group activities and meeting spaces, notes Gregory Bateson.[65] Walking around in villages and neighborhoods, you quickly notice many open-sided meeting and performance halls, temples, huts, and other spaces that bring community members together rather than dividing them. Extended-family compounds are walled off, but their front doors remain open during the day, and neighbors and visitors can always wander in. Side openings between adjacent compounds give next-door neighbors quick access, though in the past two decades I've seen many of these openings become sealed off. As people's attention turns toward their televisions and smartphones, lives are becoming more private, and neighbors are slowly becoming less neighborly.

Balinese people begin to understand space very differently from Westerners from an early age. Whereas modern Western practices enclose and divide, Balinese practices learned from infancy open, connect, and support. Infants and small children in Bali have greater freedom to go where they like, no matter how inconvenient that might be for the people caring for them. Parents and siblings give them constant physical guidance and hold them rather than putting them in restricting enclosures. They sleep on regular beds with parents and siblings, not in cribs (I don't remember ever seeing one). Bateson describes a scene of a child learning to walk. She's given a horizontal bamboo pole to steady herself. The anthropologist concludes, "The topology of this arrangement is the precise opposite of that of the play-pen of Western culture. The Western child is confined within restricting limits . . . the Balinese child is supported within a central area."[66] Perhaps the playpen, where only the baby goes, teaches that space should be compartmentalized for special, individual use (and owned, as the baby "owns" the crib).

Non-modern cultures tend to integrate stories and songs into their landscapes, defeating any possibility of them becoming compartmentalized or homogenized. Aboriginal Australian songlines, mentioned above, enact strong connections to landscape and reveal an inherently uncompartmentalized way of engaging with space. The songs and the travels weave through landscapes over great distances in a way that would be impossible in divided modern lands. Each place has a particular story, so the setting can't be homogenized. The songs are both the people's stories and the landscapes' stories. A section of a songline will impregnate a woman and become part of the life story of her child, for instance.[67] When landscape, culture, and history are so intertwined, it becomes difficult to imagine fragmentations of space such as efforts to remake nineteenth-century New England neighborhoods in Florida or huge walls such as those running along the US–Mexico border.

But Westerners need not look to other parts of the world to find less divided space; we can look to our own not-too-distant past because all the spatial division we experience is new with modern culture. Medieval houses in Europe, for instance, were built on an open-hall plan so that members of the full extended family and guests could gather. There and in other shared spaces, work and leisure time were mixed before modern ideals separated functions to different rooms and buildings.[68] Only with the advent of modernity did the spaces of common homes get divided according to separate functions—cooking, eating, weaving.[69]

Moreover, throughout the modern world spatial dissociations are far from complete and total. More American homes in recent decades, for example, include a "great room" resembling the medieval open-hall plan, where cooking, dining, and hanging out can happen together—even if the prevalence of smartphones, tablet computers, and texting means family members pay less attention to each other when gathered than they did in medieval houses. Most wilderness areas remain open and undivided, permitting free passage to hikers. Theaters and sports arenas continue to bring people together as a community. Religion also anchors its rites and ideology in space, Lefebvre observes. The names of Christian spaces—church, confessional, alter, sanctuary, tabernacle—immediately evoke religious ideology and practice. They hold meaning and resist homogenization because particular ceremonies and rites must be held in particular spaces. That these connections among ritual, space, and ideology endure is no accident: "The Christian ideology . . . has created the spaces which guarantee that it endures," writes Lefebvre.[70] Such examples seem like relics, though, as the modern perspective renders spaces increasingly generic, disposable, reproducible, and fractured.

FRACTURED SPACE

Modern spatial practices inscribe the principle of dissociation onto Earth's surface inch by inch, creating divided, homogenized, and simulated zones: continent-sized grids and anatopistic ski slopes. Freed of long-standing relationships that anchor nature and culture to place, the landscape becomes fodder for the mechanized economy to digest and then discharge in commodity form: a Bavarian fairy-tale castle rises in Florida, and an Egyptian pyramid graces the Mojave Desert. Modern spaces reflect and reinforce social divisions. The rich sequester themselves in wealthy suburbs and expensive neighborhoods in the city, the middle class fill commuter suburbs, and the poor live in inner cities and segregated suburbs. The walled, gated communities of Los Angeles and other American cities, full of nuclear-family, detached homes, vividly express the dissociations of modern spatial and social orders.

It can cost dearly to trample existing ecological and social relationships as we modernize the landscape. For instance, keeping the tiles of the Great Plains agricultural grid green requires a lot of fossil fuels to generate electricity to run irrigation pumps. Burning the fossil fuels drives global climate change, with its human toll. Meanwhile, the pumps *mine* the Ogallala Aquifer, a giant, underground freshwater sea stretching from South Dakota to Texas. That is, they draw water faster than it's replaced by natural processes. At the current rate, this water that grows many of the crops that feed the United States and other nations will eventually run out. Modern spatial divisions have human costs as well. Creating the Great Plains grid required dispossessing Native Americans from their lands, which killed many outright, broke up families and communities, and wiped out cultures. The machine-like logic of the grid has no place for them. As nuclear-family homes and privatized public zones such as shopping malls limit circulations of people, moderners slowly become more like objects of free enterprise than subjects in their own right. Modern spatial practices discipline humans and nature alike.

<hr/>

This concludes our tour of dissociation in modern thought, practice, and experience. We've seen that layers of commerce and industry obscure the harms of products we consume. Material networks connect us like enormous tentacles out to the larger world, dispersing our consequences far and wide, mixing them with those of many other people. Nature has become an abstraction, and so, too, have our social and environmental problems. Underlying all the dissociations that fragment modern experience is a philosophical heritage run through with divisive concepts about the material world and how we should relate to it.

For half a millennium Western ideas and practices have increasingly infused much of the globe, inflecting thought and industry almost everywhere (while being nourished by other cultures). At first their propagation took place by force—colonization—but now modern industry and its fruits, everything from mechanized agriculture to iPads®, Viagra®, and nuclear power plants, are the yearning of the world. Societies across the planet adopt modern dissociating ways, creating their own unique hybrids of modernity and tradition. So my friend's family in Japan sells European cars and puts *mochi* offerings on them at the holidays to appease the spirits. And my Balinese friends experience an increasingly industrialized, wage-labor economy, which gradually makes nature more distant for them, while still performing their age-old rituals, music, and dance. But modernity's dissociations prove destructive in many guises. In China rapid industrialization is sprouting skyscraper-filled cities with millions of residents ever more alienated from the landscapes they once knew, which are now being ravaged by industry.

It may be daunting to think of the great extent of environmental challenges we face. But knowing how they're related, that they're mutually advanced by an underlying organizing principle—dissociation—creates the possibility of tackling them more effectively. Better knowledge of the sources of our problems and the connections among them gives us more freedom—and, perhaps, resolve—to respond. Just as conditions haven't always been like this, they don't always have to be. Finally, let's consider ways of healing these divides in search of a healthier world.

CHAPTER 8
RECONNECTING AND HEALING A PLANET

In 2005, armed with nothing but a credit card and a web browser, you could hunt and kill "wild" animals on a ranch in Texas. A remote-controlled rifle was housed in a hunting blind with a camera on top providing real-time images of the scene over the Internet. The rifle fired with the click of a mouse, and after a kill your prey could be butchered and shipped to you. Huntable game on the ranch included sheep, hog, and antelope introduced to an enclosed area.[1] Public outcry from many corners, including groups not usually on the same side of issues, such as the US Humane Society and the National Rifle Association, resulted in a frenzy of legislation prohibiting such activities in most US states. The one known site, live-shot.com, didn't last long.

Figure 8.1. Internet Hunting: "Real Time, On-Line Hunting and Shooting Experience."[2] Advanced dissociation of a primal experience.

Why did Internet hunting provoke such ire and revulsion that the industry would be shut down before it ever got off the ground? The answer lies, I believe, in its stark combination of dissociation and violence. The leap from *in situ* hunting to Internet hunting was so great that it highlighted the latter's dramatically dissociated structure. Never mind that online hunting can fairly be seen as an extension of a series of technologies—from clubs and spears to bows and arrows to high-powered rifles—that progressively distance the hunter farther from the prey, offering increased power, safety, and anonymity. Also set aside for a moment that most Americans purchase factory-farmed meat, which likely involves more suffering for animals than being shot as they roam on a ranch.[3] The dissociations inherent in both *in situ* hunting and factory-farmed meat are less stark because they result from many stepwise technological changes over centuries.

Similarly, the dissociations limiting the experiences of US Air Force personnel in Nevada when they kill people in Pakistan and Afghanistan are less obvious because they arise from a long series of military technologies leading up to "unmanned" drones: from clubs, knives, and spears to rifles, tanks, and bombs dropped from airplanes. But the dramatic leap from a hunter in the landscape with a rifle to one in front of a computer in another state rendered the dissociations between hunter and prey naked. The demise of Internet hunting is good news because it shows that there's still something people intuitively don't like about dissociated power when it has been made visible and obvious.

Online hunting is an apt symbol of the power modern technologies and social institutions put into the hands of individuals. For virtually all of human history, people would have thought that only a god could instantly kill an animal on the other side of the planet. Indeed, modern living may be the culmination of ancient dreams of godlike power. But dreams, once realized, can bring unwelcome repercussions. As I've tried to show in this book, our impressive power to act in distant places is accompanied by an ironic powerlessness and lack of control over the outcomes that accompany our actions, such as toxic contaminations from high-tech manufacturing. Our dissociations from consequences, our lack of knowledge and direct perception of them, challenge our ability to manage how we engage with and shape the world. We may wish to stop the Internet hunt because it's easy to imagine and empathize with individual animals being shot by a remote-controlled machine, and we may be particularly repulsed when the hunter has such a casual and disconnected relationship with the prey. Yet at the same time we affect the lives of plants, animals, and people around the world with our actions every day and are hardly aware of the outcomes.

If modern life is rife with all sorts of dissociations—between people and nature, between people and the consequences of their actions, between people and the others their actions affect—how can we possibly expect to solve the problems arising from dissociation? It seems we could hardly choose a more Sisyphean task. But the fact that harmful dissociations run through our world is good news in disguise. As the many examples later in this chapter suggest, the ubiquity of harmful dissociations gives us a lot to work with, both for incremental changes and major, long-term ones. Their abundance presents many opportunities in our private and work lives for improvement—for creating a healthier, more livable, just, and ecologically attuned world.

Deep change isn't easy, especially when the nurturing techno-mother makes most of us in the modern, industrialized world extremely comfortable, if not happy. But the recognition of the problems related to our dissociations, and the possibilities for improvement, can motivate us. Accepting greater responsibility for our actions means reducing the dissociations in our lives, a process that begins by seeking to learn more about how we affect the rest of the world.

It's also important to realize that we don't need to completely restructure modern life to begin seeing benefits. Progress lies in just one person working to restore an urban stream or learning about where her trash goes. Moreover, many changes that counter dissociations and their effects are already under way. People have sensed the benefits of living in a more fully connected way, so they participate in producing more of their food or in learning about where their food is from, they get out of their cars more often and walk through neighborhoods, they form community organizations, and they seek to learn more about how their choices shape the world. The possibility that we've found an important key to unlocking a healthier, more just society should give us hope. We can't solve all these problems in our lifetimes, but we can make a good start and inspire future generations.

PERSONAL RESPONSES TO DISSOCIATED LIFE

As individuals we can begin by identifying the dissociations shaping our own lives. You might begin by thinking about where things you use are coming from, how they're made, and the consequences of their manufacture. This book provides some tips, for example, about consumer electronics. You might study food, where it comes from, and how it's grown, processed, and transported. Or perhaps you're more interested in what happens to the waste produced in your home when it's taken away. It can be rewarding to find out about these things your life depends on.

Many fascinating resources exist for learning about the larger world our lives touch. A good starting place for monitoring environmental problems and solutions is Andrew C. Revkin's Dot Earth environmental blog at the *New York Times*. The blogs Treehugger and Yale Environment 360 are also popular and informative.[4] Magazines such as *Orion* and *Resurgence* track environmental problems and their origins and solutions. Some local publications teach about nearby nature. San Francisco Bay Area residents, for instance, can learn about the region's ecological conditions in *Bay Nature* magazine. New books on the environment are being published all the time. The environmental writer Lester Brown's works are particularly readable and compelling. Television and radio programs such as Public Radio International's Living on Earth focus on environmental issues. These and many other sources offer help in tracing the path of dissociated consequences and incorporating that information into daily decision making.

As we analyze our lived dissociations, we discover that they come into play in differing ways. You might buy your food from an industrial grower in a distant country, or you might trade or share food with a neighbor. In the latter case you have direct personal connections to the producer and the land where the food was grown. In the former you have only indirect economic connections that percolate through many layers. But both cases include more dissociation than producing your own fruits and vegetables. In evaluating the consequences linked to your actions, you may decide that the environmental damages of a certain industry are acceptable but that those of another aren't. Of course, being disconnected from something or someone is relevant only when your actions can affect that thing or person in some way. If you don't consume meat, it would be meaningless to say you're dissociated from meat production.

Our choices as individuals or as actors in corporate or government roles can be informed by learning about the links between our disconnections and the impacts they create. You might simply reduce or eliminate a whole category of activity such as air travel, which contributes greatly to global warming, which in turn is destabilizing agricultural systems and increasing flooding of coastal cities. Each person and each community can choose the most relevant problems to respond to, and in the aggregate we'll be creating a better world for ourselves and our fellow travelers on the planet, human and otherwise. We don't have to completely disrupt our busy lives to move in a better direction; gradual change can be meaningful and productive and is less daunting.

But to adequately combat the effects of dissociation, we can't simply respond individually, each person seeking his own healthier path. Too much focus on personal action, overemphasized in our individualist society, may dampen more meaningful change at the societal level. The environmental sociologist Andrew Szasz argues that our personal responses to the poisons in drinking water, industrial chemicals in food, and dirty air in recent decades has been to double down to protect ourselves by buying water filters, bottled water, "natural" bedding, and "green" household products. Some of these products are effective at protecting us individually, but others, such as bottled water, which is less regulated than tap water, may not be. More important, says Szasz, focusing on these individual solutions detracts from efforts in the public sphere to actually fix these problems at their source—to stop contaminating our precious resources in the first place.[5]

Moreover, it's important to realize that just as our dissociations block us from fully appreciating the harms we create, they also make it difficult to be sure about possible solutions. Industry and the government alike can deceive us, perhaps intentionally through greenwashing (making products seem more environmentally friendly than they are), into adopting solutions that merely bury the problem or create new ones. Dissociations mask the truth, even from titans of industry and government officials. Today's engineering, manufacturing, and economic systems are so complex that it's often hard to know with any certainty the most beneficial choices. Take for example the electric car. In 2010, the US National Research Council (the "Advisors to the Nation on Science, Engineering, and Medicine") analyzed the full life-cycle energy costs of manufacturing, operating, and disposing of electric cars and found they may consume as much or more energy overall than regular gas-powered cars.[6] Mining the metals for their batteries, for instance, consumes massive amounts of energy. Yet other studies contradict this conclusion. The lesson here, I think, is that it's truly hard to know the costs of any complex technology, and we should remain skeptical of attempts to solve all our environmental problems with exciting new inventions.[7] In many cases lower-tech solutions that reduce energy and resource use—local food production, public transit, biking, organizing communities more efficiently—are better for human and environmental health.

Besides individual and community initiatives, addressing harmful dissociations systematically will require a lot of long-term political work and policy changes. I believe the key will be to adopt new principles against dissociation that can guide laws, administration, and regulation to help us reclaim our ethical agency. Such principles will favor policies that

1. inform people about more of the consequences of their consumption decisions;
2. more directly engage people with nature;
3. incorporate more of the costs of things into their prices;
4. promote local and regional goods and services;
5. advance educational policy that teaches students how everyday actions propagate out into the world;
6. put people into closer contact with, or at least give them better knowledge of, the most important consequences of their actions in their personal and professional lives;
7. limit inequalities in wealth.

A policy following these precepts might, for instance, require products to come with more information about their production. Today in the United States, food packages list ingredients, nutritional content, and country of origin. It would be nice to know even more about our food's manufacturing conditions, such as the amount of energy used, a more specific geographic location, or the average wage of the workers whose labor we're benefiting from. Such a policy could be applied to other types of products as well.

When people are more connected to nature in both work and leisure, perhaps involved in farming part time or working on ecological restoration projects, they learn important lessons about nature's needs and how we can partner with rather than dominate it. New programs could educate people about the range of real-world effects of particular consumption choices. Incorporating more costs into consumer goods and services, including costs over the full life cycle from production through disposal, acknowledges the ongoing role of market capitalism in our lives. When we strive to teach children how their living in the world affects nature and other people, they can learn to make more thoughtful decisions and have greater command over their own ethical engagements. Imagine schoolchildren regularly taking field trips to industrial farms, factories, waste dumps, and sewage treatment plants to see how the extended economy serves them and shapes their world. What about a universal sunshine law that allows people to see how things they buy are made?

When we promote locality, we cultivate an approach to living that reorients our sensitivities to the land and to our neighbors, which makes us more aware of how we affect one another (though we should not then ignore our continued effects on more distant others). More even distribution of wealth is also essential to establishing less dissociated lives. It's difficult to teach children the ethics of greater awareness of their own consequences when very wealthy people, by the sheer magnitude of their potential for consumption, can so easily overshadow the efforts of many of us. Besides, the ideal of unlimited individual wealth that underlies economic inequality depends on a notion of freedom that's built on a fantasy that denies the contributions of others.

THE FANTASY OF PURE FREEDOM

In modern Western cultures, freedom occupies an exalted position that seems beyond the reach of criticism. Yet our inherited notion of freedom must be seen in its historical context. In chapter 5, I discuss the privileged position that Plato enjoyed as a free Athenian upper-class male citizen. Plato's status apparently so enveloped him that his efforts to free his society from the incessant political struggles and strife of the time, and from what he viewed as the chaos of tumultuous nature, ironically led him to solutions that required the maintenance of a class of slaves, the promotion of elite authoritarian rulers, and the development of a conceptual system that codified his own dissociated relationship to nature in the Forms. That is, Plato's version of freedom required power, denial, and detachment over and from the humans and others on whom his life depended.

Plato's freedom was relative, contingent (dependent on outside factors), and privileged: it was intertwined with the unfreedom of others. The history of the United States manifests a similar contradiction. The colonists desired to free themselves from the colonial power of England. But they created and maintained an unfree class of African slaves and denied Native Americans the freedom to continue to pursue life on the land as their cultures had long done. The colonial and early US national economies depended both on slave labor and on the "free gift of Nature," as the colonists brought landscapes into the capitalist system of production, often rupturing the connections of Native Americans to the land in the process.[8] These major moments in the history of the West reveal that unexamined, absolute, and undialectical (immune to reasoned argument from multiple perspectives) views of freedom mask the dissociated social and political structures that enable people to wield power over the human and non-human others who sustain their lives.

Plato's desire for freedom from the chaotic and uncontrollable physical world finds a modern counterpart: the desire for total techno-scientific mastery of nature. Both forms of freedom—Plato's disavowal of physical reality and ours using science and technology to attempt to master nature—deny our dependence on the larger community of life to which we belong. They negate nature's role as an active partner in the ongoing dance of life. Both notions of freedom dissociate humans from nature by interpreting what is actually a rich, subject-to-subject, active relationship as a reduced subject-to-object one. To be better grounded, more meaningful, and more beneficial to all people and to the other-than-human world, these modern Western ideas of freedom must be adjusted to account for actual interconnections and to envision all people in reciprocal and interdependent relationships of mutual responsibility.

ASSOCIATING ETHICS

An ethic is a set of precepts or principles that guide action. Is there an ethic that will give us a more systematic way of accounting for and responding to the problems of dissociation? Actually, there are many possible ones; I call them collectively associating ethics. They share certain goals and features. Most important, they promote conditions that help people learn more about and bear greater witness to the consequences of their actions and the nature and other people their decisions affect. They operate on the shared assumption that a less dissociated world supports greater ecological and human health. When people are more in touch with the outcomes of their choices, they'll be more likely—though of course not guaranteed—to make choices aimed at healthier environments and a more just society.

More associated life takes many forms because the ecological and cultural situation of each community is unique. In my view, it's not my proper role to present an abstract, universal environmental ethic to be applied everywhere. One radical community may decide that each tree should be valued equally to each human. Another may choose to value clean air and clean water above all else. But under more associated conditions, the ethic a community adopts is more likely to lead to ecological sustainability and humane treatment of other people.

The Merriam-Webster Dictionary defines the verb *associate* as "to join as a partner, friend, or companion."[9] Associating ethics acknowledge that humankind is engulfed in the natural world and interdependent with it. That's why we should partner with nature rather than trying to dominate and control it. Partnering means cooperating with the other-than-human world—living beings, ecological "services," ecosystems, the biosphere, and the cosmos—and seeing its value and responding to its needs. Thinking of these beings as our companions in life will give them voice in our decisions and will acknowledge their role in our lives. For instance, we can seek to be more aware of our dependence on the air we breathe and the connection between the quality of that air and the actions that degrade it. We can honor the clean water we drink and be mindful of how buying products made from toxic materials can contaminate drinking water.

The processes that produce our food, clothing, appliances, and housing also support our lives. An associating ethic aims to better connect us with them as well. That may entail learning more about how the things we buy are manufactured or making more things—clothing, food, games—ourselves. Getting informed about and participating in the production and disposal of things we want and need gives us a clearer sense of what's involved and empowers us to make decisions more aligned with our values.

Adopting an associating ethic doesn't necessarily mean going back to a romanticized past of log cabins and herding sheep. Rather, we can seek to move forward to unknowable and unimagined modes of life with new technologies that don't dissociate unnecessarily. A utopian project is not required: perfect imaginary worlds might just seem out of reach anyway. Indeed, the word *Utopia*, the title of Sir Thomas More's 1516 book and the ideal society it depicts, is a pun referring to two Greek words, one meaning "good place," the other, "no place." Any model of the future could actually defeat change by serving as a mental repository where ideas for moving forward are tidily filed away and don't interfere with lived experience. Projects for reassociating can begin immediately, in everyday life, without a master plan.

Such projects can be either practical or theoretical. Because we create the world as we think it—and, as I've presented, ideas central to Western modernity provide a foundation for dissociated living—truly healing the planet will require new ideas that build on existing ones that see the cosmos in its integrated wholeness. The logical, ontological, and epistemological disintegrations in modern thought can be counteracted with new philosophy and everyday discourse that gives proper attention to context, connection, dependency, relation, similarity, and wholeness while maintaining realistic and adequate consideration of particularity and difference. In the rest of this chapter, I explore various theories and practical ideas for living more connected lives.

First, a word of clarification is due. Associating ethics might seem to favor conservative politics, at least the brand of conservatism popular in the United States. Conservatives call for power to be "devolved" from the federal government to local governments. On the surface, this position seems to work against dissociations, but conservative principles also promote individualism, which divides people economically and empowers some over others. The supremacy of the individual drove the dissociating thought of ancient Greece, infusing the West's intellectual heritage with dissociating dualistic concepts. Local politics have sometimes led to the worst human rights and environmental abuses, and the US federal government abolished slavery, racially integrated schools, established basic human rights, and instituted environmental protections, without which nature and civil rights in much of the United States would be worse off (though, of course, the federal government could do even more and has caused its own problems). A localism that turns inward to enact unjust and environmentally damaging practices would not fulfill the ideals of an associating ethic.

In her book *Environmental Culture: The Ecological Crisis of Reason*, Val Plumwood points out that the idea of autarchy—that each locale should serve and rule only itself, practicing complete self-sufficiency and autonomy, avoiding involvement with or responsibility toward the outside world—can be destructive if it means ignoring the harms its economy is perpetrating elsewhere or turning a blind eye when people or nature are abused in other parts of the world.[10] Communication and cooperation across communities and political hierarchies are necessary to avoid such problems. Any new ideas for more associated living are best applied carefully, gradually, and in full sight of actual conditions and results. And they have to be practical: clearly, we can't all be expected to grow our own food tomorrow. The examples I present here don't represent a complete solution or program for counteracting harmful dissociations. They're only a starting point and a small sampling of endless possibilities.

LIVING THE WORLD CONNECTED

To achieve a more connected world means reorganizing our lives to put us in closer contact with nature and the consequences of our actions. That starts with learning more about the world beyond, the world we affect with our everyday choices. We can look into the processes of production that we depend on or seek activities that give us more experience and time in places wilder than our usual built environments or that foster better understandings of the ecologies of our home regions. We can become more involved in governance through town hall meetings and neighborhood organizations. A less dissociated world engages citizens in political participation, increasing both local control and global responsibility. It removes the corporations, lobbyists, and money that intervene between people and their elected representatives, particularly in the United States. Reassociating means getting involved, learning about how our society works, accepting responsibility, and regaining control over how we affect our world.

A reconnected world starts with more livable, walkable, and bike-able communities where residents encounter human, domestic, and wild neighbors on a daily basis. It requires civic planning that puts people closer to their jobs and fosters community interaction with dedicated pedestrian areas and slow-traffic zones. Also needed are shared, public facilities such as community halls and better public transit. Architecture can reassociate by helping to build connections across extended families and communities rather than compartmentalizing people in nuclear-family, standalone houses.

Communities organized this way provide immediate social and environmental benefits. Less single-occupancy car travel and fewer cars mean less pollution and energy use and more dependence on human-powered and shared, public transportation. More connected communities also alleviate alienation because they foster social encounters. Neighborhoods have been organized this way throughout human history, and even many large cities today, particularly European cities such as Amsterdam and Munich, create as much community as they can by promoting walking, biking, public transit, and neighborhood centers. But we can go much farther in smaller towns by arranging them so that food is grown close by and nature isn't far away.

The work of reassociating can be rewarding in itself. Learning about food systems is empowering, for instance, and participating in producing our own food brings still greater satisfaction, as the Edible Schoolyard Project does for junior high school students in Berkeley, California. The connection and cooperation that come from reassociating practices provide their own benefits: conversation, mutual aid, and friendships. Advantages to nature and other people, combined with the feeling of working toward positive change, counteract feelings of despair and disempowerment that arise from seeing a world degrading and disintegrating.

A reassociated world is a long-term quest approached through many small and large steps. Some changes may come more quickly; others take generations. Wealthier nations and people may have more possibilities for change than poorer ones, and more benefit can be gained when they change their ways because their environmental and human toll is so much greater. In spite of the limitations we face as we try to reduce our alienation from nature and from our consequences, we can find satisfaction in knowing we're headed in a positive direction. Perhaps the examples below will inspire changes in your life.

Reconnecting with Nature

A better understanding of ecology helps in any effort to get in touch with nature. Public education in ecology and the impacts of human activity on nature could be a more prominent part of the curriculum of public schools. Possessing a basic understanding of ecology, students can study environmental literature, ethics, philosophy, and other environmental humanities and social sciences to round out their understanding of the human place in nature. If ecological curricula emphasize nearby nature, students can learn about the particular ecological needs of their local environments. They can also gain perspective on how their own actions play out in the larger world by learning how industry uses nature in their region and elsewhere through classroom study and field trips. Besides creating intellectual bonds between students and nature, ecology can teach important lessons about connection and interdependence, qualities that jump to the fore in this science that sees the natural world as a thoroughly integrated, interconnected realm, as I elaborate further on in this chapter.

Programs that introduce adults and children to the natural spaces of their home regions through talks and guided tours enhance awareness of surrounding nature in active, social ways. One example is Close to Home, founded by the San Francisco Bay Area environmentalists Cindy Spring and Sandra Lewis. Participants attend monthly presentations on a topic of natural or cultural history in the region; a week after each presentation, they go on a related day-long field trip to view flora, fauna, and geology and take part in human-nature interactions in selected locations: marshlands, estuaries, stream beds, parklands, or nearby mountains. Local naturalists and indigenous culture experts make the presentations and lead the field trips. On a Close to Home outing on Mt. Diablo east of the San Francisco Bay, I learned about the rare yellow fairy-lantern lily growing wild on its flanks and other aspects of its ecology, and so earned a deeper appreciation for a mountain that has always loomed in the distance. Programs like this one could be extended to larger segments of the population, with outreach to disadvantaged groups and others with less access to natural areas.

Projects that involve local residents with nature in an active, working capacity have even greater potential. They can deepen knowledge of natural processes and develop a community's sense of commitment to nature. Working together toward a common goal also strengthens communities. In the San Francisco East Bay, citizen groups work to "daylight" creeks that run from the hills to the bay. East Bay cities historically have paved over the creeks, placing them in culverts below streets, yards, and buildings. Some of these groups lobby cities to open the creeks again to the sky and restore indigenous vegetation to their banks. Volunteers working on ecological restoration projects learn about the region's hydrology and vegetation and gain a more intimate relationship with local nature.[11]

Urban gardening connects people with both nature and the processes of food production their lives depend on. In community gardens, participants tend plots granted by the city or reclaimed vacant lots. Sharing land and knowledge of growing techniques and conditions, along with seeds and produce, fosters communal cooperation. Gardeners get contact with living nature when they may not otherwise have the opportunity. They learn about the seasons and issues of soil quality, hydration, and other natural processes; nature-deprived inner-city residents can experience direct, productive interaction with the land. They also gain the deep satisfaction of producing some of the food they eat, from earth to plate. In Berkeley, the Spiral Gardens Community Food Security Project uses urban gardening "to create healthy sustainable communities by promoting a strong local food system and encouraging productive use of urban soil."[12] Social justice groups in Oakland, California, use urban gardening as an effective organizing tool for empowering disenfranchised and impoverished communities.[13] Similar organizations work toward food security in urban settings across the United States and elsewhere. The Detroit Black Community Food Security Network operates D-Town Farm and other projects to "build self-reliance, food security, and food justice in Detroit's black community." They pursue these goals through public policy, urban agriculture, promoting healthy eating and cooperative purchasing, and preparing young people for careers in food-related fields.[14]

Beyond gardening, there are many ways people can become better engaged—less dissociated—from the food they eat. Throughout human history and across cultures, growing, preparing, and eating food has brought families and communities together daily to enjoy and celebrate their common sustenance and Earth's bounty. As food has become increasingly industrialized over the past two centuries, imported from farther reaches of the globe, it has shifted from communally produced substance packed with meaning and sometimes love to a cold commodity disconnected from family and community with unfamiliar ingredients that might be manufactured in chemical plants, with high densities of salts, sugars, and fats to fire up inattentive taste buds. Many schoolchildren eat snack-like breakfasts while walking to school, and for a lot of American families, dinnertime has fractured into individual meals out of the microwave, arranged around busy schedules, often gulped down while distracted by television, email, and text messages. Books such as Michael Pollan's *The Omnivore's Dilemma: A Natural History of Four Meals*, Eric Schlosser's *Fast Food Nation: The Dark Side of the All-American Meal*, Christopher D. Cook's *Diet for a Dead Planet: How the Food Industry Is Killing Us*, and Wenonah Hauter's *Foodopoly: The Battle over the Future of Food and Farming in America* teach about some of these late-modern changes to our relationship with food.[15]

But preparing and eating food doesn't have to be so individualized and commodified. The "slow food" movement, which began in Italy, and authors such as Pollan and the restaurateur Alice Waters of Chez Panisse fame encourage people to become more actively and intentionally involved in food production by growing, selecting, cooking, and eating food with and for friends and family. Through Waters's Edible Schoolyard Project, Berkeley schoolchildren grow, cook, and eat their own fresh, healthy food in a large school garden and kitchen and at home. They gain a lasting appreciation of the values of community, nourishment, and land stewardship. The movement is spreading throughout the United States. It's not too late for adults to become more engaged in producing the food they eat. It can be simple and fun to buy and cook simple ingredients, and to learn more about those ingredients by talking to farmers at farmers' markets and the few remaining independent grocers. The farmer-author Wendell Berry and other food and agriculture writers give us stimulating starting points for learning more about contemporary agriculture and alternatives.[16]

When we cook our own food and share it, we reconnect ecological and social networks.[17]

We can also connect with nature through animals in our everyday environments. Companion animals and undomesticated "familiar" animals put us in relationships that extend our experience with the world beyond humans. Val Plumwood made friends with wild wombats, an Australian marsupial with powerful jaws and teeth, living in the wooded area around her house. She picked them up occasionally and let at least one of them wander inside often. Not many of us would tolerate chewed-up furniture, but even in large cities many people become familiar with critters living in their environs. Closer to home for most of us, dogs, cats, and other pets invite us to care for them with their vulnerability, and when we do so, we're bridging the gulf between us and nature. Caring for animals can be particularly rewarding when they come from rescue shelters, such as the one in Berkeley where I volunteered for years to walk and comfort homeless dogs. Our relationships with companion animals teach us about cross-species communication and empathy. Learning to value connections with animals in their own right, rather than treating them as a source of entertainment or physical beauty, can be a lesson in valuing nature as a whole.

Any journey to reconnect with nature will benefit from some inspiring travel companions who've recently appeared. David Abram, whose ideas in *The Spell of the Sensuous* I build on in previous chapters, has written a wonderful guidebook for experiencing, feeling, and understanding ourselves as the sensually attuned beings we truly are when immersed in nature: *Becoming Animal: An Earthly Cosmology*.[18] Following Abram into the forest, we can learn the poetry, magic, and thrill of opening all our senses to the flowing, vibrating rush of nature, to be fully engaged and engulfed. It's possible to once again experience not only our bodies but also our minds as integral parts of nature.

Another guide is Richard Louv, whose book *Last Child in the Woods* details the developmental and psychological consequences of "nature-deficit disorder."[19] Children growing up in the past couple of decades are suffering from their lack of contact with wild nature, but Louv doesn't merely reveal problems stemming from their nature alienation. He proposes many concrete ways of bringing nature into children's lives and bringing children into natural landscapes: building a bird bath or bat house, hiking, adopting a tree, camping, learning to track. Besides suggestions for children, Louv offers a slew of ways of getting communities, businesses, educators, and governments involved in a nature reconnection movement: transforming vacant lots into "wild zones," building new conservation organizations, creating nature-based onsite child care, greening schoolyards (many of which have been paved over), and supporting policies that provide more naturalists in parks. Each of these examples on its own may have limited impact, but practiced together they can help make nature a living presence and powerful force in children's and adults' lives.

Shifting Consumption

Truly responding to the dissociations that shape our lives will require major changes in economic systems, but we can also work for the time being within existing structures. Associating people more closely with the outcomes of their actions can be done, for instance, when prices are made to reflect environmental harms more directly, or when information about those harms is connected more closely with consumption and use choices. Many methods, including taxes on environmentally damaging purchases at the consumer level or fines and penalties at the producer or manufacturer level, can accomplish this goal. This approach is hampered, however, because price collapses various sorts of information to a single, one-dimensional quantity. Moreover, wealthier individuals and corporations are in many cases immune to price pressures. But other, more effective, forms of feedback are possible. For instance, frequent (daily or more) data about energy consumption provided by special meters in homes has been shown to reduce heating fuel consumption.[20]

Other methods associate product information with consumption choices before they're made. One new and intriguing possibility is the use of bar codes on items, together with software on smartphones, to display details about the production processes and materials of particular products—for instance, the conditions in which a chicken was raised, with photographs. Shoppers use their smartphones to scan bar codes, and the software keys into a database containing relevant environmental, labor, and other details of the product's origins.[21] One such facility is already available via web browsers and smartphone apps: goodguide.com.[22] These technologies heighten consumers' awareness of the fuller set of consequences related to things they might buy, reducing knowledge dissociation.

Greater transparency in manufacturing gives consumers more power in their choices and lets them know how goods are produced before buying them. Alta Gracia, an innovative company based in the Dominican Republic, makes logo apparel for the collegiate market. It treats its workers extraordinarily well, maintaining high health and safety standards and paying them a fair wage to lift them out of poverty. The company lets American college students use Skype® to talk to the workers onsite and keeps its factory open to monitoring by a labor rights organization. Alta Gracia may provide even better transparency than the "fair trade" certifiers, another excellent resource for people wanting more connection with the origins of their purchases.[23]

Imagine a world in which a window is always available onto the land-scapes and work places that sustain us, one that doesn't continuously conceal those spaces and the people who do the hard and dirty work we all depend on. Timothy Pachirat, the political scientist who worked in an industrial cattle slaughterhouse, coaxes his readers to consider such a world, in which "those who benefited from dirty, dangerous, and demeaning work had a visceral engagement with it. . . . In this world . . . to eat meat would be to know the killers, the killing, and the animals themselves."[24] Along the same lines, Pollan poses the idea of glass walls on abattoirs to let the public see what's going on with the animals killed for their benefit, to give us the "right to look." A handful of small meat processing plants do provide such transparency, for example, Lorentz Meats in Cannon Falls, Minnesota.[25]

But responsibility must accompany transparency. Too often, the disso-ciations organizing modern life alienate consumers and corporations from responsibility for their choices. The Extended Producer Responsibility (EPR) framework reestablishes the connections between choices and con-sequences. Corporations must assume responsibility for the goods they make, "from cradle to grave." They're not released from accountability for things after selling them—that is, they don't become dissociated from their commodities and all the consequences of creating them. Articles may be returned to manufacturers at any time, and the latter are held to envi-ronmental and other regulations regarding the re-use, re-manufacture, or disposal of returned items. The consequences of their manufacturing decisions thus remain with them. This result is sensible because producers have control over the materials and processes they use. Under the pressure of having to dispose of unwanted products, they may choose to make only those that can be disposed of or recycled more cheaply, completely, and benignly. EPR is already implemented widely in the European Union but has not yet been adopted in the United States, where it could make a big difference.

Other methods are available to mitigate harmful dissociations under capitalism in the near term. The Norwegian peace studies pioneer and Alternative Nobel Prize winner Johan Galtung proposed "exchange principles" that would stop harmful consequences from being shifted onto less powerful societies or ones with weaker environmental protections. Exchange would be allowed across similar levels of environmental regulation and production when the exchange doesn't create or intensify power imbalances or relative disadvantage between parties. The total amount of processing used to create the items for exchange should be about the same in both directions to reduce the possibility of inequitably rearranging environmental and social costs. Galtung encourages autonomy of the parties to exchange, horizontal division of labor, solidarity, participation, and other types of integration.[26]

Galtung also proposes a concept of self-reliance that encourages self-sufficiency in basic needs at local and regional levels. The economics of center-periphery—in which economically more powerful, thus "central," societies treat economically more peripheral ones as an external sector of their own economy, thus dissociating away many of their harmful consequences—is replaced by greater self-reliance, in which you "produce what you need using your own resources, internalising the challenges this involves, growing with the challenges, neither giving the most challenging tasks (positive externalities) to somebody else on whom you become dependent, nor exporting negative externalities to somebody else to whom you do damage and who may become dependent on you."[27] The goals of self-reliance are meeting basic needs while preserving "essential ecological balances"; producing what is needed using, as far as possible, locally available production factors; trading as necessary; and applying strong ethical constraints based on compassion. Trade is acceptable under self-reliance within the constraints of the exchange principles: (1) the net balance of costs and benefits, including externalities (costs that aren't counted, such as acid rain), of the parties is as equal as possible, and (2) production at least of basic needs should be carried out in such a way that the country or region is at least potentially self-reliant.[28] Applying Galtung's exchange principles and concept of self-reliance at multiple scales—from the town to the nation—could protect communities from exploitation by more powerful outside (consequentially dissociated) parties.

The Bioregional Approach

Bioregionalism resembles Galtung's ideas but goes beyond them to a more all-encompassing focus on local and regional society and ecology. It associates people more closely with nature and community by promoting sensitivity to place and alignment of economic institutions to local landscapes and ecological conditions. Inspired by Native Americans' intimate connections to the land, the poet Gary Snyder expressed a bioregionalist sensitivity to place in his 1974 book, *Turtle Island*.[29] The conception of bioregionalism developed by the movement's leading proponents Raymond Dasmann (1919–2002) and Peter Berg (1937–2011) envisions human communities that correspond to and cooperate with more-than-human biological communities defined by ecological features such as a watershed (a region in which all precipitation flows ultimately into a single river). For Dasmann and Berg "living-in-place" means "following the necessities and pleasures of life as they are uniquely presented by a particular site, and evolving ways to ensure long-term occupancy of that site." Living-in-place fosters links between "human lives, other living things, and the processes of the planet—seasons, weather, and water cycles—as revealed by the place itself." Recognition of the multiple interconnections of life on Earth and efficient resource use are integral parts of bioregionalist philosophy.[30]

In addition to factors such as climate, animal and plant geography, and natural history, Dasmann and Berg emphasize the importance of the "distinctive resonances" among the living beings within a bioregion, which might include the political, economic, and cultural systems within a clearly definable ecological zone such as a watershed or basin area. An example of a bioregion might be the Mattole River watershed in Northern California or the island of Kauai, Hawaii. Bioregionalists envision people in intimate relation with their environs, experiencing connections that lead to ideas, symbols, and life-ways that are deeply appropriate to place.[31]

Sometimes the words we choose can help to maintain connections. In Bali an everyday idiomatic expression, *desa-kala-patra*, is used to insist that people consider the multiple aspects of the contexts in which all events occur. The words translate to place, time, and social position or role, the latter deriving from the notion of an actor's part or character in a play.[32] The idea of *desa-kala-patra* evokes a basis for knowing and acting that's bound to setting. It's used often in discussions about appropriate behavior or when a group, such as a *banjar* (hamlet or neighborhood organization), works together to make difficult decisions. A less opulent ceremony held by a poor banjar might be judged equivalent to a much more lavish one held by a rich banjar because the economic surroundings of both are taken into consideration. A person or group should act based on what is appropriate to the *desa-kala-patra*. The term is an easy reminder of contextual factors that may otherwise be forgotten or ignored. In English, the simple question "What about the context?" could go a long way toward reminding people of the inevitable connections around every event.

Psychology and Connection in Daily Life

Emotions are powerful drivers of action. Feelings of being connected with human or nonhuman others produce empathy, an important source of benefiting others and inhibiting behavior that harms them, according to Ervin Staub, the pioneer of peace and violence studies. When one feels distant from others and sees them and their experiences as remote, feeling empathy is unlikely. When connectedness—with nature and other people in an extended community of life—is valued for itself, real connection and the *experience* of connection will arise. That experience is in turn a precondition for empathy and related emotions such as love and respect that provide the foundation for caring behavior.

In his work on large-scale violence against people, Staub shows how dissociating thoughts that devalue people along class, ethnic, gender, and other lines are basic elements of the conditions under which such violence originates. Besides valuing connection itself, Staub suggests specific remedies such as the creation of "crosscutting relations" and superordinate (overarching, commonly held) goals. Members of groups alienated from one another should spend time together, developing familiarity, if not caring. But proximity isn't enough; members of subgroups must live, work, and play together, and their children must go to school together. Under such conditions, genuine knowledge, empathy, and authentic relationships begin to form between formerly hostile groups. Connections of caring and cooperation emerge.[33] More Palestinians and Israelis must be in each other's lives in productive and enjoyable activities, for example. Likewise, we people alienated from nature must find more ways of spending time in and working with the natural world, not just playing there.

If empathy depends on direct and engaged relationships, the dramatic rise of email, cell phones, texting, Facebook®, and Twitter® over the past decade is cause for concern. These devices and interfaces are markedly changing how we carry out our personal and professional relationships. People feel more connected than ever because they're in more frequent contact with friends and relatives far and near. With cell phones and texting, parents now track their children's whereabouts (or supposed ones) by the hour. And they sometimes panic when they lose contact for just a short time. But we're spending more time with our gadgets and less in face-to-face contact. It's become common to see people in groups ignoring their immediate companions in favor of texting with others (or texting private messages to someone in the same group). I haven't yet noticed students texting during my classes, though I've heard the practice is common. Even some adults out for a dinner with a friend, perhaps on a date, think it's acceptable to give their attention to someone else by texting.

Sherry Turkle, a technology and society researcher at MIT with a background in psychology, has been studying the impact of these technologies on relationships for the past fifteen years. She finds that people constantly connected through electronic media are losing not only the desire but also the capacity for solitude—the ability to spend time alone without feeling lonely. The personal-growth benefits of being alone in reflective contemplation can't be realized some other way. Many people today experience an unnerving loss when separated from their devices, a feeling sometimes bordering on panic.

Most important to this study of dissociation, Turkle's findings confirm that people become more aggressive when communicating through electronic media. Face-to-face communications suppress hostile tendencies and help people work through relationship difficulties.[34] Sites such as Facebook and Twitter have been credited with speeding democratic reforms in various nations. But as an entire global generation from Bali to Zaire, from Albuquerque to St. Petersburg, grows up linked in to "tweets" and texts and focused on synthesizing online personalities on Facebook, we must consider what the ubiquitous use of electronics in social relating means for the health and stability of relationships at the personal level and between communities as well.

A Model for Relating to Artifice and Nature

In the philosophical novel *Zen and the Art of Motorcycle Maintenance: An Inquiry into Values*, Robert Pirsig's unusually thoughtful relationships with artifice take center stage. I first read this mind-expanding book in a college philosophy of science course.[35] Pirsig indicts the modern pathological obsession with quantity. *Quality* and direct, intimate relations with a smaller set of things are the best foundations of values. He laments the disconnections between people and the objects in their lives; outside experts and institutions intervene. Our dependencies on them, he shows, are often exaggerated in the common imagination: it's possible to be more aware, deliberate, and engaged in the world. Pirsig offers a beautifully nuanced accounting of his relationships of immediacy with his motorcycle, which he meticulously maintains, and with the surrounding social and natural environments that it carries him through. His patient, plodding, appreciative approach to the machine that he depends on brings him into a reciprocity of sorts, in which his nurturing of the motorcycle is matched by its steadfast service to him. Caring for the machine, keeping it maintained and running smoothly, brings it into partnership with him, just as doing the same with nature makes us partners with nature.

Pirsig's connection with his motorcycle is an inspiring model for relating with all sorts of artificial, natural, and hybrid nonhuman and human others and the cosmos as a whole. The relation is the thing, and participation in a mutually nurturing one is central to being—to life in a more livable world. When we relate more intentionally and thoughtfully with the human-made objects in our lives, we get equal or greater satisfaction from fewer of them. It can be liberating to have less clutter and possessions to manage, as I found when traveling internationally for a couple of years with no more than I could carry. Deeper appreciation of our artifice means recognizing the productive nature that's an inherent part of it. As I walk over the oak floor of my home, I often look down and admire its beautiful grain and think of the trees, harvested in the 1920s, from which it came. I value it more deeply because I think of its origins in growing, living nature. The image of Pirsig on his motorcycle with his son, wind in his hair, sights all around, and scents filling his nose, evokes a deliberate and thoughtful partnership with artifice—not a generic, taken-for-granted one—that enhances rather than detracts from encounters with nature.

THINKING THE WORLD CONNECTED

> In the beginning is the relation.
> —Martin Buber, *I and Thou*, 1937*

> When we try to pick out anything by itself, we find it hitched to everything else in the universe.
> —John Muir, *My First Summer in the Sierra*, 1911†

To meaningfully reassociate the phenomenal world of nature and experience, we must reconceive the world as whole, interconnected, and interrelated. Objects don't precede relations but emerge together with them. Such ideas are fortunately not new, though they're certainly not in the mainstream of modern thought. They come from thinkers in various scholarly fields, religions, and cultures, from physics, philosophy, systems theory, and environmental theory to Buddhism. Environmental philosophers and other scholars have long recognized the problems of fractured conceptions of the world. As I've discussed, the modern philosophy of mechanism helped to turn people away from nature and view it as a passive, inert realm to be exploited. Environmental philosophers, nature writers, and others have responded to the modern divorce from the natural world with various ideas about reestablishing our vital connections with it. These ideas hold in common a focus on connection, particularity, immediacy, the body and use of the senses, respect, and deep relation.

We can start by seeing ourselves more clearly as part of the larger natural world. Theodore Roszak launched the promising field of ecopsychology twenty years ago with the goal of awakening an inherent sense of reciprocity with nature that he believed is innate in each of us. At the core of our minds resides an "ecological unconscious," he writes, a living record of cosmic evolution. The key is to access that core. Roszak reminds us that a truly comprehensive, universal understanding of our identities must acknowledge that "salt remnants of ancient oceans flow through our veins, ashes of expired stars rekindle in our genetic chemistry."[36] Recovering the newborn's enchanted sense of the world—through practices such as walking into a forest, grassland, or other wild area with our senses fully open and our thinking adult minds quieted, to just sit there and absorb the natural wonder—will help mend our fundamental alienation from nature. Eventually, our ethic may expand to a planetary one built on the "synergistic interplay between planetary and personal well-being."[37]

The Self in Nature

In the late 1970s, a decade before Roszak published the first ecopsychology text, Gregory Bateson challenged the inherited Western duality between cognition and the natural world in his book *Mind and Nature: A Necessary Unity*. He showed that the criteria of mind are hardly unique to humans, as Descartes insisted; indeed, they pervade all of nature. Bateson observed that mental faculties can be characterized as collections of interacting parts or processes involving circular or more complex chains of determination—a cause becomes an effect, which then becomes a cause leading to additional effects, and so on. In the other-than-human world, we see this phenomenon when wolves hunt caribou and kill the slowest or weakest individuals. The faster, stronger caribou reproduce, passing on those traits to their young. Likewise, the swifter, cleverer wolves eat more reliably and thus are more likely to reproduce. The causal relationship between these caribou and wolf characteristics is circular, expressing a certain kind of mental activity, in Bateson's view. Thought and evolution are both mental processes; mind belongs to both humans and nature more broadly.[38]

As a major contributor to cybernetics and system theory, Bateson attempted to bring together various fields of science and social science under a single system of knowledge. His epistemology cut across many disciplines, and he himself worked in linguistics, semiotics, psychiatry, and other fields in addition to anthropology. Bateson's thinking was wholly integrative. "The relationship comes first," he wrote, "it precedes."[39] He turned Aristotelianism on its head. Whereas Aristotle saw the behavior of objects as an outcome of attributes possessed by them, Bateson taught that "'things' are produced, are seen as separate from other 'things,' and are made 'real' by their internal relations and by their behavior *in relationship* with other things and with the speaker [emphasis added]."[40] Internal relationships are inherent parts of the objects in relation. The relationship between you and your mother is embedded in both you and her as qualities of both. You carry around inside you your own relationship with nature. External relationships only describe interactions among things.[41]

This approach to thinking about relationships undergirds ecopsychology. The psychologist Andy Fisher takes an intensely theoretical yet passionate and practicable approach in *Radical Ecopsychology: Psychology in the Service of Life* and his other writings and teachings. Like Bateson, Fisher sees our essential relationships in the natural world as "interrelationships"—relationships that are internal. All things in the cosmos and in nature are not simply connected, but they all imply one another by their very existence. The existence of each is refracted in the existence of all others. "There are no self-contained or self-identical entities but only fields of mutually informing relations," Fisher writes, so "all things imply or contain their past and their future in the present." Through this concept of interrelationship, ecopsychology constructs "a psychology of internal relations that includes more-than-human reality" and thus gives meaning and depth to nature and human nature through their mutual interrelatedness.[42]

Ecopsychologists say our disconnected modern experiences of the world are reflected in our peculiarly detached egos. When we hyper-separate ourselves from the rest of nature, our egos experience that separation as a lack or void. Ego's experience of the engulfing void, "this nonsolidity, groundlessness," produces anxiety in us, writes Fisher.[43] Another ecopsychologist, David W. Kidner, writes that our egos' lack of relatedness frustrates our innate relational tendencies and thus "causes the pathological turning inward of the libido which Freud labeled narcissism." And capitalist economics have a stake in our detached egos. When our dissociated world frustrates our innate "erotic" tendencies toward relationality, our egos go into "survivalist" mode. They become colonists, feeding off the world. As Sigmund Freud taught, ego wants to incorporate the object into the self by devouring it. Through this line of reasoning, ecopsychologists insightfully link our dissociated experiences to detached egos that attempt to satisfy themselves through consumption.[44]

Fisher prescribes a "melting of the ego" and an opening of the self toward all of reality.[45] If we can begin to see and experience ourselves *in* nature, continuous with the rest of nature, the void—the divide experienced at the deepest levels of our selves—will begin to heal. But this doesn't quite mean *identifying* with nonhuman nature, as deep ecologists espouse. We need not replace a split with nature with a merging. The trick instead is to understand what distinguishes us within nature while rejecting both the removal of humans from the natural world we've inherited from the West's intellectual past and the undifferentiated identification with nature proposed as its antidote.[46] For Kidner this means defining and experiencing selfhood not in "contradistinction" with the world, but through "resonance" with it: perceiving and participating in it in a "more-than-rational" way that liberates facets of our selves, including emotional ones, normally confined in modern experience. An enhanced self-awareness arises in combination with an enriched awareness of the natural world.[47]

Abundant sensory engagement with the pulsating natural world can both restore our lives and enable proper responses to nature's needs. Knowledge "gained directly through the senses and in the gut," writes the ecopsychologist Laura Sewall, puts "flesh and bones" on abstractions such as climate statistics.[48] It can "inform us so deeply that denial would be a personal betrayal."[49] To heal our ailing environments, Sewall believes, "our hearts must be triggered, and our bodies and emotions engaged."[50] It's essential but not enough to turn our attention to the positive in nature; we must also focus on what's most troubling. Looking courageously and resolutely at ecological distress is the only way to reconfigure our synapses to attune to the "environmental signals needed for serious course correcting."[51] We "drivers of ecosystem health or demise" must "become both deeply informed by the condition of organic systems and saturated by the generous beauty of the natural order."[52] That is, we must develop both intimate, sense-based knowledge and a deep, embodied appreciation of what we're protecting. This work won't be easy, Sewall warns: "It is cognitively and emotionally expensive to sustain attention on our fractured state of affairs."[53] But the health of the environment and human communities depends on it.

Ecopsychology's descriptions of a fully integrated yet differentiated psyche, or self, recall the sense of self experienced in non-modern cultures. As I've explained, the Canaque, the Zuñi, the Loritja, and other less modern people experienced themselves as more fully interrelated socially and ecologically. The self reflects and contains the other, whether human, animal, or landscape. People have likely conceived and experienced themselves in similar ways throughout human history and across the vast majority of cultures that have ever existed. It's a mode of life and experience that probably still reverberates in our bones and may not be as difficult to recover as we might first expect.

But in modern society, materialist and scientific explanations hold more sway for many people than psychological and anthropological ones. I urge them to take to heart Roszak's reminder that the most basic elements of our flesh and bones were manufactured in stars billions of years ago. All the substances in our bodies were once in other organisms and before that in the seas and soils. Even as you live, nutrients and energy enter and leave your body as food, air, water, sunlight, and waste as it constantly reconstructs itself in an active and ongoing homeostatic process.

This flow of materials and energy into and out of you, and among all organisms, is the focus of ecology, which arose in the nineteenth and twentieth centuries following the discoveries and theories of Charles Darwin and other biologists. Ecology sees interaction and connection as primary aspects of the world. Ecologists study, for instance, how the populations of one species in a biological community affect those of another through predation, helping behavior, nutrient exchange, competition over resources, and so on. Connection is so important in ecology that the very identity of each organism or other element of the natural world depends on how it's connected to the rest: "Each entity is constituted by a particular and unique junction in a matrix of relationships."[54] A species can be thought of as a model of the environment in which its ancestors evolved, in which they lived, and to which they adapted over eons.[55]

Ecology links inorganic and organic materials and processes into seamless wholes. The ecosystem, a core concept, brings together parts into a unity that's ontologically fundamental: the community of living organisms, nonliving materials, and energy in a region, all interacting. The leaf was part of a plant, harvesting sunlight and manufacturing the energy it needed to grow and sustain itself. Now it's food for decomposers—fungi and microorganisms that break it down into component nutrients that will return to the soil to be taken up by other plants—or perhaps the same one. The identity of the leaf is built on these functional relationships.

As the modern West's first true science of connection, ecology is our greatest defense against dissociating thought. By revealing lived relationships, it informs us of the effects and limits of human participation in the natural world and the fuller significance of modern environmental problems. Consider the eradication of wolves from Yellowstone Park, where they were almost wiped out in the 1920s. Ecological degradation followed. Without wolves, their prey, such as elk, overgraze and destroy vegetation that other animals such as beaver thrive on.[56] By building dams, beavers create wetland habitats where many species live. The effects of the beavers' work ripple outward to sustain a rich, productive ecology, illustrating the "trophic cascade" model I mention in the introduction. When wolves were reintroduced to Yellowstone in the mid-1990s, previously degraded areas along streams once again flourished.[57] Understanding ecology can lead us to see ourselves as part of an interconnected community of life that also includes the air, soils, plants, and animals. Seeing others in nature as part of our community engages our moral sentiments and sympathies for them.[58]

Starting from the concept of biotic community—the soil, water, sunlight, air, and organisms in a place all interacting together, creating and re-creating life—the ecologist Aldo Leopold (1887–1948) constructed a new environmental ethic in the 1940s, one that's ecocentric (places the biotic community, not people, at the center of concern) and holistic. One day Leopold shot a wolf to promote more deer (the wolf's prey) and thus more game hunting in a forest he was managing. Inspired by the "fierce green fire" dying in the wolf's eyes with her last breaths, Leopold suddenly saw—and felt—the whole interconnected and interdependent biotic community, not just deer hunters, as deserving care and protection, worthy of moral consideration. His land ethic, which has inspired generations of conservationists and environmentalists, says that "a thing is right when it tends to preserve the integrity, stability, and beauty of the biotic community. It is wrong when it tends otherwise."[59] Although scholars argue over whether a biotic community is precisely any of those things, the land ethic and other "nonanthropocentric" (not human-centered) ethics give us a model for incorporating a more integrative view of the world into our choices.

But it's not only ecosystems and biotic communities that comprise many organisms. Recent findings in biology show that each one of us humans isn't merely an individual organism but is rather a collection of trillions of organisms working together in symbiosis, including over ten thousand different species of microbes: bacteria, yeasts, and others. Your body includes about ten times as many nonhuman cells as human ones. The nonhuman ones are substantial—in total they weigh about 2 to 5 pounds (.9 to 2.3 kilograms). And they aren't merely hitchhikers: many are essential for survival. In the gut, bacteria break down nutrients our stomach acids can't. Some appear to detoxify carcinogens we eat. On our skin, they ward off microbes that can devour our flesh (consider this fact before bombarding them with anti-bacterial soap). Scientists have been learning that the combinations of microbial species differ between people and among locations on a single body, yet they perform functions in the aggregate that no single species could perform alone. Our personal microbes may even affect our behavior.[60] So at the most intimate level of your own body, you yourself are a community of living organisms. The boundaries of the human self are thus decidedly porous: we're all members of the larger community of life and each of us *is* a community of life.

Philosophers and Physicists Countering Dissociation

Some twentieth-century physicists and philosophers, like indigenous storytellers, have laid a lot of groundwork for rethinking the cosmos and nature in ways that give as much weight to connection as to objects. Recall that Alfred North Whitehead's metaphysics resolutely discards dissociating structures. In his cosmology, the whole of reality is an organism in the sense that the essence of every existing thing depends not merely on its components but also on the particular pattern or structure in which those components are composed. An example of organismic relationships can be seen in the human digestive system, which conveys nutrition to the blood, which delivers it to the rest of the body, which can then work to get more food. Nature, wrote Whitehead, is process and is composed of moving *patterns* whose movement is essential to their being. These patterns are organized into *events* or *occasions*, each instance of which binds space and time together. Mind is an inherent part of all nature, not something external (the idea Bateson later developed).

Whitehead cautioned that breaking down a complex occasion into its component events does reveal the individual parts, but it also disintegrates structure and thus obscures essential characteristics of the occasion being observed and analyzed.[61] If a company buries a load of toxic chemicals in a vacant lot somewhere, you might see it as the work of a corrupt manager, perhaps combined with the economic demand for cheaper products. But that would ignore the larger picture of a technological society where thousands of tons of toxic chemicals are produced every year, where we demand high-tech products with such chemicals in them, where free-market ideals resist tighter regulation, and so on.

A complementary approach to Whitehead's was developed by one of his contemporaries, the statesman, general, philosopher, amateur botanist, and fourth prime minister of South Africa, Jan Christiaan Smuts (1870–1950), in his famous 1926 book, *Holism and Evolution*.[62] Devoting himself to the cause of peace, Smuts played a major role in establishing the League of Nations and later the United Nations. His philosophy as well as his politics exhibited commitment to the idea of interconnection. Smuts promoted a conception of holism that was innovative and has remained influential, though its impact is felt most strongly outside of academia. He offered it as an alternative to the mechanistic models of nature predominant in Western natural science. Wholes and wholeness are not confined to the biological domain but describe the entire cosmos, including inorganic substances. The fundamental characteristic of holism is a unity of parts,

which is so close and intense as to be more than the sum of its parts; which not only gives a particular conformation or structure to the parts but so relates and determines them in their synthesis that their functions are altered; the synthesis affects and determines the parts, so that they function towards the "whole"; and the whole and the parts therefore reciprocally influence and determine each other, and appear more or less to merge their individual characters: the whole is in the parts and the parts are in the whole, and this synthesis of whole and parts is reflected in the holistic character of the functions of the parts as well as of the whole.[63]

The science of ecology deeply influenced this holistic cosmology, a radical departure from the dissociative modern cosmology of mechanism. Holism integrates wholes temporally in a flux of time, implicitly challenging Aristotle's law of identity, which presupposes a sort of stasis: "Holism is a process of creative synthesis; the resulting wholes are not static, but dynamic, evolutionary, creative."[64]

The quantum physicist David Bohm (1917–1992) proposed a new, highly integrating theory of physics. His early work took an unusual turn when his graduate research at UC Berkeley became useful to the top-secret Manhattan Project, which was developing the atomic bombs to be dropped on Japan during World War II (Bohm's UC Berkeley mentor, Robert Oppenheimer, was leading the project). Because Bohm didn't have the proper security clearance, he was prevented from writing or defending his own dissertation. As his career developed and he became a renowned theoretical physicist, Bohm criticized the mechanistic cosmology inherited by mainstream science, in which the world is reduced as far as possible to a basic set of elements whose "fundamental natures" are independent of one another and whose forces of interaction don't affect their "inner natures." The success of the program of mechanism doesn't prove its truth, he thought, and adhering to it threatens our very existence.

Albert Einstein's (1879–1955) relativity theory was a start at producing a nonmechanistic physics. It provides the basis for a cosmological view of "unbroken wholeness or flowing wholeness," according to Bohm. Quantum theory, which implies that the physical outcome of an event depends on how and whether it's observed, leads ontology (the nature and relations of being) away from mechanism by insisting on context dependence and the existence of internal relationships among parts and between the parts and the whole. Bohm argued for a new, "postmodern" physics that doesn't separate matter and consciousness and therefore doesn't separate facts, meaning, and value.[65] The separation of truth and virtue, he wrote, "is part of the reason we are in our present desperate situation,"[66] alluding to the environmental and political crises of the twentieth century.

Bohm's theory of physics envisions an unbroken movement of enfolding and unfolding: the *holomovement*. The holomovement is ontologically primary—nothing else is more real—while the apparently discrete objects perceived in our everyday experience are secondary phenomena, derivations of the holomovement (but not unreal, as Plato claimed). Bohm's concept captures the flux of the natural world and the relationships among things at scales from the macroscopic (cars, planets, trees) to the subatomic (protons, electrons). "The whole universe is *actively* enfolded to some degree in each of the parts"—all parts are enfolded into each other.[67] This concept rejects the mechanistic assumption that the relation of parts is external to them: all parts and the universe are *internally* related to each other. That's the "fundamental truth" of the implicate order—what lies beneath the perceptible world. External relatedness is a secondary, derivative truth called the explicate, or unfolded, order—the physical world as it presents itself to our senses and observations at any given moment. Bohm saw this approach as a counter to the traditional fragmentary thinking that gives rise to a reality that's "constantly breaking up into disorderly, disharmonious, and destructive partial activities."[68]

Whitehead, Smuts, Bohm, Einstein, and other philosophers and physicists have driven major changes in scientific theory and models of nature in this late-modern period. Scientific models have for over a century now begun to reintegrate spatial and temporal aspects of matter, reconceiving it as activity rather than as collections of inert particles. Ecology further shifts attention to interrelations and to aggregates of organisms and environments at multiple scales—organisms, watersheds, and ecosystems. Although vestiges of it remain, the modern mechanistic, atomistic paradigm is now theoretically defunct. The image of nature as a vast machine is slowly giving way to the image of nature as a vast organism, or something resembling one, and relations have become more central in many fields of study.[69]

A Response to Dissolution

Confronted with a difficulty, the prosaic man gets busy . . . he changes, modifies and manipulates; he institutes projects and programs. . . . And he does all this without ever asking a single fundamental ques-

tion, without ever attending to such basic things as the aims, underlying assumptions, values, or justification of what he is dealing with and what he is doing. Therefore all his busyness—restless, nerve-racking, and exhausting—is at bottom only a tinkering with and an acceleration of what already exists. . . . Since he reduces everything in the world to a problem, his awareness is extremely superficial and narrow.
—George W. Morgan, *The Human Predicament: Dissolution and Wholeness*, 1968‡

The Brown University philosopher George W. Morgan developed an extensive, multilayered answer to conceptual and lived dissociations in the middle of the twentieth century. Morgan's neglected gem (though it was lauded by the historian Lewis Mumford) *The Human Predicament: Dissolution and Wholeness* is a broad-reaching treatise on the modern fragmentation of knowledge and subjectivity.[70] It's based on the concept of "dissolution," similar to dissociation. Dissolution stands for many kinds of divisions manifest in modern life, though Morgan focuses primarily on matters of epistemology (knowledge creation) and subjectivation (the social processes by which individuals are formed). Dissolutive modern scientific methods, Morgan writes, contrast with those of history and philosophy, which are more integrative. Science excludes certain ways of knowing, ultimately neglecting the very purpose of knowledge.[71] The subjective modes of exploration in history, philosophy, and art, in which symbol, narrative, interpretation, alternative modes of representation, and uncertainty are tolerated and even embraced, allow for fuller, truer understandings. In these disciplines, reflection in any one subject reaches out to other subjects.[72] The stepwise advance, clear-cut boundaries, and explicitness that predominate in the methods of modern science have come to predominate everyday thinking, and they push aside fuller ways of knowing the world that can be found in poetry and art. Morgan doesn't discount scientific ways of knowing entirely but merely seeks to balance them.

Science, industry, and everyday life can learn from the more organic, integral, non-piecemeal structures of the arts and philosophy, says Morgan. The limits to clear-cut explication need to be understood so that flexible classifications, comparisons, and analogies are embraced. Like David Abram, Morgan prescribes modes of relation that open up a person to another, human or not, in reciprocal relation. He believes you must open yourself to the impact of nature on yourself, cherishing the shape, color, and scent of the fruits of Earth. "The 'man' who responds to nature cares for it, not like one who merely wants to maximize its usefulness, but like one who is concerned for its being, respects its otherness, and assumes responsibility for what it will become. . . . In response and responsibility there is a unity between 'man' as 'man' and nature as nature." As the philosophical thought of another person can't be well understood unless it's met receptively, so are many other aspects of the others we encounter lost to us without adequate receptivity. The impersonality and detachment exalted by science for truth-seeking are anathema to full knowing. It's in a mutually subjective encounter that self and other are most fully revealed.[73]

Morgan attributes many human problems to dissolution. One is social alienation, the "loss of 'man' by 'man,'" at the root of which is the "prosaic mentality"—matter-of-fact thinking that lacks imagination, feeling, and poetic imagination—to which relation remains unknown.[74] Instrumentality, using other people and nature as means to your own ends, is the driving force of the prosaic mentality, and it reduces and annihilates relation. "Responding to a person," writes Morgan, "is the contrary of approaching him with the prosaic interest of gaining mastery or control over him, either by explaining him or by using him." Like Martin Buber, whom I discuss below, Morgan believes that reducing others to means negates true relation. Similar problems are at play in human relations to nature: "'Man' must stand in a relation of response to nature." Both humans and nature must be recognized in their concrete uniqueness. The relation of people to the natural world requires that we don't reduce it to abstract properties.[75] In accord with my above criticism of popular naive notions of freedom, Morgan writes that freedom isn't found in an individual's separation from the world but rather in her capacity to respect it, apprehend it, respond to it, be responsible for it, and "judge and decide and engage in action." The capacity of freedom, a power of the self to be in the world, is inseparably linked to wholeness.[76]

Environmental Scholars Bridging the Divide

David Abram's ideas about "realignment with reality" represent the most elaborate philosophical envisioning of a more intimate, direct, and sensuous engagement with nature. He develops the phenomenological idea of perception as a reciprocal, "intersubjective" process that sees life and action embedded in the more-than-human world. He contrasts these ideas with mechanistic assumptions that view the nonhuman world as dead and passive and see the human mind as disjointed from the human body and the rest of nature. Speech and human gesture, in truth, emerge out of nature: the human mind is "induced by the . . . participations between human body and animate earth," as Abram eloquently reveals.[77] His approach, grounded in the phenomenological theories of the philosophers Maurice Merleau-Ponty and Edmund Husserl (1859–1938), meets natural others not as objects but as subjects who participate in our world and in our perceptions of them. It sees the "sensuous, perceptual life-world" as primary; thoughts about that life-world are derivative and dependent.

Abram teaches that relating to others in the more-than-human world requires the full capacity of our human senses. Interactions of less modern peoples with an alive and spirited nature demonstrate this mode of relating: the songlines that traverse landscapes of Aboriginal Australians, the "world that watches" of the Koyukon, the Balinese ritual offerings to spirits inhabiting the world around them. Abram explains that increased interaction with artifice also reduces sensuous engagement with nature because compared with organic things artifice lacks variation and richness of form. Modern society, he writes, is extremely self-reflexive (it references, imitates, and reflects itself) and sensually impoverished as a result of our extensive and continual involvement with artifice.

Eloquently evoking the power of landscape connections, Abram says that forcing non-modern peoples out of their landscapes, as European colonists did in much of the world, is "to dislodge them from the very ground of coherence . . . to force them out of their mind."[78] For us moderners, sensuously reengaging with landscapes that are seen and felt to be alive will bring about richer and deeper environmental ethics: "A new environmental ethic . . . will come into existence not primarily through the logical elucidation of new philosophical principles and legislative structures, but through a renewed attentiveness to this perceptual dimension that underlies all our logics, through a rejuvenation of our carnal, sensorial empathy with the living land that sustains us."[79]

Edward S. Reed, the ecological psychologist, advocates for direct sensory experience from an equally broad perspective. In *The Necessity of Experience* he argues that to properly know the world and be fully developed persons, we must elevate primary experience and realize that secondhand information is slowly turning our minds into machines. A critical first step is to reverse the disparagement of experience that Descartes and other influential Western thinkers have handed down to us. We must learn through our senses and become less dependent on secondhand, mediated communications and information in all our interactions, including those with nature.[80] Like Abram and the phenomenologists, Reed says that direct perception through the senses creates a vital, fuller sense of the world: "To perceive something is to enter into a rapport with it as a meaningful object, place, or event. Perception in this sense is not merely mental; it is an act that best serves to unify mind and body; it is an achievement of the whole person who is looking, listening, scrutinizing, and discerning."[81]

Accordingly, Reed rails against the analytic philosophy popular in American universities, which limits itself to analyzing language and argument: "Philosophers ... need to spend more of their time thinking about the places within which people find themselves (schools, workplaces, in front of the television) and less time dealing with abstractions."[82] Nonmodern cultures tend to elevate primary, firsthand experience. Children learn from direct observation of the activities of adults, from which they're not barred: "In most traditional, rural cultures, children as young as three may be found on the fringes of adult activities, beginning to learn for themselves what adult situations are like."[83] One of the things that always impresses me about Bali is how immersed children are in the thick of any community activity. During music rehearsals, they sit on the laps of their parents, who are playing. When rehearsal is over, they start experimenting with the instruments. It's also common to see Balinese toddlers learning to dance while being held, before they can stand on their own. Children often play in the rice fields while their parents farm, and they make offerings or ceremonial structures alongside adults.

Anthony Weston is another environmental philosopher who, like

Abram, sees a need to establish deeper relations with the more-than-human world through renewed sensuality. Weston laments what he calls the "desolation" of modern life—the loss of social relationships with other animals and the loss of sensuous contact with nature.[84] The senses play a fascinating role in relation, as illustrated by the function of smell in inter-human relationships and the corresponding suppression of smells required by the rise of individualism in the mid and late eighteenth centuries. "Coming to our senses" and establishing "transhuman etiquettes" are necessary for humans to truly know and relate to nature and to reduce our harms to the natural world. Major disruptions sometimes bring us to our senses, if momentarily: in the aftermath of Hurricane Gloria in New York City and Long Island in 1985, power outages immediately stimulated greater sociality and an attention to the moon and other nature. Language and other human capacities develop through relation with nature: Earth "sings through us." The processes of bringing nonhumans into the realm of sociability and opening up our human sensibility to the world beyond us are vital to the future of humankind.[85]

Depth and intentionality in relationships was the focus of the Jewish philosopher Martin Buber (1878–1965) in his classic work *I and Thou*.[86] Buber championed a version of relation akin to Abram's and other phenomenologists' views of intersubjectivity. "Relation is reciprocity," he writes. Each relationship with an other brings to us unique moments of understanding. Each other to whom we're related creates something unique in each of us. We're educated by children and animals, and "we live in the currents of universal reciprocity."[87] Buber urges us to be in relation with the *whole* other and argues that this is a necessity of love.[88] Instrumentalizing others, making them a means to accomplish our own ends, is antithetical to true relation. Human and nonhuman others should be approached with the respect and depth implied by the term *thou*, traditionally used to refer to God.[89]

The idea that we need engaged, immediate, sensuous relationships with nature is supported also by ecofeminists in their call for particularistic (not universal and generic) relationships of caring with human and nonhuman others. Caring for particular others requires direct contact with and knowledge of them, not merely an abstract concern for whole categories of others. Criticizing environmental ethics based on rationalist, universalist views (in which logical processes produce universally applicable rules of behavior) and ethical extensionist views (in which moral consideration given to humans is extended to other categories of nature), Plumwood writes, "The feminist suspicion is that no abstract morality can be well founded that is not grounded in sound particularistic relations to others in personal life, the area which brings together in concrete form the intellectual with the emotional, the sensuous and the bodily."[90]

In *Earthcare: Women and the Environment*, Carolyn Merchant writes that we need a directness of relation to nature that women's close associations with it have historically enabled. Her proposed "partnership ethic" brings human and nonhuman stakeholders (the latter through various forms of representation) into direct deliberation over environmental concerns.[91] Research by these and other feminist scholars bolsters the idea that close, particular relations are necessary for limiting the extent of modern humanity's destructiveness.

James Lovelock's Gaia theory embraces relation, too, though in a more conceptual sense. It sees the whole created by the interconnections of nature—the entire Earth's biosphere or the whole planet—as an organism.[92] Gaia theory emphasizes the flows among the biophysical and geophysical elements that produce Earth's atmosphere and regulate rainfall, temperature, and other large-scale dynamics on the planet. In concert, they produce the delicate conditions under which life as we know it flourishes. These processes interact and respond to one another, producing a whole—living Earth—the complexity and coherence of which give it the appearance of a single large organism comprising many smaller organisms and processes. Each individual organism is constituted by a concerted interplay of other organisms, materials, and processes, both over evolutionary time and in the present.[93]

Links among Connecting Theories

The ideas of these diverse thinkers support and complement each other well. Morgan's emphasis on direct respectful relation, achieved through an opening up to the other, echoes Buber's ideas about relationship and is made more concrete and sensuous by Abram. The caring attentiveness to the particularity of the other that Morgan says is vital to relation, in addition to his stance of openness and responsiveness to nature, agree well with ecofeminists' focus on particularity in caring for nature. Whitehead's notions of patterns, process, and occasions support Smuts's notion of dynamic wholes and Bohm's concept of the holomovement. And they concur well with Morgan's thinking on the essential wholeness that pervades the world and that must be acknowledged in our approaches to knowing and being. Morgan's conclusion that modern ways are too "clear-cut" harmonizes with "sociative logics," which integrate sometimes messy and intractable context into reasoning. These alternative systems of thought together form a powerful foundation for rethinking the world in a less dissociated way. They're fertile intellectual soil for a less dissociated life-world.

We modern people touch nature and other people across the globe every day. Conversely, our own lives are touched by institutions and individuals to whom we remain invisible and anonymous others. Circulating about the planet, our consequences become abstractions, like the laws and ethics needed to navigate this dissociated world. These characteristics lie at the heart of our modern lives. Meanwhile, the large-scale reconfigurations necessary to better associate actions with spheres of connected empathy seem almost impossible today. Such a program of change would entail a radical departure from life as we know it.

Yet a master plan is neither necessary nor desirable to bring about worlds that are better associated, more just, and healthier for humans and the rest of nature. Just as destructive material dissociations have proliferated throughout the modern world with the support of dissociating principles, so too can associating ones be used to guide reconfigurations that are less harmful to others and ultimately to ourselves. Such changes can occur at any scale and in all spheres of life: in social and inter-species relationships; in activities that educate about and engage with local ecology or food production; in political organizations that favor local control and disfavor remote control by global institutions; in international politics that promote environmental and social justice through local empowerment; and in global economic, organizational, and institutional changes that reduce exploitation of distant people and nature for private gain. We'll do well to remember, though, that while creating a more associated life-world puts us in closer touch with the harms we create, as well as our positive influences on the world, it won't solve all problems and ensure happiness. And it will take guts to face the world we're transforming with open eyes and an open heart.

It may seem futile to alter our own life choices by increments in the face of the large scale of our problems, but gradual change by many people, alongside more systematic political change, can have a major impact. Consider how much our lives and economy have been transformed in just the last few decades by the adoption of computers and smartphones. Moreover, small-scale changes that seek to keep consequences more immediate or reduce dissociations of knowledge can themselves lead to new, more associated experiences and ways of thinking that could be infectious. Your personal project to start growing more of your own food in a community garden may inspire neighbors to do the same. That could lead to a stronger social network in your community and more knowledge about where food comes from. And getting out of your car just one day a week to bicycle, carpool, or use public transit, meeting neighbors along the way, adds up quickly if many people do it. The road to a new world is paved with small decisions as well as large ones.

For anyone reasonably familiar with the extent and depth of modernity's environmental problems, a radical theory such as this that calls for major change seems appropriate and necessary. This book tries to match the enormous scope of global environmental crisis with a view of it not merely as a collection of difficult problems to be solved by better scientifically based techno-managerial and policy solutions but rather as a fundamental problem of relation. Social and environmental harms don't result simply from people and corporations doing evil deeds. Instead, they arise from conditions that promote harmful choices by us all. Deeper, more immediate and authentic connections will provide the context for more caring behavior and more livable worlds.

ACKNOWLEDGMENTS

All our work takes place within social and natural contexts that influence and contribute to it. This book is no exception. Although I'm fully responsible for it (including the errors), help has come from many directions to make it possible.

First, I owe much to all the labors of distant and unseen nature: the growth of forests, the cleansing of waters, the production of oxygen. It usually goes unnoticed, but the nonhuman world supplies me with much of what I need to sustain myself and be productive. I feel gratitude also for all the anonymous people who work most closely with nature, whose efforts are usually taken for granted, and for other laborers doing the dirty and difficult jobs that drive this economy that supplies me.

This book could not have been completed without the intellectual ground laid by many inspirational and often courageous authors on whose work I build in these pages. Various scholars also devoted themselves to guiding and improving my efforts more directly. Carolyn Merchant tirelessly devoted her time and energy to help me hone my thoughts and writing early on. She, Nancy Peluso, and Kaiping Peng gave steadfast encouragement and support, introduced me to important existing work, and read innumerable early drafts. Without them, there would be no book.

Many others reviewed and commented on early drafts, including Shana Cohen, Todd Trattner, and Elisa Cooper. Carol Manahan and Kurt Spreyer provided thoughtful responses and sharp editing skills to drafts of sections early on and near the end, too. Stefania Pandolfo, Isha Ray, Ignacio Chapela, and Ian Boal engaged me in provocative conversations on these questions. Richard Norgaard gave me abundant moral support and inspiration. The great environmental philosopher Val Plumwood, who has unfortunately since passed from this world, added her wisdom to these ideas and gave them the gift of her blessing. My friend Brian Dahmen contributed numerous psychological insights and dedicated long hours writing detailed comments and pulling together extensive background material on the concept of dissociation in psychology. Our discussions of the subject on long hikes helped me to clarify and articulate my thoughts. Warm thanks also to Nestar Russell for heaps of last-minute encouragement and support; he quickly read through many drafts of the chapter on psychology, suggesting many helpful ideas, corrections, insights, and supporting references.

I was fortunate to have my friend Marianna Cherry, an astonishingly brilliant editor (and ace gamelan player), review much of the manuscript. She engaged passionately with these ideas and taught me priceless lessons about writing in the process. The editor Elizabeth Bernstein applied her excellent craft to parts, too. Rachel Hall deserves special thanks for diligently reading the entire manuscript, every word, with a keen and thoughtful yet gentle eye. Other friends and colleagues read chapters and shared their wisdom: Susanna Benningfield, Jim Silver, David Rowland, Elissa Gaynor, and Justin Park, who also connected me with the literary agent Elizabeth Kracht, who in turn introduced the project to her agency head, Kimberley Cameron, who became my agent and has given me her unwavering faith.

The following people helped in various ways, from providing advice and encouragement over the long haul to referring me to important sources: David Harris (who discovered the obscure yet important book *The Human Predicament*), Yukari Yoshida, David Buxbaum, Daniele Rossdeutscher, Markus Storzer, Irene Wibawa, Graeme Vanderstoel, Samantha Chen, Andy Drake, and David Abram. Many friends in Bali taught me many fascinating things about their culture. Among them are Anak Agung Anom Putra, I Gusti Ketut Lidur, I Gusti Ngurah Sukra, I Gusti Made Suparta, Ayu Sukmawati, Anak Agung Rai Srinah. Richard Stallman referred me to Greg Egan's novel *Diaspora*. The artist Sean O'Dell found resonance between his art and my writing, read early drafts, and offered valuable comments and support. My college students deserve thanks for their enthusiasm, curiosity, and passion over the years and for their interest in this work. They've been a valuable sounding board for many of these ideas. My uncle Richard Pouliot helped with genealogical sources. Special thanks go to my parents for their enduring support, care, and encouragement.

Thank you to the people at Prometheus Books for your hard work: Steven L. Mitchell, Melissa Shofner, Julia DeGraf, Mariel Bard, Catherine Roberts-Abel, Richard Snyder, and Mark Hall, among others.

Wikipedia has been helpful on many occasions to search for new sources of information and to confirm items of general knowledge. Several organizations provided funding for this work in its earliest stages: the UC Berkeley Center for Southeast Asian Studies, the US Department of Education Foreign Language and Area Studies Program, and the National Recycling Coalition. Finally, I'm grateful for the recognition, excitement, appreciation, playfulness, and affection of the many homeless dogs at Berkeley's city animal shelter who afforded many delightful respites during years of research and writing.

I welcome comments and ideas at http://invisiblenature.com.

NOTES

INTRODUCTION

1. James A. Estes et al., "Trophic Downgrading of Planet Earth," *Science* 333, no. 6040 (2011).

2. "Tobacco Farmers to Access Data through Cellphones," *Citizen* (Dar es Salaam), June 13, 2011, http://www.thecitizen.co.tz/business/13-local-business/11884-tobacco -farmers-to-access-data-through-cellphones.html (accessed December 28, 2012).

3. James Gustave Speth, *The Bridge at the Edge of the World: Capitalism, the Environment, and Crossing from Crisis to Sustainability* (New Haven, CT: Yale University Press, 2008).

4. Judith Shapiro writes, "The 1958–60 Great Leap Forward . . . failed to reach its goals, decimated China's forests, and caused widespread starvation." Judith Shapiro, *Mao's War against Nature: Politics and the Environment in Revolutionary China*, Studies in Environment and History (Cambridge; New York: Cambridge University Press, 2001), p. 2.

5. Jared M. Diamond, *Collapse: How Societies Choose to Fail or Succeed* (New York: Viking, 2005).

6. Kari Marie Norgaard, *Living in Denial: Climate Change, Emotions, and Everyday Life* (Cambridge, MA: MIT Press, 2011).

7. David Abram, *The Spell of the Sensuous: Perception and Language in a More-Than-Human World*, 1st ed. (New York: Pantheon Books, 1996).

8. Carolyn Merchant, *The Death of Nature: Women, Ecology, and the Scientific Revolution*, 1st ed. (San Francisco: Harper & Row, 1980).

9. Thomas H. Birch, "The Incarceration of Wildness: Wilderness Areas as Prisons," in *The Great New Wilderness Debate*, ed. J. Baird Callicott and Michael P. Nelson (Athens: University of Georgia Press, 1998).

10. Mark David Spence, *Dispossessing the Wilderness: Indian Removal and the Making of the National Parks* (New York: Oxford University Press, 1999).

11. This refers to the Renaissance architect and engineer Filippo Brunelleschi (1377–1446) and his piazzas, large plazas paved in rectangular stone. The piazza metaphor comes from Neil Smith, *Uneven Development: Nature, Capital, and the Production of Space* (New York: Blackwell, 1984), p. 178.

12. On garbage and its mysterious life, see Elizabeth Royte, *Garbage Land: On the Secret Trail of Trash*, 1st ed. (New York: Little, Brown, 2005); Heather Rogers, *Gone Tomorrow: The Hidden Life of Garbage* (New York: New Press: Distributed by Norton, 2005); Edward Humes, *Garbology: Our Dirty Love Affair with Trash* (New York: Avery, 2012).

CHAPTER 1: THE BANALITY OF EVERYDAY DESTRUCTION

1. Millennium Ecosystem Assessment Program, *Ecosystems and Human Well-Being: Synthesis*, Millennium Ecosystem Assessment Series (Washington, DC: Island Press, 2005), pp. v, ii.

2. The MA is anthropocentric (human centered, concerned only about humans) throughout. It evaluates human-made changes to nature almost entirely with regard to their effects on humans, both benefits and harms. It doesn't explicitly value the well-being of other-than-human nature. The MA represents nature and natural processes as "ecosystem services," such as the wood produced by forests. This implies an instrumental ("What's in it for me?") relationship that many environmental ethicists blame for humanity's destructiveness toward nature, so the MA may replicate the conditions that have led to the harms it reports. Nevertheless, it provides a valuable accounting of the scale and varieties of our transformations of the planet.

3. Millennium Ecosystem Assessment Program, *Ecosystems and Human Well-Being: Synthesis*, pp. 1–2.

4. Ibid., p. 1.

5. Ibid., p. 2.

6. Ibid., pp. 2–3.

7. The latest four-year update of the report is in draft form at this time but isn't expected to undergo major changes. US Global Change Research Program, "Third National Climate Assessment (Draft)" (Washington, DC: US Global Change Research Program, 2013), p. 3, http://ncadac.globalchange.gov/download/NCAJan11-2013 -publicreviewdraft-fulldraft.pdf (accessed February 1, 2013).

8. Ibid.

9. Ibid., pp. 3, 8–10.

10. DARA and the Climate Vulnerable Forum, "Climate Vulnerability Monitor, 2nd Edition: A Guide to the Cold Calculus of a Hot Planet" (Madrid: DARA International, 2012), pp. 17, 157.

11. Ibid.

12. Ian Stirling and Andrew E. Derocher, "Effects of Climate Warming on Polar Bears: A Review of the Evidence," *Global Change Biology* 18, no. 9 (2012).

13. Millennium Ecosystem Assessment Program, *Ecosystems and Human Well-Being: Synthesis*, pp. 2–4.

14. Ibid., p. 1.

15. ICF International, "Electronics Waste Management in the United States through 2009," ed. US Environmental Protection Agency Office of Resource Conservation and Recovery (US Environmental Protection Agency, 2011), http://www.epa.gov/osw/conserve/materials/ecycling/docs/fullbaselinereport2011.pdf (accessed October 16, 2012).

16. Eric D. Williams, Robert U. Ayres, and Miriam Heller, "The 1.7 Kilogram Microchip: Energy and Material Use in the Production of Semiconductor Devices," *Environmental Science & Technology* 36, no. 24 (2002).

17. Ibid., p. 5507.

18. Ibid.

19. Ibid.

20. Christopher D. Cook and A. Clay Thompson, "Silicon Hell: The Computer Industry Prides Itself on a 'Clean' Image—But It's Actually Doing Horrible Damage to Its Workers and the Environment," *San Francisco Bay Guardian*, April 26, 2000.

21. Myron Harrison, "Semiconductor Manufacturing Hazards," in *Hazardous Materials Toxicology: Clinical Principles of Environmental Health*, ed. John B. Sullivan and Gary R. Krieger (Baltimore: Williams & Wilkins, 1992), p. 475.

22. Cook and Thompson, "Silicon Hell."

23. Jim Fisher, "Poison Valley, Parts 1 and 2," Salon.com, http://archive.salon.com/tech/feature/2001/07/30/almaden1/print.html and http://archive.salon.com/tech/feature/2001/07/31/almaden2/index.html (both accessed June 21, 2011).

24. Ken Geiser, *Materials Matter: Toward a Sustainable Materials Policy* (Cambridge, MA: MIT Press, 2001). Geiser recommends the precautionary approach, used widely in Europe, which requires new chemicals to be subject to toxicity tests before use.

25. Fisher, "Poison Valley, Parts 1 and 2."

26. Benjamin Pimentel, "Ex-IBM Workers Lose Toxics Case: Jury Says Company Not Responsible for Cancer They Claimed Was Caused by Chemicals," *San Francisco Chronicle*, February 27, 2004, http://www.sfgate.com/business/article/Ex-IBM-workers-lose-toxics-case-Jury-says-2817137.php (accessed April 22, 2012).

27. Cook and Thompson, "Silicon Hell."

28. Benjamin Pimentel, "Big Blue Settles Lawsuits; Sick Employees Blamed Chemicals," *San Francisco Chronicle*, June 24, 2004.

29. Cook and Thompson, "Silicon Hell."

30. John D. Boice Jr. et al., "Cancer Mortality among US Workers Employed in Semiconductor Wafer Fabrication," *Journal of Occupational and Environmental Medicine* 52, no. 11 (2010).

31. Harris Pastides et al., "Spontaneous Abortion and General Illness Symptoms among Semiconductor Manufacturers," *J Occup Med* 30, no. 7 (1988).

32. Adolfo Correa et al., "Ethylene Glycol Ethers and Risks of Spontaneous Abortion and Subfertility," *Am J Epidemiol* 143, no. 7 (1996).

33. Cook and Thompson, "Silicon Hell."

34. Harrison, "Semiconductor Manufacturing Hazards," p. 490.

35. Ibid.

36. Fisher, "Poison Valley, Parts 1 and 2."

37. Ibid.

38. Ibid.

39. Ibid.

40. Harrison, "Semiconductor Manufacturing Hazards," p. 477.

41. Ibid.

42. William Wan, "To Analysts, Foxconn Brawl Is No Anomaly," *Washington Post*, September 26, 2012; Charles Duhigg, David Barboza, and Gu Huini, "In China, Human Costs Are Built into an iPad," *New York Times*, January 26, 2012, http://www.nytimes .com/2012/01/26/business/ieconomy-apples-ipad-and-the-human-costs-for-workers-in -china.html?smid=pl-share (accessed November 14, 2012).

43. Clifford Coonan, "Workers Threaten Mass Suicide at Company That Supplies Apple; Fresh Labour Dispute at Foxconn Factory Turns Spotlight on Working Conditions in China," *Independent* (London), January 12, 2012, http://www.independent.co.uk/news/ world/asia/workers-threaten-mass-suicide-at-company-that-supplies-apple-6288160. html (accessed December 28, 2012).

44. Fiona Tam, "Foxconn Factories Are Labour Camps: Report," *South China Morning Post* (Hong Kong) October 11, 2010, http://www.scmp.com/article/727143/foxconn -factories-are-labour-camps-report (accessed September 19, 2012).

45. Wan, "To Analysts, Foxconn Brawl Is No Anomaly."

46. Cook and Thompson, "Silicon Hell"; Roger Papler and California Regional Water Quality Control Board–San Francisco Bay Region, "Third Five-Year Review: Hewlett-Packard (620–640 Page Mill Road) Superfund Site; Palo Alto, Santa Clara County, California," ed. Stephen A. Hill and Kathleen Salyer (2010), http://yosemite.epa.gov/r9/ sfund/r9sfdocw.nsf/3dc283e6c5d6056f88257426007417a2/351973b3feacb7a8882577af00 73b60e!OpenDocument (accessed October 10, 2012).

47. Cook and Thompson, "Silicon Hell."

48. Fisher, "Poison Valley, Parts 1 and 2."

49. Ibid.

50. Cook and Thompson, "Silicon Hell."

51. Ibid.

52. "SVTC Eco-Map Family," Silicon Valley Toxics Coalition, http://www.map cruzin.com/svtc_ecomaps (accessed December 28, 2012).

53. ICF International, "Electronics Waste Management in the United States through 2009."

54. "Management of Electronic Waste in the United States: Approach Two" (United

States Environmental Protection Agency, 2007), p. 22, http://www.epa.gov/wastes/conserve/materials/ecycling/docs/app-2.pdf (accessed December 28, 2012).

55. Jim Puckett et al., "Exporting Harm: The High-Tech Trashing of Asia" (Seattle: Basel Action Network, Silicon Valley Toxics Coalition, 2002), pp. 9–10, http://www.ban.org/E-waste/technotrashfinalcomp.pdf (accessed September 12, 2012).

56. Ibid.

57. Ibid., p. 3.

58. Kejing Zhang, "Rough Times in Guiyu—E-Waste Recycling in China Has Become a Serious Threat to Human Health and the Environment. Now the Authorities Are Stepping In," *Recycling m@gazine*, 2007, http://www.recyclingmagazin.de/epaper/rm0005/default.asp?ID=8 (accessed October 8, 2012).

59. Puckett et al., "Exporting Harm."

60. Sheila Davis and Ted Smith, "Corporate Strategies for Electronics Recycling: A Tale of Two Systems" (Silicon Valley Toxics Coalition; Computer TakeBack Campaign! 2003), http://svtc.org/wp-content/uploads/prison_final.pdf (accessed December 28, 2012).

61. Puckett et al., "Exporting Harm." The figure is based on an estimate of 80 percent of recycled computers moving offshore.

62. Ibid., pp. 9–10.

63. Ibid., p. 14.

64. Ibid., p. 1.

65. Ibid., p. 16; Zhang, "Rough Times in Guiyu."

66. Puckett et al., "Exporting Harm," pp. 17–18.

67. Lin Peng Xia Huo, Xijin Xu, Liangkai Zheng, Bo Qiu, Zongli Qi, Bao Zhang, Dai Han, and Zhongxian Piao, "Elevated Blood Lead Levels of Children in Guiyu, an Electronic Waste Recycling Town in China," *Environmental Health Perspectives* 115, no. 7 (2007).

68. Puckett et al., "Exporting Harm," pp. 20–22.

69. Ibid., p. 22.

70. Ibid., p. 16.

71. Ibid., p. 22.

72. Davis and Smith, "Corporate Strategies for Electronics Recycling."

73. Ibid., p. 13.

74. Ibid.

75. For historical context, René Descartes, one of the fathers of modernity, died around that same time, in 1650. Various genealogical sources refer to Charles Pouliot's immigration to New France. A copy of the August 8, 1664, contract between him and Charles de Lauzon, Seigneur of Charny, for the construction of a windmill on Lauzon's seigneurie on the Île d'Orléans, still exists.

76. Charles's parents-in-law emigrated earlier, from France to Montréal, New France, and apparently were the first French people married in that town. From various genea-

logical records, some summarized in Thomas John Laforest and Gérard Lebel, *Our French-Canadian Ancestors*, vol. 2 (Palm Harbor, FL: LISI Press, 1984). Additional genealogical information from Bibliothèque et Archives Nationales du Québec.

77. Jim Vallette, "Larry Summers' War against the Earth," Global Policy Forum, http://www.globalpolicy.org/socecon/envronmt/summers.htm (accessed February 4, 2013).

78. Duhigg, Barboza, and Huini, "In China, Human Costs Are Built into an iPad."

79. Rob Nixon, *Slow Violence and the Environmentalism of the Poor* (Cambridge, MA: Harvard University Press, 2011), p. 2.

80. David N. Pellow and Lisa Sun-Hee Park, *The Silicon Valley of Dreams: Environmental Injustice, Immigrant Workers, and the High-Tech Global Economy* (Critical America) (New York: New York University Press, 2002); Cook and Thompson, "Silicon Hell." Also, a disproportionately high proportion of semiconductor worker injuries and illnesses (9.3 percent) come from toxic exposures.

81. The German sociologist Niklas Luhmann writes of the difficulty of attributing environmental effects to particular decisions: "We know . . . that important ecological conditions for life can be changed by the employment of technology and its products, with the prospect of grave harm. But we can hardly ascribe this problem to individual decisions, because the extremely complex mesh of causes of numerous factors and the longevity of these [ecological] trends do not allow such an attribution." Niklas Luhmann, *Observations on Modernity* (Writing Science) (Stanford, CA: Stanford University Press, 1998), pp. 73–74.

82. Luhmann discerned the differences between confidence and trust: "The normal case is that of confidence. You're confident that your expectations will not be disappointed: that politicians will try to avoid war, that cars will not break down or suddenly leave the street and hit you on your Sunday afternoon walk. You can't live without forming expectations with respect to contingent events and you have to neglect, more or less, the possibility of disappointment. You neglect this because it is a very rare possibility, but also because you do not know what else to do." Niklas Luhmann, "Familiarity, Confidence, Trust: Problems and Alternatives," in *Trust: Making and Breaking Cooperative Relations*, ed. Diego Gambetta (New York: Blackwell, 1990), p. 97.

83. Cook and Thompson, "Silicon Hell."

84. Fisher, "Poison Valley, Parts 1 and 2."

85. Ibid.

86. US National Research Council, *Hidden Costs of Energy: Unpriced Consequences of Energy Production and Use* (Washington, DC: National Academies Press, 2010).

87. National Institute for Occupational Safety and Health Education and Information Division, "Methyl Iodide," US Centers for Disease Control and Prevention, http://www.cdc.gov/niosh/npg/npgd0420.html (accessed December 4, 2012).

88. Ed Kashi and Michael Watts, *Curse of the Black Gold: 50 Years of Oil in the Niger Delta* (Brooklyn, NY: PowerHouse Books, 2008).

CHAPTER 2: SENSE AND CONNECTION

1. Chris Ware, "Thanksgiving.com," *New Yorker*, November 27, 2000, cover illustration.

2. This departure from tradition would not be as great as that of Thanksgiving.com. On January 1, 1997, staying with Japanese friends outside of Nagoya, I saw they had purchased from a vendor an elaborate New Years' holiday meal, presented in a beautiful set of lacquered boxes, even though the mother of the family is an accomplished cook. A daughter in the family informed me that this practice is quite common because of the difficulty of making the traditional meal. Most important was that the family still ate the holiday meal together. In the same vein, it's acceptable for Balinese families to purchase complex religious offerings required for rituals.

3. Max Horkheimer and Theodor W. Adorno, *Dialectic of Enlightenment* (New York: Seabury Press, 1972), p. 25.

4. George Orwell, *Nineteen Eighty-Four, a Novel* (London: Secker & Warburg, 1949); Mary Wollstonecraft Shelley, *Frankenstein; or, the Modern Prometheus* (London: Printed for Lackington, Hughes, Harding, Mavor, & Jones, 1818); Aldous Huxley, *Brave New World, a Novel* (London: Chatto & Windus, 1932).

5. Plumwood writes, "Bad news from below is not registered well by any of liberal democracy's information systems, hardly at all by the market, and often poorly by liberal democratic, electoral, and administrative systems. Yet it is precisely this bad news from below that has to be heard if many crucial forms of ecological damage are to be socially registered and opened to political action." Val Plumwood, *Environmental Culture: The Ecological Crisis of Reason* (New York: Routledge, 2001), p. 86.

Niklas Luhmann explains the limitations of social systems to apprehend and respond to ecological conditions through the concept of "resonance capacity," a system's ability to register information from other systems. Modern social systems can't "communicate" well enough with ecological systems. The complexity of the latter isn't matched by the resonance capacities of the former. Communication is a key aspect of ecological functioning. One part of the system must know what other parts are doing; the fox knows and feels when its prey is getting scarcer. The communicability of social and ecological systems is a key to ecological and social health, says Luhmann. Niklas Luhmann, *Ecological Communication* (Chicago: University of Chicago Press, 1989), p. 15.

6. David Abram, *The Spell of the Sensuous: Perception and Language in a More-Than-Human World*, 1st ed. (New York: Pantheon Books, 1996).

7. Merleau-Ponty's major works on phenomenology are Maurice Merleau-Ponty, *The Primacy of Perception, and Other Essays on Phenomenological Psychology, the Philosophy of Art, History, and Politics*, Northwestern University Studies in Phenomenology & Existential Philosophy (Evanston, IL: Northwestern University Press, 1964); Maurice Merleau-Ponty, *Phenomenology of Perception* (London; New York: Routledge, 1962).

302   **NOTES**

8. Following Abram and the Czech-German philosopher Edmund Husserl (1859–1938), in this study I use the term *life-world* to mean "the world of our immediately lived experience, as we live it, prior to our thoughts about it . . . that which is present to us in our everyday tasks and enjoyments—reality as it engages us before being analyzed by our theories and our science." Abram, *Spell of the Sensuous*, p. 40.

9. Anthony Giddens, *The Consequences of Modernity* (Stanford, CA: Stanford University Press, 1990), p. 137. I follow Giddens in expanding the term *phenomenology* beyond its conventional meaning in philosophy to embrace a broader study of daily experience. Husserl is considered the father of phenomenology, though the term was used previously. He described phenomenology as the study of intuitive experience—that which presents itself to us in conscious experience before analysis and cognition. See Edmund Husserl, *Cartesian Meditations* (The Hague: M. Nijhoff, 1965). Phenomenology was taken up by the philosophers Maurice Merleau-Ponty, Martin Heidegger, and others. David Abram builds on their work in his environmental philosophy. He urges us to turn toward the world as it's experienced in its felt immediacy. Rather than working to explain the world, his objective in phenomenology is to "describe as closely as possible the way the world makes itself evident to awareness, the way things first arise in our direct, sensorial experience." Abram, *Spell of the Sensuous*, p. 35.

To support his broader definition of phenomenology, Giddens summons two interesting images arising from prevalent phenomenologies of modernity. One is that of Max Weber, who wrote of a contemporary everyday experience that retains its color and spontaneity, "but only on the perimeter of the 'steel-hard' cage of bureaucratic rationality." The second, proposed by Karl Marx, sees modernity as a monster. The monster, though, is an unfinished project and one that can be partially tamed. Giddens suggests a third image: "The juggernaut—a runaway engine of enormous power which, collectively as human beings, we can drive to some extent but which also threatens to rush out of our control and which could render itself asunder. . . . The ride . . . can often be exhilarating and charged with hopeful anticipation. But, so long as the institutions of modernity endure, we shall never be able to control completely either the path or the pace of the journey. In turn, we shall never be able to feel entirely secure." Giddens, *Consequences of Modernity*, pp. 137–39.

10. I use the term *modernity* in a way roughly compatible with Anthony Giddens's thinking on modernity. He writes of "conditions of modernity," which allows for a nuanced conception of modernity not as a particular time period or even a period experienced in particular cultures but rather as a set of characteristic yet changing conditions. Some of the broad features that define modernity include a pronounced pace and scope of change and distinctive, highly developed institutions. Examples of the latter include the nation-state, the wholesale dependence of production on inanimate power sources, and the commodification of products and labor. Giddens, *Consequences of Modernity*, pp. 6, 14–15, 34.

11. Alex H. Taylor et al., "Complex Cognition and Behavioural Innovation in New Caledonian Crows," *Proceedings of the Royal Society B–Biological Sciences* 277, no. 1694 (2010).

12. Carolyn Merchant, *Major Problems in American Environmental History: Documents and Essays*, 2nd ed., Major Problems in American History Series (Boston: Houghton Mifflin, 2005), p. 71.

13. Kate Soper, *What Is Nature?: Culture, Politics and the Non-Human* (Oxford; Cambridge, MA: Blackwell, 1995), p. 249.

14. By *negentropic*, I mean háving the characteristic of decreasing entropy. Processes that increase the order within a system are negentropic. Physicist Erwin Schrödinger introduced the concept of negative entropy, later shortened to negentropy, to identify the thermodynamic characteristics of life. Living systems are highly ordered (low entropy), and they export entropy. Conversely, they import negentropy. Erwin Schrödinger, *What Is Life?: The Physical Aspect of the Living Cell; with, Mind and Matter: & Autobiographical Sketches*, Canto ed. (Cambridge; New York: Cambridge University Press, 1992), p. 70.

Modern thinkers often negate other-than-human contributions to productivity. John Locke's notion of the creation of value out of valueless nature through the addition of human labor, which underlies his notion of property and was an effective justification for the European transformation of the New World, denies the value and productivity of nonhuman nature and indigenous people. John Locke and Peter Laslett, *Two Treatises of Government; a Critical Edition with an Introduction and Apparatus Criticus*, 2nd ed. (Cambridge: Cambridge University Press, 1970), "The Second Treatise," chap. 5, "Of Property," sects. 25–51, pp. 303–20.

15. I mean the industrial revolution in Europe from the middle of the eighteenth to the early nineteenth century CE. It was sparked by the introduction in England of steam power, machine tools, and heavy-duty manufacturing machinery.

16. The British historian Edward P. Thompson's studies of the enclosures of forest and agrarian commons legitimated by the 1723 "Black Act" reveals how the law was used to protect the economic elite in the guise of safeguarding the national interest. Edward P. Thompson, *Whigs and Hunters: The Origin of the Black Act*, 1st American ed. (New York: Pantheon Books, 1975). Janet Neeson elaborated on the history of the removal of public access to and management of common lands. Janet M. Neeson, *Commoners: Common Right, Enclosure and Social Change in England, 1700–1820* (Cambridge; New York: Cambridge University Press, 1993).

17. About the intertwining of time and labor discipline with industrial capitalism and the use of mechanical clocks to regulate labor and divide the day, see Edward P. Thompson, "Time, Work-Discipline and Industrial Capitalism," in *Customs in Common* (New York: New Press: Distributed by W. W. Norton, 1991). With the rise of industrial capitalism, time became currency to be measured, divided, and parceled out, and the "enclosure [of commons] and agricultural improvement were both, in some sense, concerned with the efficient husbandry of the time of the labour-force." Ibid., pp. 359, 380.

18. Domestic dogs and cats are good models of *being*. They know how to lie around

with their senses open to the environment. By contrast we modern humans are usually pre-occupied with *doing*.

19. "Stress in America: Our Health at Risk," American Psychological Association (American Psychological Association, 2012), pp. 15–16, http://www.apa.org/news/press/releases/stress/2011/final-2011.pdf (accessed December 17, 2012).

20. Giddens, *Consequences of Modernity*, p. 146.

21. Ibid., p. 139.

22. Ibid., p. 146.

23. Ibid., pp. 22, 26, 83–88.

24. Ibid., pp. 27–28.

25. Niklas Luhmann, "Familiarity, Confidence, Trust: Problems and Alternatives," in *Trust: Making and Breaking Cooperative Relations*, ed. Diego Gambetta (New York: Blackwell, 1990), p. 97.

26. Greg Egan, *Diaspora: A Novel* (New York: HarperPrism, 1999).

27. Egan's use of the term *polis* throughout *Diaspora* hints at the underlying Platonic order of the fantasy world of his novel. The polis was the primary sociopolitical organization of Plato's time: the city-state of ancient Greece. Over the centuries culminating in the classical period in Greece, the power and self-contained might of the polis became a central preoccupation of Greek citizens, a phenomenon that Egan echoes in the autonomy and self-sufficiency of his polises.

Egan's use of the term *citizen* to refer to the occupants of his polises is also interesting. As I write in chapter 4, citizenship in the Greek polises was an elite position reserved for Greek-born males. Women, aliens, slaves, and others didn't have the political rights conferred by citizenship. Egan reflects the elite status of citizens in his novel. The lower, inferior, order of "fleshers" are those who remain associated with the body and physicality, as were women, slaves, and workers in ancient Greece. Through these oppositions, *Diaspora* reproduces both Plato's disparagement of physicality, the senses, and people closer to nature and his valorization of the work of the mind. But Egan adds a modern libertarian twist: the fleshers are welcome to leave their bodies and become citizens in a polis.

28. Plato, *The Collected Dialogues of Plato: Including the Letters*, ed. Edith Hamilton and Huntington Cairns (New York: Pantheon Books, 1961), *Phaedo*, 82d–83b, p. 66.

29. As reflected in *Diaspora*, many people who are interested in cloning today are concerned with immortality. In *Diaspora*, an entire polis clones itself to ensure survival—partly to guarantee that at least one copy survives and partly to increase available mental resources for developing a solution to the impending cosmic disaster. Today, many people interested in cloning want to defeat illness and death in several ways: by cloning oneself when death is near; by cloning a deceased loved one (human or pet) to bring them back to life; or by cloning a person to create a source of replacement organs or tissues (which is also done using natural reproduction). While these uses of cloning attempt to thwart natural cyclical

processes of life and death or decay in favor of a linear, eternal conception of time, the first two purposes are built naively on a simplistic understanding of biological determination. It assumes not only that the self is identical with the biological self and that one's mind will be reproduced along with the body but also that a body is entirely genetically determined. The latter ignores environmental and other biological factors in the development of a body. The former ignores the role of the senses and experience in the development of mind and self.

The phenomenology of Maurice Merleau-Ponty (1908–1961) contradicts the ideas of disembodiment of mind and the mechanical conception of mind. He anchored knowledge in perception—in lived, bodily experience—making it the foundation of "all rationality, all value, and all existence." Merleau-Ponty, *Primacy of Perception*, pp. 13, 27.

30. A Micmac elder responded to condescending French opinions of his peoples' uncivilized, poor, earth-bound conditions in a speech delivered to a group of French settlers, recorded by French missionary Chrestien LeClercq in the 1670s:

> Thou reproachest us, very inappropriately, that our country is a little hell in contrast with France, which thou comparest to a terrestrial paradise, inasmuch as it yields thee every kind of provision in abundance. Thou sayest of us also that we are the most miserable and most unhappy of men, living without religion . . . without social order . . . like beasts in our woods and our forests. . . . I beg thee now to believe that, all miserable as we seem in thine eyes, we consider ourselves much happier than thou in this, that we are content with the little that we have; and believe also once for all, I pray, that thou deceivest thyself greatly if thou thinkest to persuade us that thy country is better than ours. For if France, as thou sayest, is a little terrestrial paradise, art thou sensible to leave it? . . . We believe, further, that you are also incomparably poorer than we, and that you are only simple journeymen, valets, servants, and slaves, all masters and grand captains though you may appear. . . . Which of the two is the wisest and happiest—he who labors without ceasing and only obtains, and that with great trouble, enough to live on, or he who rests in comfort and finds all that he needs in the pleasure of hunting and fishing? . . . Learn now, my brother, once and for all, because I must open to thee my heart: there is no Indian who doesn't consider himself infinitely more happy and more powerful than the French.

Chrestien Le Clercq, William Francis Ganong, and Champlain Society, *New Relation of Gaspesia: With the Customs and Religion of the Gaspesian Indians*, Publications of the Champlain Society; 5 (Toronto: Champlain Society, 1910), pp. 104–106. Quoted in Colin G. Calloway, *The World Turned Upside Down: Indian Voices from Early America*, Bedford Series in History and Culture (Boston: St. Martin's Press, 1994), pp. 50–52.

31. Abram, *Spell of the Sensuous*, p. 257.

32. Richard Louv, *Last Child in the Woods: Saving Our Children from Nature-Deficit Disorder*, updated and expanded. ed. (Chapel Hill, NC: Algonquin Books, 2008).

33. Luhmann, *Ecological Communication*, p. 15.

34. The style of relation with artifice I want to evoke here is influenced by Robert M. Pirsig, *Zen and the Art of Motorcycle Maintenance: An Inquiry into Values* (Toronto; New York: Bantam Books, 1981).

35. Club Méditerranée (Club Med®) was founded in 1950 to provide, famously, "an antidote to civilization." Appeals to sensual pleasure have been a key part of its sales literature. Nigel Evans, David J. Campbell, and George Stonehouse, *Strategic Management for Travel and Tourism* (Oxford: Butterworth-Heinemann, 2003), p. 57.

36. *Islands*, July/August 2003.

37. Alice Leuchtag, "Merchants of Flesh: International Prostitution and the War on Women's Rights," *Humanist* 55, no. 2 (1995). Even sex tours that procure sex with children have been offered via the Internet and other media: Andy Ho, "Time to Throw the Book at Child Sex Tour Organisers," *Straits Times* (Singapore), May 6, 2005; Sara Wilson, "Internet Is Blamed for Increase in Sex-Tours," *Scotsman* (Edinburgh), July 3, 1996.

38. The introduction to the spa at the landmark Claremont Hotel in Berkeley is typical of spa literature: "Welcome to Spa Claremont, where tranquil elegance encourages you to lose yourself in a therapeutic environment that embraces and caresses the senses . . . wellness professionals bring their worlds of treatments to the spa that overlooks San Francisco and its soothing Bay." "Flying Dutchmen Travel," http://www.flyingdutchmentravel.com/family_vacations/golf/spa/spa-california.html (accessed February 4, 2013).

39. *Islands*, July/August 2003, p. 13.

40. Ibid., p. 81.

41. *Islands*, December 2012, back cover.

42. For an example of guided tours to the summit of Mount Everest, see Alpine Ascents International, "Mount Everest Guided Climbing Expedition with Alpine Ascents International," http://www.alpineascents.com/everest.asp (accessed September 7, 2012). As of September 2012 the company had escorted almost two hundred climbers to the summit of Everest.

43. Regarding the separation of leisure nature from work nature and how leisure nature obscures environmental degradation, see Richard White, "Are You an Environmentalist, or Do You Work for a Living?" in *Uncommon Ground: Rethinking the Human Place in Nature*, ed. William Cronon (New York: W. W. Norton, 1996).

44. For an account of the removal of American Indians in the creation of the US national parks, see Mark David Spence, *Dispossessing the Wilderness: Indian Removal and the Making of the National Parks* (New York: Oxford University Press, 1999). For more on the problems of the concept of wilderness (and some of its benefits), see William Cronon,

"The Trouble with Wilderness; or, Getting Back to the Wrong Nature," in *Uncommon Ground: Rethinking the Human Place in Nature*, ed. William Cronon (New York: W. W. Norton, 1996).

CHAPTER 3: FROM DISSOCIATION TO DESTRUCTION THROUGH THE PSYCHE

1. Henri F. Ellenberger, *The Discovery of the Unconscious; the History and Evolution of Dynamic Psychiatry* (New York: Basic Books, 1970). Discussions with the psychologist Brian A. Dahmen contributed a lot to my thinking for this section. This chapter is based on Kenneth Worthy, "Modern Institutions, Phenomenal Dissociations, and Destructiveness toward Humans and the Environment," *Organization & Environment* 21, no. 2 (2008).

2. Roberto Lewis-Fernández, "Culture and Dissociation: A Comparison of Ataque De Nervios among Puerto Ricans and Possession Syndrome in India," in *Dissociation: Culture, Mind, and Body*, ed. David Spiegel (Washington, DC: American Psychiatric Press, 1994), pp. 5, 9.

3. Breuer was the pioneer of the cathartic method, which cured his famous patient, "Anna O."; he documented the process in collaboration with Freud, who later built upon the technique in the development of psychoanalysis. Josef Breuer and Sigmund Freud, *Studies on Hysteria*, 2nd ed. (New York: Basic Books, 1957).

4. Lewis-Fernández, "Culture and Dissociation," p. 9.

5. David Spiegel, *Dissociation: Culture, Mind, and Body*, 1st ed. (Washington, DC: American Psychiatric Press, 1994), p. xiii; Marlene Steinberg, "Systematizing Dissociation: Symptomatology and Diagnostic Assessment," in Spiegel, *Dissociation*.

6. Task Force on DSM-IV American Psychiatric Association, *Diagnostic and Statistical Manual of Mental Disorders: DSM-IV*, 4th ed. (Washington, DC: American Psychiatric Association, 1994).

7. Steinberg, "Systematizing Dissociation."

8. Ibid.

9. Israel Orbach, "The Role of the Body Experience in Self-Destruction," *Clinical Child Psychology & Psychiatry* 1, no. 4 (1996): 608–11, 615. The distorted childhood caregiving that may result in lack of self-care includes lack of taming of self-aggression, lack of representational learning, and lack of attunement to needs. Experiences that may result in self-destruction are internalized and symbolized hate, distorted experience of pain and pleasure, and dissociation and detachment from the body. Ibid., p. 614.

10. Stanley Milgram, *Obedience to Authority: An Experimental View*, 1st ed. (New York: Harper & Row, 1974), p. 1. Together with Hannah Arendt's writings on the Holocaust, particularly *Eichmann in Jerusalem: A Report on the Banality of Evil*, Milgram's findings

dispelled the belief that the Holocaust had been the result mainly of a peculiar, perhaps momentary, characteristic of the collective German psyche. The experiments showed in contrast that situational, social features were more relevant to understanding how so many people could have supported and carried out the German Nazi attempt at exterminating European Jews. Hannah Arendt, *Eichmann in Jerusalem; a Report on the Banality of Evil* (New York: Viking Press, 1963).

11. Thomas Blass, "Understanding Behavior in the Milgram Obedience Experiment: The Role of Personality, Situations, and Their Interactions," *Journal of Personality & Social Psychology* 60, no. 3 (1991): 408. Thomas Blass and Arthur G. Miller have written extensively on the influence of Milgram's obedience experiments. See Thomas Blass, "The Milgram Paradigm after 35 Years: Some Things We Now Know About Obedience to Authority," *Journal of Applied Social Psychology* 29, no. 5 (1999); Thomas Blass, *Obedience to Authority: Current Perspectives on the Milgram Paradigm* (Mahwah, NJ: Lawrence Erlbaum Associates, 2000); Arthur G. Miller, Barry E. Collins, and Diana E. Brief, "Perspectives on Obedience to Authority: The Legacy of the Milgram Experiments," *Journal of Social Issues* 51, no. 3 (1995); Arthur G. Miller, *The Obedience Experiments: A Case Study of Controversy in Social Science* (New York: Praeger, 1986); Milgram, *Obedience to Authority*, p. 207.

12. As quoted in Harold Takooshian, "How Stanley Milgram Taught About Obedience," in Blass, *Obedience to Authority*, p. 10.

13. Miller, Collins, and Brief, "Perspectives on Obedience to Authority," introduction.

14. In this chapter, I use both *subject* and *participant* to denote people being studied in psychological experiments, following the original authors. The term *subject* has gradually been replaced by the term *participant* in psychology, perhaps because of the former's colonialist and imperialist overtones.

15. Milgram, *Obedience to Authority*, pp. 3–4, 20.

16. This describes Experiment 5: A New Base-Line Condition. Ibid., pp. 3–4, 34, 55–57.

17. Blass, "Understanding Behavior in the Milgram Obedience Experiment," p. 398.

18. Milgram, *Obedience to Authority*, pp. 56, 60–61.

19. Stanley Milgram, "Behavioral-Study of Obedience," *Journal of Abnormal and Social Psychology* 67, no. 4 (1963): 375; Milgram, *Obedience to Authority*, p. 30.

20. For instance, the obedience experiments have "played a central and enriching role" in a number of controversies, such as those over research ethics, the social psychology of experimentation, and deception versus role-playing. Blass, "Understanding Behavior in the Milgram Obedience Experiment," p. 398.

21. Miller, Collins, and Brief, "Perspectives on Obedience to Authority"; Blass, "Understanding Behavior in the Milgram Obedience Experiment," p. 398; Miller, *Obedience Experiments*, pp. 88–89.

22. Milgram, "Behavioral-Study of Obedience," p. 375; Blass, "Understanding

Behavior in the Milgram Obedience Experiment," pp. 398–99. Subjects were observed sweating, trembling, stuttering, biting their lips, groaning, digging their fingernails into their flesh, and smiling and laughing nervously. In four of the experimental conditions, fifteen subjects experienced full-blown, uncontrollable seizures. Stanley Milgram, "Some Conditions of Obedience and Disobedience to Authority," *Human Relations* 18, no. 1 (1965): 68.

Some people might question whether it's proper to use the results of the obedience studies due to the ethical questions surrounding them. Evaluating the propriety of Milgram's experiments is a matter that can fill volumes; there certainly was at least momentary suffering for many of the subjects, though Milgram claimed that follow-up studies showed that subjects didn't suffer long-term consequences of participation in the experiments, and very few subjects (1.3 percent) stated that they were sorry to have participated. Ibid., p. 58; Stanley Milgram, "Issues in the Study of Obedience: A Reply to Baumrind," *American Psychologist* 19, no. 11 (1964). As far as I'm concerned, referring to the obedience results (and those of the other experiments cited here) causes no substantial further harm and may lead to important benefits, though it's not my intention to condone the methods.

23. Blass, "Understanding Behavior in the Milgram Obedience Experiment," pp. 398–99.

24. This was Experiment 11: Subject Free to Choose Shock Level. Milgram, *Obedience to Authority*, pp. 61, 70–72.

25. Milgram, "Behavioral-Study of Obedience," p. 375.

26. Alan C. Elms, *Social Psychology and Social Relevance* (Boston: Little, Brown, 1972), pp. 130–31.

27. Miller and Blass are among the psychologists who continue to find the Milgram experiments of value and who survey their continued influence in psychology and other fields. Miller, Collins, and Brief, "Perspectives on Obedience to Authority"; Miller, *Obedience Experiments*; Blass, *Obedience to Authority*; Blass, "Understanding Behavior in the Milgram Obedience Experiment," p. 398.

28. Milgram, "Some Conditions of Obedience and Disobedience to Authority," p. 74.

29. Blass, *Obedience to Authority*, p. 59; Thomas Blass, "A Cross-Cultural Comparison of Studies of Obedience Using the Milgram Paradigm: A Review," *Social and Personality Psychology Compass* 6, no. 2 (2012).

30. Blass, "Milgram Paradigm after 35 Years," p. 969.

31. Charles L. Sheridan and Richard G. King, "Obedience to Authority with an Authentic Victim," *Proceedings of the Annual Convention of the American Psychological Association* 7, no. 1 (1972). The complete obedience of the female participants may be related to the youth of the participants, who were in their late teens. You may wonder about the detachment required for the experimenters themselves when they designed an experiment in which a puppy suffered like this.

32. This experiment was conducted by Carney Landis and associates at the University of Minnesota to study physiological expressions of emotion. The job of decapitation was often awkward and prolonged due to the stress and internal conflict experienced by the participants. Peter V. Butler, "Destructive Obedience in 1924: Landis'"Studies of Emotional Reactions' as a Prototype of the Milgram Paradigrm," *Irish Journal of Psychology* 19, no. 2–3 (1998): 242.

33. Milgram, *Obedience to Authority*, pp. 32–34.

34. Ibid., pp. 34–36.

35. Ibid., p. 35.

36. Milgram, "Some Conditions of Obedience and Disobedience to Authority," p. 62.

37. Ibid., p. 61.

38. Thomas Blass analyzed the Milgram data and showed that, although there actually is no statistical significance to the differences in obedience rates between the remote and voice-feedback variations or between the proximity and touch-proximity variations, all the other differences were significant. So there's an inverse relation between proximity and obedience. Blass, "Understanding Behavior in the Milgram Obedience Experiment," p. 401.

One of the most interesting experiments following the Milgram obedience model is the one performed in the late 1960s by Harvey A. Tilker of the City University of New York. Tilker investigated subject responsibility and victim feedback by manipulating them through several experimental conditions. Responsibility was varied between No Responsibility, Ambiguous Responsibility, and Total Responsibility. Feedback was varied between No Feedback, Auditory Feedback, and Auditory-visual Feedback. Tilker found that total accepted responsibility for another person's well-being and maximum feedback from that person regarding his or her condition are major determinants of socially responsible behavior. Harvey A. Tilker, "Socially Responsible Behavior as a Function of Observer Responsibility and Victim Feedback," *Journal of Personality and Social Psychology* 14, no. 2 (1970): 99. These results match previous results showing that feedback from a victim reduces the intensity of aggression directed toward the victim. Arnold H. Buss, "Instrumentality of Aggression Feedback and Frustration as Determinants of Physical Aggression," *Journal of Personality and Social Psychology* 3, no. 2 (1966). When responsibility was diffused, as it is when there are multiple witnesses to an emergency, speed of assistance or action was reduced. John M. Darley and Bibb Latané, "Bystander Intervention in Emergencies: Diffusion of Responsibility," *Journal of Personality & Social Psychology* 8, no. 4, pt. 1 (1968). Maximum feedback and perceived responsibility for outcomes can reduce harms.

39. Milgram, "Some Conditions of Obedience and Disobedience to Authority."

40. Dave Grossman, *On Killing: The Psychological Cost of Learning to Kill in War and Society* (New York: Little, Brown, 2009), p. 118.

41. Ibid., pp. 117, 308.

42. Milgram, "Some Conditions of Obedience and Disobedience to Authority," p. 63.

43. Milgram, *Obedience to Authority*, pp. 37–38.

44. Milgram, "Some Conditions of Obedience and Disobedience to Authority," p. 61.

45. Milgram, *Obedience to Authority*, pp. 38–39.

46. Milgram, "Some Conditions of Obedience and Disobedience to Authority," p. 65.

47. François Rochat and Andre Modigliani, "Authority: Obedience, Defiance, and Identification in Experimental and Historic Contexts," in *A New Outline of Social Psychology*, ed. Martin Gold and Elizabeth Douvan (Washington, DC: American Psychological Association, 1997), pp. 235, 237, 238; Nestar J. C. Russell and Robert J. Gregory, "Spinning an Organizational 'Web of Obligation'? Moral Choice in Stanley Milgram's 'Obedience' Experiments," *American Review of Public Administration* 41, no. 5 (2011): 500–501.

48. Milgram, *Obedience to Authority*, pp. 39–40.

49. Ibid., pp. 143–48, 154–57.

50. Ibid., pp. 132–34, 153–64. Blass believes that the theoretical component of Milgram's book was its weakest section: Thomas Blass, *The Man Who Shocked the World: The Life and Legacy of Stanley Milgram*, 1st ed. (New York: Basic Books, 2004). Nevertheless, it's the most comprehensive framework available for understanding the obedience results.

51. Milgram, *Obedience to Authority*, p. 149.

52. Ibid., pp. 149–52.

53. Ibid., p. 152.

54. George Monbiot, "On the 12th Day of Christmas . . . Your Gift Will Just Be Junk," *Guardian* (London), December 10, 2012, http://gu.com/p/3cdnx (accessed February 4, 2013).

55. Milgram, *Obedience to Authority*, p. 121.

56. Milgram, "Some Conditions of Obedience and Disobedience to Authority," p. 66.

57. Ibid., p. 63.

58. Milgram, *Obedience to Authority*, p. 122.

59. Terrance O'Connor, "Therapy for a Dying Planet," in *Ecopsychology: Restoring the Earth, Healing the Mind*, ed. Theodore Roszak, Mary E. Gomes, and Allen D. Kanner (San Francisco: Sierra Club Books, 1995).

60. Milgram, *Obedience to Authority*, p. 149.

61. Jonathan L. Freedman and Scott C. Fraser, "Compliance without Pressure: The Foot-in-the-Door Technique," *Journal of Personality and Social Psychology* 4, no. 2 (1966).

62. Similarly, it may be that corporate executives sometimes resist switching to new, more environmentally benign practices out of fear of implying that past behavior was improper or even illegal, thereby opening their companies to lawsuits or punitive action by government agencies.

63. Ervin Staub, *The Roots of Evil: The Origins of Genocide and Other Group Violence* (Cambridge; New York: Cambridge University Press, 1989), p. 85.

64. Milgram, *Obedience to Authority*, p. 9.

65. Staub, *Roots of Evil*, pp. 33–34, 79, 86.

66. Milgram, *Obedience to Authority*, p. 10.

67. Ibid., pp. 121–22.

68. Ibid., p. 122.

69. Wesley Kilham and Leon Mann in Australia directly tested and compared obedience levels of transmitters (the role of conveying the command) and executants (the role of pressing the levers). In their transmitter condition, 54 percent of their subjects were fully obedient, as compared with 28 percent of subjects in the executant condition. In other words, subjects were twice as likely to be destructively obedient when they didn't have to directly press the lever than when they instructed another person to do so. Wesley Kilham and Leon Mann, "Level of Destructive Obedience as a Function of Transmitter and Executant Roles in Milgram Obedience Paradigm," *Journal of Personality and Social Psychology* 29, no. 5 (1974): 700. Kilham and Mann included a study of gender effects. They found that females were much more defiant than males in the executant condition, but in the transmitter condition, they were only slightly more defiant than were males. Ibid., p. 699.

70. Russell and Gregory, "Spinning an Organizational 'Web of Obligation'?" p. 512. They take the concept of "zone of indifference" from Chester Irving Barnard, *The Functions of the Executive* (Cambridge, MA: Harvard University Press, 1938).

71. Wim H. J. Meeus and Quinten A. W. Raaijmakers, "Obedience in Modern Society: The Utrecht Studies," *Journal of Social Issues* 51, no. 3 (1995). The original article reporting the experiment was Wim H. J. Meeus and Quinten A. W. Raaijmakers, "Administrative Obedience: Carrying Out Orders to Use Psychological-Administrative Violence," *European Journal of Social Psychology* 16, no. 4 (1986).

72. Meeus and Raaijmakers, "Obedience in Modern Society."

73. Craig Haney, Curtis Banks, and Philip Zimbardo, "Interpersonal Dynamics in a Simulated Prison," *International Journal of Criminology & Penology* 1, no. 1 (1973): 80–81.

74. Ibid.

75. Janice T. Gibson and Mika Haritos-Fatouros, "The Education of a Torturer," in *Readings in Social Psychology: General, Classic, and Contemporary Selections*, ed. Wayne A. Lesko (Needham Heights, MA: Allyn & Bacon, 1991), pp. 247–49.

76. Staub, *The Roots of Evil*, pp. 28–29, 83.

77. Kilham and Mann, "Level of Destructive Obedience," p. 696.

78. John Sabini and Maury Silver, *Moralities of Everyday Life* (Oxford; New York: Oxford University Press, 1982), pp. 65–66.

79. Thomas Blass, "Attribution of Responsibility and Trust in the Milgram Obedience Experiment," *Journal of Applied Social Psychology* 26, no. 17 (1996): 1532.

80. Milgram, *Obedience to Authority*, p. 70.

81. Haney, Banks, and Zimbardo, "Interpersonal Dynamics in a Simulated Prison," p. 88.

82. Milgram, *Obedience to Authority*, p. 146.

83. Bibb Latané and John M. Darley, "Group Inhibition of Bystander Intervention in Emergencies," in *Readings in Social Psychology: The Art and Science of Research*, ed. Steven Fein, Steven Spencer, and Sharon S. Brehm (Boston: Houghton Mifflin, 1996), pp. 135–36.

84. Milgram, "Behavioral-Study of Obedience," p. 377.

85. Richard Louv, *Last Child in the Woods: Saving Our Children from Nature-Deficit Disorder*, updated and expanded. ed. (Chapel Hill, NC: Algonquin Books, 2008).

86. Discussions of the psychic effects of environmental degradation in ecopsychology include the following: Paul Shepard, "Nature and Madness," in *Ecopsychology: Restoring the Earth, Healing the Mind*, ed. Theodore Roszak, Mary E. Gomes, and Allen D. Kanner (San Francisco: Sierra Club Books, 1995); Chellis Glendinning, "Technology, Trauma, and the Wild," in *Ecopsychology: Restoring the Earth, Healing the Mind*, ed. Theodore Roszak, Mary E. Gomes, and Allen D. Kanner (San Francisco: Sierra Club Books, 1995); O'Connor, "Therapy for a Dying Planet."

87. Roszak, Gomes, and Kanner, *Ecopsychology*; Joanna Macy, "Working through Environmental Despair," in Roszak, Gomes, and Kanner, *Ecopsychology*; Allen D. Kanner and Mary E. Gomes, "The All-Consuming Self," in Roszak, Gomes, and Kanner, *Ecopsychology*; O'Connor, "Therapy for a Dying Planet."

88. Theodore Roszak, *The Voice of the Earth* (New York: Simon & Schuster, 1992), p. 14. In his Epilogue, Roszak proposes a set of principles for ecopsychology. Ibid., p. 319.

89. Ibid., p. 18.

90. Laura Sewall, "Beauty and the Brain," in *Ecopsychology: Science, Totems, and the Technological Species*, ed. Peter H. Hasbach and Patricia H. Kahn (Cambridge, MA: MIT Press, 2012), p. 271.

91. Ibid., p. 276.

92. Ibid., pp. 268, 269, 271, 274, 276, 277.

93. Haney, Banks, and Zimbardo, "Interpersonal Dynamics in a Simulated Prison," pp. 80–81.

94. Looking at prosocial, helping behavior, the psychologist Robert V. Levine found that the highest population densities in cities can lead to feelings of alienation, anonymity, and social isolation. In his experimental procedure carried out in a large group of American cities, these factors were found to reduce helping behavior. Robert V. Levine, "Cities with Heart," in Lesko, *Readings in Social Psychology*.

95. In a well-known report, Robert Kraut and associates found that Internet use was associated with elevated depression. Robert Kraut et al., "Internet Paradox: A Social Technology That Reduces Social Involvement and Psychological Well-Being?" *American Psychologist* 53, no. 9 (1998). These findings were contested by various authors. Kraut and associates later published a follow-up report stating that the effects originally noted dissipated over time. Judith S. Shapiro, "Loneliness: Paradox or Artifact?" *American Psychologist*

54, no. 9 (1999); Robert Kraut et al., "Internet Paradox Revisited," *Journal of Social Issues* 58, no. 1 (2002). Any study of detrimental psychic effects from Internet use is made difficult by the many possible types of use, the various ways those uses complement non-Internet life for different people, and the particular psychology that individuals bring to the situation. At least one study finds that the overall effect of Internet use is positive for well-connected people and negative for those already isolated or suffering from psychological problems. M. Wirion et al., "Is the Internet a Pathological Communication Tool or Not? Where the Weakest Are the Most Vulnerable," *Annales Medico-Psychologiques* 162, no. 6 (2004).

Various studies carried out since the original Kraut report relate Internet use and psychological harms. (1) Boys with higher levels of Internet usage have higher levels of social anxiety and less mature identity statuses: Dennis Mazalin and Susan Moore, "Internet Use, Identity Development, and Social Anxiety among Young Adults," *Behaviour Change* 21, no. 2 (2004). (2) Higher levels of Internet use result in higher levels of "emotional loneliness" and decreased social well-being: Eric J. Moody, "Internet Use and Its Relationship to Loneliness," *CyberPsychology & Behavior* 4, no. 3 (2001). (3) More time spent playing online computer games leads to decreased quality of interpersonal relationships and increased social anxiety: Shao-Kang Lo, Chih-Chien Wang, and Wenchang Fang, "Physical Interpersonal Relationships and Social Anxiety among Online Game Players," *CyberPsychology & Behavior* 8, no. 1 (2005).

96. Milgram, "Some Conditions of Obedience and Disobedience to Authority," p. 65.

CHAPTER 4: DISSOCIATIONS IN WESTERN PSYCHES

1. Jonathan Lewis, "Shifting Nature," in *China from the Inside* (BBC/Granada/PBS: 2007).

2. Edward Twitchell Hall, *Beyond Culture* (Garden City, NY: Anchor Books, 1981), p. 7.

3. Richard E. Nisbett et al., "Culture and Systems of Thought: Holistic Versus Analytic Cognition," *Psychological Review* (April 2001); Richard E. Nisbett, *The Geography of Thought: How Asians and Westerners Think Differently—and Why* (New York: Free Press, 2003).

4. Nisbett et al., "Culture and Systems of Thought," p. 295.

5. Ibid., p. 291.

6. Ibid., pp. 296–97.

7. Ibid.

8. Kaiping Peng, Daniel R. Ames, and Eric D. Knowles, "Culture and Human Inference: Perspectives from Three Traditions," in *The Handbook of Culture & Psychology*, ed. David Ricky Matsumoto (New York: Oxford University Press, 2001), p. 250; Li-Jun Ji, Kaiping Peng, and Richard E. Nisbett, "Culture, Control, and Perception of Relationships in the Environment," *Journal of Personality and Social Psychology* 78, no. 5 (2000).

9. Nisbett et al., "Culture and Systems of Thought," p. 298.

10. Ziva Kunda, *Social Cognition: Making Sense of People* (Cambridge, MA: MIT Press, 1999), p. 532.

11. Nisbett et al., "Culture and Systems of Thought," p. 298; Michael W. Morris and Kaiping Peng, "Culture and Cause: American and Chinese Attributions for Social and Physical Events," *Journal of Personality and Social Psychology* 67, no. 6 (1994).

12. Incheol Choi and Richard E. Nisbett, "Situational Salience and Cultural Differences in the Correspondence Bias and Actor-Observer Bias," *Personality and Social Psychology Bulletin* 24, no. 9 (1998).

13. Nisbett et al., "Culture and Systems of Thought," p. 299.

14. Ibid.; Peng, Ames, and Knowles, "Culture and Human Inference," p. 252.

15. Nisbett et al., "Culture and Systems of Thought," p. 298.

16. Ibid., p. 300.

17. Peng, Ames, and Knowles, "Culture and Human Inference."

18. Nisbett et al., "Culture and Systems of Thought," pp. 300–301.

19. Ibid., p. 301.

20. Ibid., p. 302.

21. Ibid.

22. Ibid., p. 303.

23. Gus Lubin, "Satellite Pictures of the Empty Chinese Cities Where Home Prices Are Crashing," *Business Insider*, December 10, 2011, http://www.businessinsider.com/china-ghost-cities-2011-11 (accessed January 29, 2013).

24. Nisbett et al., "Culture and Systems of Thought," p. 303.

25. Ibid.

26. Ibid., pp. 304–305.

27. Ibid., p. 304.

28. Donald J. Munro, *Individualism and Holism: Studies in Confucian and Taoist Values* (Ann Arbor: Center for Chinese Studies, University of Michigan, 1985), p. 4.

29. Ibid., p. 23.

30. Ibid., p. 2.

31. Hall, *Beyond Culture*.

32. Ibid., pp. 34, 35.

33. Ibid., pp. 23, 35, 113.

34. Ibid., p. 9.

35. Ibid., p. 187.

36. Ibid., pp. 98–99.

37. Margaret S. Mahler, *Infantile Psychosis and Early Contributions*, 2 vols., vol. I, "The Selected Papers of Margaret S. Mahler, M.D.," ed. Marjorie Harley and Annemarie Weil (New York: J. Aronson, 1979), p. 223; Margaret S. Mahler, Fred Pine, and Anni Bergman,

The Psychological Birth of the Human Infant: Symbiosis and Individuation (New York: Basic Books, 1975), pp. 3, 41–51. "Separation consists of the child's emergence from a symbiotic fusion with the mother, and individuation consists of those achievements marking the child's assumption of his own individual characteristics." Ibid., p. 4.

38. Peng, Ames, and Knowles, "Culture and Human Inference," p. 248.

39. Hazel Rose Markus and Shinobu Kitayama, "Culture and the Self: Implications for Cognition, Emotion, and Motivation," in *The Self in Social Psychology*, ed. Roy F. Baumeister (Philadelphia: Psychology Press/Taylor & Francis, 1999).

40. Peng, Ames, and Knowles, "Culture and Human Inference," p. 248.

41. Vincent Crapanzano, "Preface," in Maurice Leenhardt, *Do Kamo: Person and Myth in the Melanesian World* (Chicago: University of Chicago Press, 1979), pp. vii–xiv. Throughout this section, my references to the Canaque and to Mead and Bateson's Balinese studies aren't meant to imply societies that are static and homogeneous. Descriptions that account for the changes and variety within these societies could fill volumes.

42. Marcel Mauss, "A Category of the Human Mind: The Notion of the Person; the Notion of the Self," in *The Category of the Person: Anthropology, Philosophy, History*, ed. Michael Carrithers, Steven Collins, and Steven Lukes (Cambridge; New York: Cambridge University Press, 1985), p. 14.

43. Ibid., p. 16.

44. Ibid., pp. 14, 18.

45. Ibid., p. 21.

46. Ibid., pp. 3, 18, 20.

47. Ibid., p. 12.

48. Ibid., pp. 5, 7.

49. Ibid., p. 18.

50. Leenhardt, *Do Kamo*, p. 24.

51. Ibid., p. 61.

52. Crapanzano, "Preface," in Leenhardt, *Do Kamo*," pp. xxii, xxiv.

53. Gregory Bateson and Margaret Mead, *Balinese Character, a Photographic Analysis* (New York: New York Academy of Sciences, 1942), p. 8.

54. Ibid., p. 9.

55. Ibid., p. 12.

56. Self-enhancement reasoning is more prevalent in more individualistic cultures, and self-criticism reasoning is more prevalent in more collectivist cultures. Self-criticism shows less need to build up the self over others. It probably also reflects more dependence on fitting in and performing within one's position in society to attain satisfaction. For a helpful synthesis of related research, including North American self-enhancement bias, see Kunda, *Social Cognition*, pp. 533–49.

57. Crapanzano, "Preface," in Leenhardt, *Do Kamo*," p. xii.

58. Leenhardt, *Do Kamo*, p. 32.

59. Crapanzano, "Preface," in Leenhardt, *Do Kamo*, p. xxiii.

60. Mauss, "A Category of the Human Mind," p. 11.

61. Ibid., p. 5.

62. Traditional names in Bali signify social context, though it recently has become more popular to omit caste-based naming conventions, particularly among lower-caste people.

63. Mauss, "A Category of the Human Mind," p. 10.

64. Ibid., p. 8.

65. David Malo, *Hawaiian Antiquities (Moolelo Hawaii)*, trans. Dr. N. B. Emerson (Honolulu: Hawaiian Gazette, 1903), pp. 24–28.

66. J. Baird Callicott, *Earth's Insights: A Survey of Ecological Ethics from the Mediterranean Basin to the Australian Outback* (Berkeley: University of California Press, 1994), pp. 110–19.

67. Leenhardt, *Do Kamo*, pp. 24–26.

68. Bateson and Mead, *Balinese Character*, p. 14.

69. Ibid., plate 65. I've seen Balinese dance teachers teach this way many times, guiding their pupil's body with their hands and legs, though the technique is used less often and less forcefully when teaching foreigners.

70. Ibid., plate 63.

71. Crapanzano, "Preface," in Leenhardt, *Do Kamo*, p. xxiv.

72. Leenhardt, *Do Kamo*, p. 41.

73. Ibid., pp. 34, 35.

74. Crapanzano, "Preface," in Leenhardt, *Do Kamo*, p. xxiv.

75. Leenhardt, *Do Kamo*, p. 84.

76. Gary Snyder, *The Practice of the Wild: Essays* (San Francisco: North Point Press, 1990), p. 82.

77. Leenhardt, *Do Kamo*, p. 127.

78. Bateson and Mead, *Balinese Character*, p. 3.

79. Some Balinese do seek solitude regularly. Among them are traditional healers, "balians," who seek it as an ascetic or spiritual practice. Others just like it. But it's less common for Balinese than for Americans to enjoy being alone.

80. Bateson and Mead, *Balinese Character*, p. 4.

81. Ibid., p. 15.

82. Ibid., plate 1.

83. Ibid., plate 15.

84. Leenhardt, *Do Kamo*, p. 39.

85. Ibid., p. 143.

86. Ibid., pp. 39–40.

87. Rebecca Solnit, *River of Shadows: Eadweard Muybridge and the Technological Wild*

West (New York: Viking, 2003), pp. 111–20. In his detailed accounting of Indian removals from the West in the making of the US National Parks, Mark David Spence writes that "two versions of a tragic story about Crow warriors who took a raft over Yellowstone Falls and committed suicide instead of surrendering to pursuing troops of the US army suggest how painful the loss of the Yellowstone area must have been for some individuals." Mark David Spence, *Dispossessing the Wilderness: Indian Removal and the Making of the National Parks* (New York: Oxford University Press, 1999), p. 155, n. 59.

88. Crapanzano, "Preface," in Leenhardt, *Do Kamo*, p. xxv. Crapanzano also admits that "Leenhardt's attitude toward individuation was not without ambivalence. He recognizes as one of the hazards of full individuation the loss of mythic modes of apperception and communal (participational) existence." Ibid.

Henri Lefebvre expresses a view of plenitude opposite to Leenhardt's. Discussing alienation, he associates "traditional everyday life" with a certain "human plenitude," which is lost in the historical process of social alienation and modernization. He says that traditional everyday life is based on non-separation and an absence of differentiation in the cosmic order. Henri Lefebvre and Michel Trebitsch, *Critique of Everyday Life*, vol. 1 (London; New York: Verso, 1991), p. xxiii.

CHAPTER 5: ANCIENT TRACES OF DISSOCIATION

1. J. Baird Callicott, *Earth's Insights: A Survey of Ecological Ethics from the Mediterranean Basin to the Australian Outback* (Berkeley: University of California Press, 1994), p. 83.

2. Robin George Collingwood, *The Idea of Nature* (New York: Oxford University Press, 1960; reprint, 1967), pp. 1–9.

3. Edith Hamilton, *The Greek Way* (New York: Time, 1963), pp. 314–15.

4. Samuel Sambursky, *The Physical World of the Greeks* (London: Routledge, 1960), pp. 3, 223, 244.

5. Alvin Ward Gouldner, *Enter Plato; Classical Greece and the Origins of Social Theory* (New York: Basic Books, 1965), p. 340.

6. Ibid., p. 9.

7. Ibid., p. 74.

8. Ibid., p. 13.

9. Ibid., pp. 79–80.

10. Ibid., p. 46.

11. Ibid., p. 8.

12. Ibid.

13. Ibid., p. 99.

14. Ibid., p. 70.

15. Ibid., p. 71.

16. Ibid.

17. Ibid., pp. 66–67.

18. Ibid.

19. Robert K. Logan, *The Alphabet Effect: The Impact of the Phonetic Alphabet on the Development of Western Civilization*, 1st ed. (New York: Morrow, 1986).

20. Some of these ideas originate in the Toronto School of communication theory, established by Harold Innis and Marshall McLuhan. Logan absorbs and develops their theories. See also David Abram, *The Spell of the Sensuous: Perception and Language in a More-Than-Human World*, 1st ed. (New York: Pantheon Books, 1996). Leonard Shlain argues that the subjugation of women and alienation from nature are consequences of the rise of alphabetic writing. Leonard Shlain, *The Alphabet Versus the Goddess: The Conflict between Word and Image* (New York: Viking, 1998).

21. Logan, *Alphabet Effect*, p. 23.

22. Ibid.

23. Ibid., p. 105.

24. Plato, *The Collected Dialogues of Plato: Including the Letters*, Phaedrus, 275d–76, pp. 521.

25. Logan, *Alphabet Effect*, pp. 38–39.

26. Gouldner, *Enter Plato*; Sambursky, *Physical World of the Greeks*.

27. This discussion roughly follows Callicott, *Earth's Insights*, pp. 28–29.

28. Joseph Needham, "Poverties and Triumphs of the Chinese Scientific Tradition," in *The "Racial" Economy of Science toward a Democratic Future*, ed. Sandra G. Harding, *Race, Gender, and Science* (Bloomington: Indiana University Press, 1993), p. 30; Geoffrey Ernest Richard Lloyd, *Adversaries and Authorities: Investigations into Ancient Greek and Chinese Science* (Cambridge; New York: Cambridge University Press, 1996), p. 227.

29. Needham, "Poverties and Triumphs," p. 30; Lloyd, *Adversaries and Authorities*, p. 227.

30. Gouldner, *Enter Plato*, pp. 172–73.

31. Ibid., pp. 171–73.

32. Ibid.

33. Ibid., p. 174.

34. For Plato, philosophy tries to set the soul free: "She [philosophy] points out that observation by means of the eyes and ears and all the other senses is entirely deceptive, and she urges the soul to refrain from using them unless it is necessary to do so, and encourages it to collect and concentrate itself by itself, trusting nothing but its own independent judgment upon objects considered in themselves . . . because such objects are sensible and visible but what the soul itself sees is intelligible and invisible." Plato, *The Collected Dialogues of Plato: Including the Letters*, Phaedo, 83a–b, p. 66.

35. Gouldner, *Enter Plato*, pp. 327–30. About Plato's disdain for received, social

wisdom (versus reason), Gouldner writes, "The culminating achievement of Greek philosophy is, at the same time, the crowning Greek treachery—Plato's condemnation and betrayal of classical poetry and science. This Athenian triumph marks the ebb of the Ionian enlightenment." Ibid., p. 134.

36. Ibid., pp. 378–79.

37. Benjamin Farrington, *Greek Science; Its Meaning for Us* (Harmondsworth, Middlesex: Penguin Books, 1949), pp. 146–47.

38. Geoffrey Ernest Richard Lloyd, *Science and Morality in Greco-Roman Antiquity: An Inaugural Lecture* (Cambridge; New York: Cambridge University Press, 1985), p. 25.

39. Gouldner, *Enter Plato*, pp. 216–17, 228, 241. Plato justified slavery in two ways: (1) the slave is diminished in the highest human quality—reason (repeated frequently by Plato when he draws oppositions between reason and women, slaves, and nature) and (2) it's ordained and natural, being an expression within the society of the hierarchical relationships said to be characteristic of the universe as a whole (Plato's commitment to social hierarchy runs through his *Republic* and other writings).

40. Ibid., pp. 333–40.

41. Plato frequently associates women and slaves together in a lower order of nature, one that can't summon reason to rule over the emotions: "The mob of motley appetites and pleasures and pains one would find chiefly in children and women and slaves and in the base rabble of those who are free men in name." Plato, *The Collected Dialogues of Plato: Including the Letters, Republic*, 431c, p. 673. He associates the female with a formless void onto which male reason and the logos imprint form, ibid., *Timaeus*, 50b–51b, pp. 1177–78. He associates the female also with a baser part of the body, ibid., *Timaeus*, 70a, p. 1193. Women are inferior to men in most capacities, ibid., *Republic*, 455c–56, p. 694. And they descended from men "who were cowards or who led unrighteous lives," ibid., *Timaeus*, 90e–91a , pp. 1209–10.

42. Gouldner, *Enter Plato*, pp. 333–40.

43. John Burnet and Philip Whalen, *Early Greek Philosophy* (New York: Meridian Books, 1957). Heraclitus, *Fragments*, 41–42, p. 136.

44. We know this through Aristotle. Aristotle, *Selected Works*, trans. Hippocrates George Apostle and Lloyd P. Gerson, 3rd ed. (Grinnell, IA: Peripatetic Press, 1991), *Metaphysics, Book IV*, chap. 5, 1010a10–20, p. 375.

45. For example, in the *Phaedo* Plato writes of the perceptible world, "Did we not say some time ago that when the soul uses the instrumentality of the body for any inquiry, whether through sight or hearing or any other sense—because using the body implies using the senses—it is drawn away by the body into the realm of the variable, and loses its way and becomes confused and dizzy, as though it were fuddled, through contact with things of a similar nature?" Plato, *The Collected Dialogues of Plato: Including the Letters, Phaedo*, 79c, p. 62. By "similar nature" Plato means that the physical world is fuddled and dizzy.

46. Empedocles, "Purifications," in *An Introduction to Early Greek Philosophy; the Chief Fragments and Ancient Testimony, with Connecting Commentary*, ed. John Mansley Robinson (Boston: Houghton Mifflin, 1968), p. 152.

47. Collingwood, *Idea of Nature*, pp. 66–67.

48. Plato's well-known theory of Forms or Ideas emerges in numerous parts of his work. It's one of the Forms' central features that they're static and unchanging. The Forms (particularly Beauty) and the love of the beauty of the Forms are discussed in the *Symposium*. Plato, *The Collected Dialogues of Plato: Including the Letters*, *Symposium*. "True reality consists in certain intelligible and bodiless forms. . . . In the class of argument they shatter and pulverize those bodies which their opponents wield, and what those others allege to be true reality they call, not real being, but a sort of moving process of becoming." Ibid., *Sophist*, 246a–c, p. 990. "So to proceed downward to the conclusion, making no use whatever of any object of sense but only of pure ideas moving on through ideas to ideas and ending with ideas." Ibid., *Republic*, book VI, 511b–c, p. 746.

49. Collingwood, *Idea of Nature*, p. 56.

50. For example, see Plato, *The Collected Dialogues of Plato: Including the Letters*, *Phaedo*, 78d, pp. 61–62. Here, Plato has Socrates discuss the invariant, eternal quality of absolute reality, absolute equality, and absolute beauty.

51. "And having been created in this way, the world has been framed in the likeness of that which is apprehended by reason and mind and is unchangeable, and must therefore of necessity, if this is admitted, be a copy of something." Ibid., *Timaeus*, 29a–c, p. 1162.

52. Collingwood, *Idea of Nature*, p. 56.

53. Gouldner, *Enter Plato*, p. 372.

54. Plato is notably concerned with classification in the *Sophist* and (particularly) in the *Statesman*, where he spends pages traversing hierarchies of categories to reveal the definition of the true statesman. Plato, *The Collected Dialogues of Plato: Including the Letters*, *Statesman*, 263b, p. 1027.

55. Gouldner, *Enter Plato*, pp. 198–200.

56. Alexander P. D. Mourelatos, "Plato's Science—His View and Ours of His," in *Science and Philosophy in Classical Greece*, ed. Alan C. Bowen and Institute for Research in Classical Philosophy and Science, *Sources and Studies in the History and Philosophy of Classical Science / Institute for Research in Classical Philosophy and Science* (New York: Garland, 1991), p. 29.

57. See the commentary and text in Plato, *The Collected Dialogues of Plato: Including the Letters*, *Euthyphro*, 10d–11, pp. 169, 179.

58. Gouldner, *Enter Plato*, pp. 362, 370.

59. Ibid., pp. 256–57.

60. Val Plumwood, *Feminism and the Mastery of Nature* (London; New York: Routledge, 1993), p. 69.

61. Plato taught that the soul also exists before birth. Knowledge and wisdom are gained by *recollection* of knowledge and wisdom that existed in the soul before birth. See Plato, *The Collected Dialogues of Plato: Including the Letters*, *Phaedo*, 76c, p. 59. Plato can see the soul as transcendent and preexisting birth because his dissociating ontology lets him regard objects as primary and relations such as social relations, which enable the transmittal of knowledge and wisdom, as secondary or nonexistent. Received cultural wisdom, biologically inherited capabilities, and other capacities have been transfigured by Plato into capacities of the preexisting soul, which becomes associated with the body only at birth, carrying all its inherent characteristics and knowledge with it. Plato saw humans as transcending specific cultures and histories.

62. Collingwood, *Idea of Nature*, p. 6.

63. Plato, *The Collected Dialogues of Plato: Including the Letters*, *Phaedo*, 65a–e, p. 48, and 67d–e, p. 50.

64. Ibid., *Phaedo*, 80e–81a, p. 64. Elsewhere, Plato makes clear that bodily pleasure, along with other senses and emotions, must be expelled from the self to ensure entrance into this intellectual heaven. He writes, "There is one way, then, in which a man can be free from all anxiety about the fate of his soul—if in life he has abandoned bodily pleasures and adornments . . . and has devoted himself to the pleasures of acquiring knowledge." Ibid., *Phaedo*, 114d–e, p. 95.

65. Plumwood, *Feminism and the Mastery of Nature*, p. 92.

66. Plato implies that death is a healing of the "illness" of life in the *Phaedo* when he has Socrates order Crito to offer a cock to Asclepius, Greek god of medicine and healing, in a ritual marking recovery from illness. These were Socrates's last words according to Plato. Plato, *The Collected Dialogues of Plato: Including the Letters*, *Phaedo*, 81a, 118a, pp. 64, 98.

67. Ibid., 81a, p. 64.

68. Ibid., 116e–117, p. 96.

69. Socrates famously chose to remain subject to Athens' legal system even though he didn't agree with its findings in his case. He could easily have saved his life by leaving Athens, which friends urged him to do. Ibid., *Crito*, p. 28.

70. This follows in part Plumwood on dualism. See, for instance, Plumwood, *Feminism and the Mastery of Nature*, p. 93.

71. According to Plato, the emotions are born of the inessential bodily existence of humans and are a nuisance. He writes, "Besides, the body fills us with loves and desires and fears and all sorts of fancies and a great deal of nonsense, with the result that we literally never get an opportunity to think at all about anything." Plato, *The Collected Dialogues of Plato: Including the Letters*, *Phaedo*, 66c, p. 49.

72. Plato is suspicious of love. He doesn't call on people to love one another. Gouldner, *Enter Plato*, p. 244. Instead, he espouses love of the Forms. He even expels emotionality from art. In the *Ion*, Socrates disputes the possibility of a balance between the emotions and

the intellect in art. He says art is independent of the emotions and rather belongs to the realm of knowledge. Emotions would cause disorder, while the intellect can cause order in art. Accordingly, poets are not artists because they can't compose until reason has left them. Plato is fighting what he perceives as an Athenian shift toward emotionality. See the commentary at Plato, *The Collected Dialogues of Plato: Including the Letters*, p. 215.

73. Plato, *The Collected Dialogues of Plato: Including the Letters*, *Phaedo*, 60a, p. 43.

74. Ibid., *Phaedo*, 116a–b, p. 96.

75. Ibid., 117c–e, p. 97.

76. Plumwood, *Feminism and the Mastery of Nature*, p. 42.

77. Ibid., pp. 42, 45, 47–59.

78. Ibid.

79. Plato, *The Collected Dialogues of Plato: Including the Letters*, *Phaedrus*, 230d, p. 479. This passage contrasts with the preceding one, in which Socrates expresses considerable sensual delight in the spot that Phaedrus has chosen for them beside a stream:

> Upon my word, a delightful resting place, with this tall, spreading plane, and a lovely shade from the high branches of the *agnos*. Now that it's in full flower, it will make the place ever so fragrant. And what a lovely stream under the plane tree, and how cool to the feet! Judging by the statuettes and images I should say it's consecrated to Achelous and some of the nymphs. And then too, isn't the freshness of the air most welcome and pleasant, and the shrill summery music of the cicada choir! And as a crowning delight the grass, thick enough on a gentle slope to rest your head on most comfortably.

Ibid., 230b–c, pp. 478–79.
It's on hearing these words that Phaedrus wonders why Socrates doesn't leave town more often. Plato disparages sensual enjoyment frequently enough elsewhere. Perhaps this then marks a difference between Socrates and Plato, or perhaps Plato's point is that such enjoyment is temporary, as it's overshadowed throughout the rest of the dialogue by intellectual pursuits.

80. Ibid., *Republic*, 431c, p. 673; ibid., *Timaeus*, 50b–51b, pp. 1177–78; ibid., *Republic*, 455c–56, p. 694.

81. On the body as prison, Plato wrote, "Every seeker after wisdom knows that up to the time when philosophy takes it over his soul is a helpless prisoner, chained hand and foot in the body, compelled to view reality not directly but only through its prison bars.... Well, philosophy takes over the soul in this condition and by gentle persuasion tries to set it free. She points out that observation by means of the eyes and ears and all the other senses is entirely deceptive." Ibid., *Phaedo*, 82d–83b, p. 66.

82. Mourning and related emotions are experienced by other social animals besides

humans. Marc Bekoff, *Minding Animals: Awareness, Emotions, and Heart* (Oxford; New York: Oxford University Press, 2002), p. 113.

83. Aristotle, *Selected Works*, *Metaphysics*, *Book XIII*, chap. 10, 1087a1–5, p. 424.

84. Ibid.

85. Ibid., 1086b–1087a25, pp. 423–24.

86. Ibid., book IX, chap. 8, pp. 404–407.

87. The Academy sought to influence society, to "bridge the policy needs of ancient times with the philosophy, knowledge, and social theory of the period." Gouldner says the Academy "is in its own way the RAND Corporation of antiquity. It is, however, in business for itself and searches for a client worthy of it." Gouldner, *Enter Plato*, p. 157.

88. Aristotle, *Selected Works*, *Metaphysics*, book I, chap. 1, 980a, p. 334.

89. See Aristotle's discussions of animals. Ibid., *History of Animals*, *Parts of Animals*, *Generation of Animals*.

90. There are many examples in the passages at ibid., *Nichomean Ethics*, book VII, chap. 4–5, pp. 512–13.

91. Callicott, *Earth's Insights*, p. 29.

92. Sambursky, *Physical World of the Greeks*, p. 89.

93. Kate Soper, *What Is Nature?: Culture, Politics and the Non-Human* (Oxford; Cambridge, MA: Blackwell, 1995), p. 53.

94. Geoffrey Ernest Richard Lloyd, *Ancient Worlds, Modern Reflections: Philosophical Perspectives on Greek and Chinese Science and Culture* (Oxford; New York: Clarendon Press; Oxford University Press, 2004), p. 39.

95. "Sociative" logics are a family of old and new logics in which the validity of an implicational formula requires some genuine connection of reasoning or content between the antecedent and the consequent. They can't be completely unrelated. Richard Sylvan, *Sociative Logics and Their Applications: Essays*, ed. Dominic Hyde and Graham Priest (Aldershot; Burlington, VT: Ashgate, 2000), pp. 53, 54; Andre Fuhrmann, "Sociative Logics and Their Applications: Essays (Review)," review of Richard Sylvan, *Sociative Logics and Their Applications: Essays*, *Philosophical Quarterly* 53, no. 210 (2003): 138.

96. Aristotle, *Selected Works*, *Physics*, book I, chap. 5, 188a25–30, p. 172.

97. Ibid., *Metaphysics*, book I, chap. 5, 986a20–30, p. 343.

98. Ibid., *Generation and Destruction*, book II, chap. 2, 330a25, p. 245. Sambursky believes that the progress of Western physics was impeded by Aristotle's commitment to the fundamental status of opposites. "The history of physics has proved that all this theory of absolutely opposed qualities, even when presented with dialectical brilliance in the form of thesis and antithesis, leads nowhere." Sambursky, *Physical World of the Greeks*, pp. 89–91.

99. Aristotle, *Selected Works*, *Physics*, book II, chap. 1, 192b30–193a5, pp. 181–82.

100. Ibid., *Metaphysics*, book I, chap. 9, 990b15–20, p. 351.

101. Ibid., *Physics*, book II, chap. 9, 200a1–15, p. 196.

102. Aristotle, *The Basic Works of Aristotle*, ed. Richard Peter McKeon (New York: Random House, 1941), p. 366; ibid., *Physics*, book VIII, chap. 4, 255b5–20.

103. Aristotle, *Selected Works*, *Physics*, book III, chap. 3, 202a10–20, pp. 200–201.

104. Lloyd, *Adversaries and Authorities*, p. 138.

105. Aristotle, *Selected Works*, *Metaphysics*, book I, chap. 5, 986a22–30, p. 343.

106. Lloyd, *Adversaries and Authorities*, p. 138. Aristotle, *The Basic Works of Aristotle*, *On the Generation of Animals*, 732a1ff.

107. "So since the intellect is divine relative to a man, the life according to this intellect, too, will be divine relative to human life. . . . Happiness, then, would be a kind of contemplation." Aristotle, *Selected Works*, *Nicomachean Ethics*, book X, chap. 7–8, 1177b30, 1178b20–25, 1178b30–33, pp. 549–51.

108. Ray Kurzweil, *The Singularity Is Near: When Humans Transcend Biology* (New York: Viking, 2005), p. 136.

CHAPTER 6: MODERN TRACES OF DISSOCIATION

1. Robin George Collingwood, *The Idea of Nature* (New York: Oxford University Press, 1960; reprint, 1967), pp. 1–9.

2. Eustace Mandeville Wetenhall Tillyard, *The Elizabethan World Picture* (New York: Vintage Books, 1959), p. 46; Collingwood, *Idea of Nature*, pp. 93–94; Galileo Galilei, "Il Saggiatore" (The Assayer), vol. 6, p. 232, in *Le Opere di Galileo Galilei*, ed. Antonio Favaro, 20 vols. (Firenze: Le Monnier, 1890), transl. and quoted in Douglas M. Jesseph, "Galileo, Hobbes, and the Book of Nature," *Perspectives on Science* 12, no. 2 (2004): 202.

3. Collingwood, *Idea of Nature*, p. 94.

4. Ibid., pp. 94–96; Carolyn Merchant, *The Death of Nature: Women, Ecology, and the Scientific Revolution*, 1st ed. (San Francisco: Harper & Row, 1980).

5. Collingwood, *Idea of Nature*, pp. 5, 8.

6. Samuel Sambursky, *The Physical World of the Greeks* (London: Routledge, 1960), p. 243.

7. Collingwood, *Idea of Nature*, p. 96.

8. Merchant, *Death of Nature*.

9. Benjamin Farrington and Francis Bacon, "Thoughts and Conclusions on the Interpretation of Nature or a Science Productive of Works" in *The Philosophy of Francis Bacon; an Essay on Its Development from 1603 to 1609, with New Translations of Fundamental Texts* (Chicago: University of Chicago Press, 1964), pp. 93, 99. Bacon wrote, "I come in very truth, leading to you Nature with all her children to bind her to your service and make her your slave." Ibid., p. 62, in "The Masculine Birth of Time." The idea of sexual and rape imagery of nature is explored in Merchant, *The Death of Nature*, pp. 170, 171.

10. This discussion follows Collingwood, *Idea of Nature*, pp. 96–98.

11. Ibid.

12. Ibid., pp. 98–100.

13. Galileo wrote, "Philosophy is written in this vast book, which lies continuously open before our eyes (I mean the universe). But it cannot be understood unless you have first learned to understand the language and recognize the characters in which it is written. It is written in the language of mathematics, and the characters are triangles, circles, and other geometrical figures. Without such means, it is impossible for us humans to understand a word of it, and to be without them is to wander around in vain through a dark labyrinth." Galilei, "Il Saggiatore." See also Galileo Galilei, *Discoveries and Opinions of Galileo: Including the Starry Messenger (1610), Letter to the Grand Duchess Christina (1615), Excerpts from Letters on Sunspots (1613), the Assayer (1623)*, ed. Stillman Drake (New York: Anchor Books, 1957), p. 238.

14. Collingwood, *Idea of Nature*, pp. 100–103.

15. Edward Twitchell Hall, *Beyond Culture* (Garden City, NY: Anchor Books, 1981), p. 25. "Extension transference . . . this common intellectual maneuver in which the extension is confused with or takes the place of the process extended." Ibid. Extension transference occurs, for instance, when symbols are mistaken for the things symbolized (as in worshipping idols), when written language is valued above lived, spoken language (for example, when African American children are told their street language is inferior to what they're to learn in school), when methodology takes precedence over empirical data (as when the paradigms of "hard science" are applied to the social world), and more generally when artifice is considered the "real" world, not the culturally and historically specific extensions of a particular culture. We become preoccupied with material goods, and they take the place of relationships with people and nature. Hall sees extension transference as a main source of alienation from self and heritage worldwide when modern systems supplant traditional lived cultures. Ibid., pp. 29–30.

Extension transference was at work when Plato said the Forms are more real than the tangible world. Likewise, it happened when the mechanistic view of nature, arising from early modern peoples' experience with machines (which are human extensions), became reified in modern cosmology, which reenvisioned nature as a machine.

16. Merchant, *Death of Nature*, p. 228.

17. For an overview of the report and the ensuing scandal, see Kenneth A. Worthy et al., "Agricultural Biotechnology Science Compromised: The Case of Quist and Chapela," in *Controversies in Science and Technology: From Maize to Menopause*, ed. Daniel Lee Kleinman, Abby J. Kinchy, and Jo Handelsman, *Science and Technology in Society* (Madison: University of Wisconsin Press, 2005).

18. René Descartes, *Meditations on First Philosophy*, 2nd ed., trans. Laurence Julien Lafleur, Library of Liberal Arts (New York: Bobbs-Merrill, 1960), p. 80.

19. Ibid., p. 14.

20. Ibid., p. 74.

21. Ibid., p. 73.

22. Val Plumwood, *Feminism and the Mastery of Nature* (London; New York: Routledge, 1993), pp. 73, 112.

23. Descartes, *Meditations on First Philosophy*, p. 78.

24. Plumwood, *Feminism and the Mastery of Nature*, p. 114.

25. Edward S. Reed, *The Necessity of Experience* (New Haven, CT: Yale University Press, 1996), p. 57.

26. Ibid., p. 8.

27. Ibid., pp. 1, 2, 57.

28. Ibid., p. 52.

29. Ibid., pp. 52, 57–59, 70, 91.

30. Michael Pollan describes the full process of industrial meat production in Michael Pollan, "Power Steer," *New York Times Magazine*, March 31, 2002.

31. Timothy Pachirat, *Every Twelve Seconds: Industrialized Slaughter and the Politics of Sight*, Yale Agrarian Studies Series (New Haven, CT: Yale University Press, 2011), pp. 238–40.

32. Ibid.

33. Plumwood, *Feminism and the Mastery of Nature*.

34. Ibid.

35. Ibid., p. 93.

36. Ibid., p. 42.

37. Ibid., pp. 70–71.

38. Ibid., p. 45.

39. Ibid., p. 69.

40. Ibid., p. 117.

41. Ibid.

42. Thomas Jefferson, *Notes on the State of Virginia* (London: J. Stockdale, 1787), pp. 232, 239.

43. Mike Davis, *Late Victorian Holocausts: El Niño Famines and the Making of the Third World* (London and New York: Verso, 2001), p. 11.

44. Anne McClintock, *Imperial Leather: Race, Gender, and Sexuality in the Colonial Contest* (New York: Routledge, 1995), pp. 52–53.

45. Plato, *The Collected Dialogues of Plato: Including the Letters*, Timaeus, 76d–e, 90e–91a, pp. 1198, 1209–10. Gouldner writes also that reason was confined to the city's citizens who have full legal rights, which would not include women. Alvin Ward Gouldner, *Enter Plato; Classical Greece and the Origins of Social Theory* (New York: Basic Books, 1965), p. 340.

46. Gouldner writes, "It is because the master views the slave primarily in relation to

his, the master's, ends—and because these differ from the slave's own ends—that the slave's behavior seems unpredictable and disorderly to the master." The slave is merely a tool to the master. Gouldner, *Enter Plato*, pp. 352–53.

47. Plumwood, *Feminism and the Mastery of Nature*, pp. 47–55.

48. Collingwood, *Idea of Nature*, p. 128.

49. Alfred North Whitehead, *Process and Reality: An Essay in Cosmology*, ed. David Ray Griffin and Donald W. Sherburne, corrected ed. (New York: Free Press, 1978), p. 18.

50. Not because it has organs.

51. Whitehead, *Process and Reality*, pp. 18, 35.

52. "The philosophy of organism abolishes the detached mind." Ibid., p. 56. For more on the historical-philosophical context of Whitehead's ideas, see Collingwood, *Idea of Nature*, pp. 158–68, 174.

53. Collingwood, *Idea of Nature*, pp. 133–41.

54. Ibid., pp. 136–40.

55. Merchant, *Death of Nature*, pp. xvi–xxiv.

56. Lovelock's first book on the hypothesis is James Lovelock, *Gaia: A New Look at Life on Earth* (Oxford: Oxford University Press, 1995). For a more recent, popularized version, see James Lovelock, *Gaia: The Practical Science of Planetary Medicine* (Oxford; New York: Oxford University Press, 2000).

57. Collingwood, *Idea of Nature*, pp. 22–23.

58. Jane Rissler and Margaret G. Mellon, *The Ecological Risks of Engineered Crops* (Cambridge, MA: MIT Press, 1996).

59. Lily E. Kay, *Who Wrote the Book of Life?: A History of the Genetic Code*, Writing Science (Stanford, CA: Stanford University Press, 2000), p. 1; Carolyn Merchant, *Reinventing Eden: The Fate of Nature in Western Culture* (New York: Routledge, 2003), pp. 170–71.

CHAPTER 7: MODERN SPACES

1. J. Donald Hughes, *An Environmental History of the World: Humankind's Changing Role in the Community of Life*, 2nd ed. (London; New York: Routledge, 2009), p. 137.

2. Mike Davis, *Late Victorian Holocausts: El Niño Famines and the Making of the Third World* (London and New York: Verso, 2001).

3. Hughes, *Environmental History of the World*, p. 104.

4. Henri Lefebvre, *The Production of Space* (Oxford, UK; Cambridge, MA: Blackwell, 1991).

5. Ibid., p. 371.

6. William Cronon, *Changes in the Land: Indians, Colonists, and the Ecology of New England*, 1st ed. (New York: Hill and Wang, 1983).

7. See, for example, the problems discussed in Brian Tokar, "Monsanto: A Checkered History," *Ecologist* 28, no. 5 (1998): 260.

8. Judith Shapiro, *Mao's War against Nature: Politics and the Environment in Revolutionary China*, Studies in Environment and History (Cambridge; New York: Cambridge University Press, 2001).

9. Lefebvre, *Production of Space*, p. 49.

10. Ibid., pp. 110, 156.

11. Bruno Latour, *Science in Action: How to Follow Scientists and Engineers through Society* (Cambridge, MA: Harvard University Press, 1987), p. 228.

12. Neil Smith, *Uneven Development: Nature, Capital, and the Production of Space* (New York: Blackwell, 1984), p. 178.

13. Hughes, *Environmental History of the World*, p. 14.

14. Rebecca Solnit, *Wanderlust: A History of Walking* (New York: Viking, 2000), p. 68.

15. Ibid., p. 266.

16. John Winthrop wrote,

> The whole earth is the lords Garden & he hath given it to the sonnes of men, with a generall Condition, Gen: 1, 28 [Genesis 1:28]. Increase & multiply, replenish the earth & subdue it . . . why, then, should we stand hear striveing for places of habitation, (many men [sometimes] spending as much labor & cost to recover or keep . . . a Acre or two of land as would procure them many hundred as good or better in an other country) and in the mean tyme suffer a whole Continent as fruitfull & convenient for the use of man to lie [empty and unimproved].

John Winthrop, "Conclusions for the Plantation in New England," in *Old South Leaflets, No. 50. (1629)* (Boston: Directors of the Old South Work, 1897), pp. 1–2, 4–5. Quoted in Carolyn Merchant, *Major Problems in American Environmental History: Documents and Essays*, 2nd ed., Major Problems in American History Series (Boston: Houghton Mifflin, 2005), p. 71.

17. Smith, *Uneven Development*, p. 178.

18. Ibid., p. 135.

19. Theodore Steinberg, *Down to Earth: Nature's Role in American History* (Oxford; New York: Oxford University Press, 2002), pp. 131–32.

20. Janet M. Neeson, *Commoners: Common Right, Enclosure and Social Change in England, 1700–1820* (Cambridge; New York: Cambridge University Press, 1993).

21. David McNally, *Against the Market: Political Economy, Market Socialism, and the Marxist Critique* (London: Verso, 1993), p. 19.

22. Ibid.

23. Lisa W. Foderaro, "Privately Owned Park, Open to the Public, May Make Its Own Rules," *New York Times*, October 13, 2011, http://www.nytimes.com/2011/10/14/

nyregion/zuccotti-park-is-privately-owned-but-open-to-the-public.html (accessed January 21, 2013).

24. Solnit, *Wanderlust*, p. 253.

25. Robert Fishman, *Bourgeois Utopias: The Rise and Fall of Suburbia* (New York: Basic Books, 1987), p. 34.

26. Sharon Zukin, *Landscapes of Power: From Detroit to Disney World* (Berkeley: University of California Press, 1991), p. 54.

27. Mike Davis, *City of Quartz: Excavating the Future in Los Angeles*, Haymarket Series (London: Pimlico, 1998); Richard Walker, "Landscape and City Life: Four Ecologies of Residence in the San Francisco Bay Area," *Ecumene* 2, no. 1 (1995): 33.

28. Fishman, *Bourgeois Utopias*, p. 151.

29. Davis, *City of Quartz*, pp. 223–60.

30. Lefebvre, *Production of Space*, p. 386.

31. Ibid., p. 366.

32. Solnit, *Wanderlust*, p. 229.

33. Ibid., pp. 230–31.

34. Jack Ralph Kloppenburg, *First the Seed: The Political Economy of Plant Biotechnology, 1492–2000* (Cambridge; New York: Cambridge University Press, 1988).

35. Zukin, *Landscapes of Power*, p. 47. Brunelleschi is famed for, among other achievements, his stunning architectural and engineering triumph, the massive dome of the cathedral (Il Duomo) in Firenze, Italy.

36. Ibid., p. 43.

37. Ibid., p. 39.

38. Ibid., p. 52.

39. Elizabeth Wilson, *The Sphinx in the City: Urban Life, the Control of Disorder, and Women* (London: Virago, 1991), p. 7.

40. Fishman, *Bourgeois Utopias*, p. 25.

41. Lefebvre, *Production of Space*, p. 364.

42. Solnit, *Wanderlust*, p. 264.

43. Richard Peet and Michael Watts, *Liberation Ecologies: Environment, Development, Social Movements* (London; New York: Routledge, 1996), p. 1.

44. Peter Linebaugh and Marcus Rediker, "The Many-Headed Hydra: Sailors, Slaves and the Atlantic Working Class in the Eighteenth Century," in *Gone to Croatan: Origins of North American Dropout Culture*, ed. Ronald B. Sakolsky and James Koehnline (Brooklyn; Edinburgh: Autonomedia, 1993), p. 146.

45. Eric Robert Wolf, *Europe and the People without History* (Berkeley: University of California Press, 1982), p. 265.

46. S. Giedion, *Mechanization Takes Command: A Contribution to Anonymous History* (New York: Norton, 1969), p. 218.

47. In an intensive one-week survey of Internet sales, the International Fund for Animal Welfare "found over 9,000 wild animal products and specimens and live wild animals for sale, predominantly from species protected by law." International Fund for Animal Welfare, "Caught in the Web: Wildlife Trade on the Internet" (London: International Fund for Animal Welfare, 2005), http://www.ifaw.org/sites/default/files/Report%202005%20Caught%20in%20the%20web%20UK.pdf (accessed February 4, 2013).

48. Wolfgang Schivelbusch, *Disenchanted Night: The Industrialization of Light in the Nineteenth Century* (Berkeley: University of California Press, 1988).

49. Ibid., p. 6.

50. Ibid., pp. 25–30.

51. Ibid., p. 15.

52. Ibid., p. 44.

53. Ibid., p. 28.

54. Ibid., p. 29.

55. Ibid., p. 38.

56. Ibid., pp. 52–64.

57. According to the engineering firm that built it, the beam is produced by forty-five Xenon 7,000-watt lighting fixtures and twelve airport strobe lights. "The G-Force I.E.C. 'Beam of Luxor' History," G-Force International Entertainment Corporation, http://gforceiec.com/luxor_beam.php (accessed January 22, 2013).

58. Las Vegas is famous for very large reconstructions of landmark monuments, such as St. Mark's Square (Venezia's Piazza San Marco), Egyptian pyramids (at the Luxor Hotel), and "A signature of the Las Vegas skyline, the replica Eiffel Tower at Paris Las Vegas is an exact reproduction of one of Europe's most famous landmarks, rendered meticulously at one-half scale."

59. Jeremy Atiyah, "Ocean Dome: The Best Thing About This Most Perfect of Beaches Is That One Visit Is Sufficient for a Lifetime," *Independent* (London), July 16, 2000; Martin Parr, "What the Photographer Saw," *Independent* (London), February 18, 2001.

60. Jean Baudrillard, *Simulacra and Simulation* (Ann Arbor: University of Michigan Press, 1994), p. 1ff.

61. Celebration Company, "Welcome to Celebration, Florida," http://celebrationfl.com/index.html (accessed February 4, 2013).

62. Colin Coyle, "Celebration: The Town That Mickey Built," *Sunday Times* (London), July 18, 2004; Celebration Company, "Welcome to Celebration, Florida."

63. Lefebvre, *Production of Space*, p. 123.

64. Ibid., p. 154.

65. Gregory Bateson and Margaret Mead, *Balinese Character, a Photographic Analysis* (New York: New York Academy of Sciences, 1942), p. 56. Many of Bateson and Mead's observations of Balinese culture stand the test of time and direct observations there.

66. Ibid., p. 88.

67. David Abram, *The Spell of the Sensuous: Perception and Language in a More-Than-Human World*, 1st ed. (New York: Pantheon Books, 1996), p. 167.

68. Fishman, *Bourgeois Utopias*, p. 32.

69. Richard Harris, *Discovering Timber-Framed Buildings* (Aylesbury, England: Shire Publications, 1978), p. 31.

70. Lefebvre, *Production of Space*, p. 44.

CHAPTER 8: RECONNECTING AND HEALING A PLANET

1. Lynda Gledhill, "Point, Click and Shoot; Live Animals Can Now Be Hunted on the Web, and State Lawmaker Wants Practice Stopped," *San Francisco Chronicle*, March 10, 2005, http://www.sfgate.com/news/article/Point-click-and-shoot-Live-animals-can-now-be-2724455.php (accessed February 3, 2013). The hunting website was live-shot.com. To justify and promote the site, its creator, John Lockwood, claimed to be helping people with physical disabilities. Internet hunting is now banned in most states, including Texas and California. Steve Levin, "Internet Hunting Comes under Fire: Legislators Here and Elsewhere Aim to Ban It," *Pittsburgh Post-Gazette*, June 10, 2005, http://www.post-gazette.com/stories/local/uncategorized/internet-hunting-comes-under-fire-586493 (accessed February 5, 2013); Matt Weiser, "State Commission Outlaws Internet Hunting, Fishing," *Sacramento Bee*, August 20, 2005.

2. Live-shot.com, "Live-Shot: Real Time, On-Line, Hunting and Shooting Experience," http://www.live-shot.com/demo.shtml (accessed October 7, 2005). The site is no longer accessible.

3. Michael Pollan traces the life of a factory-farmed steer in Michael Pollan, "Power Steer," *New York Times Magazine*, March 31, 2002.

4. Andrew Revkin, Dot Earth, *New York Times*, http://dotearth.blogs.nytimes.com (accessed February 5, 2013); Treehugger, http://www.treehugger.com (accessed February 5, 2013); Yale Environment 360: Opinion, Analysis, Reporting & Debate, Yale University, http://e360.yale.edu (accessed February 5, 2013).

5. Andrew Szasz, *Shopping Our Way to Safety: How We Changed from Protecting the Environment to Protecting Ourselves* (Minneapolis: University of Minnesota Press, 2007).

6. US National Research Council, *Hidden Costs of Energy: Unpriced Consequences of Energy Production and Use* (Washington, DC: National Academies Press, 2010), p. 200.

7. Skepticism about technical solutions to environmental problems is pursued extensively in Ozzie Zehner, *Green Illusions: The Dirty Secrets of Clean Energy and the Future of Environmentalism* (*Our Sustainable Future*) (Lincoln: University of Nebraska Press, 2012).

8. Karl Marx and Friedrich Engels, *Marx and Engels on Ecology*, ed. Howard L. Parsons (Westport, CT: Greenwood Press, 1977), p. 171.

9. Entry for *associate*, *Merriam-Webster Dictionary*, 2013 edition, http://www.merriam-webster.com (accessed January 4, 2013).

10. Val Plumwood, *Environmental Culture: The Ecological Crisis of Reason* (New York: Routledge, 2001), pp. 75–76.

11. The San Francisco East Bay is home to several creek-advocating organizations that arrange for monitoring, daylighting, and restoration of local creeks: Friends of Alhambra Creek in Martinez, Friends of Sausal Creek in Oakland, Friends of Baxter Creek in El Cerrito and Richmond, and Friends of Five Creeks. "Bay Area Creek Restoration Updates," Ecology Center of Berkeley, California, http://www.ecologycenter.org/erc/creeks/creekreport.html (accessed February 4, 2013); Friends of Five Creeks, http://www.fivecreeks.org (accessed February 4, 2013). A local real estate developer, Ecocity Builders, also promotes daylighting of creeks: Ecocity-Builders, http://www.ecocitybuilders.org (accessed February 4, 2013).

12. Spiral Gardens Community Food Security Project, http://www.spiralgardens.org (accessed November 30, 2012).

13. One example is Planting Justice, which "is democratizing access to affordable, nutritious food by empowering disenfranchised urban residents with the skills, resources, and inspiration to maximize food production, economic opportunities, and environmental sustainability in our neighborhoods." Planting Justice: Grow Food. Grow Jobs. Grow Community, http://plantingjustice.org (accessed November 30, 2012).

14. Detroit Black Community Food Security Network, http://detroitblackfoodsecurity.org (accessed January 3, 2013).

15. Michael Pollan, *The Omnivore's Dilemma: A Natural History of Four Meals* (New York: Penguin Press, 2006); Eric Schlosser, *Fast Food Nation: The Dark Side of the All-American Meal* (Boston: Houghton Mifflin, 2001); Christopher D. Cook, *Diet for a Dead Planet: How the Food Industry Is Killing Us* (New York: New Press, 2004); Wenonah Hauter, *Foodopoly: The Battle over the Future of Food and Farming in America* (New York: New Press, 2012).

16. Wendell Berry, *Bringing It to the Table: On Farming and Food* (Berkeley, CA: Counterpoint: Distributed by Publishers Group West, 2009).

17. Pollan builds on these themes in his book, Michael Pollan, *Cooked: A Natural History of Transformation* (New York: Penguin Press, 2013).

18. David Abram, *Becoming Animal: An Earthly Cosmology*, 1st ed. (New York: Pantheon Books, 2010).

19. Richard Louv, *Last Child in the Woods: Saving Our Children from Nature-Deficit Disorder*, updated and expanded. ed. (Chapel Hill, NC: Algonquin Books, 2008).

20. Gerald T. Gardner and Paul C. Stern, *Environmental Problems and Human Behavior* (Boston: Allyn & Bacon, 1996), p. 83. "Smart Meters" installed by utilities such as Pacific Gas & Electric Company also give more updated feedback to energy consumers, but many people fear they increase electromagnetic radiation as well. Felicity Barringer,

"New Electricity Meters Stir Fears," *New York Times*, January 30, 2011, http://www.nytimes.com/2011/01/31/science/earth/31meters.html (accessed February 5, 2013). They also provide users only monthly feedback unless they log onto a website, which few likely do.

21. Nathanael Johnson, "The 4th Annual Year in Ideas: The Augmented Bar Code," *New York Times*, December 12, 2004, http://select.nytimes.com/gst/abstract.html?res=F009 10F73E550C718DDDAB0994DC404482&smid=pl-share (accessed December 28, 2012).

22. GoodGuide, http://www.goodguide.com (accessed February 6, 2013).

23. Dara O'Rourke, *Shopping for Good* (Boston Review Books) (Cambridge, MA: MIT Press, 2012), pp. 15–16.

24. Timothy Pachirat, *Every Twelve Seconds: Industrialized Slaughter and the Politics of Sight*, Yale Agrarian Studies Series (New Haven, CT: Yale University Press, 2011), pp. 240, 242.

25. Pollan, *Omnivore's Dilemma*, pp. 332–33.

26. Johan Galtung, *Development, Environment, and Technology: Towards a Technology for Self-Reliance: Study*, ed. Secretariat of the United Nations Conference on Trade and Development (New York: United Nations, 1979), p. 44 and throughout; Johan Galtung, "Towards a New Economics: On the Theory and Practice of Self-Reliance," in *The Living Economy: A New Economics in the Making*, ed. Paul Ekins (London; New York: Routledge, 1986), p. 102.

27. Galtung, "Towards a New Economics," pp. 100–103.

28. Ibid.

29. Gary Snyder, *Turtle Island* (New York: New Directions, 1974).

30. Raymond Dasmann and Peter Berg, "Reinhabiting California," *Ecologist* 7, no. 10 (1977): 399–410.

31. See also Michael Vincent McGinnis, ed., *Bioregionalism* (London; New York: Routledge, 1999). Bioregionalist ideas may help guide us to a less dissociated world. But the pure autarchic (politically insulated) version of bioregionalism eliminates dissociation by removing the gap between production and consumption: a bioregion can only consume what it can (sustainably) produce. The resulting autarchy reproduces at the community level the atomism and independence that liberalism produces at the individual level, according to Val Plumwood in Plumwood, *Environmental Culture*, pp. 78, 79, 249.

32. Petrus Josephus Zoetmulder, Stuart Owen Robson, and Koninklijk Instituut voor Taal- Land- en Volkenkunde (Netherlands), *Old Javanese-English Dictionary* ('s-Gravenhage: Nijhoff, 1982), pp. "deśa," "kāla VI," "pātra I." The three terms descend from Sanskrit. They also have other, more specific, senses in everyday usage. *Desa* refers to "village"; *kala* can refer to "era," "period," or "age"; and *patra* refers to "leaf" or "ornament."

33. Ervin Staub, *The Roots of Evil: The Origins of Genocide and Other Group Violence* (Cambridge; New York: Cambridge University Press, 1989), pp. 240, 274; Ervin Staub, "Individual and Societal (Group) Values in a Motivational Perspective and Their Role in

Benevolence and Harmdoing," in *Social and Moral Values: Individual and Societal Perspectives*, ed. Nancy Eisenberg, Janusz Reykowski, and Ervin Staub (Hillsdale, NJ: Lawrence Erlbaum Associates, 1989), pp. 58, 59.

34. Sherry Turkle, *Alone Together: Why We Expect More from Technology and Less from Each Other* (New York: Basic Books, 2012).

35. Robert M. Pirsig, *Zen and the Art of Motorcycle Maintenance: An Inquiry into Values* (Toronto; New York: Bantam Books, 1981). Pirsig's far-reaching philosophical exploration goes well beyond the idea of more intentional relations with the things around us. It includes novel critical views of inherited Western philosophy through the lens of Eastern philosophies, which Pirsig studied in India. He later published a synthesis of his philosophy, his "metaphysics of quality," in Robert M. Pirsig, *Lila: An Inquiry into Morals* (New York: Bantam Books, 1992).

36. Theodore Roszak, *The Voice of the Earth* (New York: Simon & Schuster, 1992), p. 319.

37. Ibid., pp. 320–21.

38. Gregory Bateson, *Mind and Nature: A Necessary Unity* (New York: Bantam Books, 1979), pp. 97, 98, 161.

39. Ibid., p. 143.

40. Ibid., p. 64.

41. Ibid., p. 143.

42. Andy Fisher, "What Is Ecopsychology? A Radical View," in *Ecopsychology: Science, Totems, and the Technological Species*, ed. Peter H. Kahn Jr. and Patricia Hasbach (Cambridge, MA: MIT Press, 2012), p. 92.

43. Andy Fisher, *Radical Ecopsychology: Psychology in the Service of Life*, SUNY Series in Radical Social and Political Theory (Albany: State University of New York Press, 2002), p. 97.

44. Ibid., pp. 71, 97, 114, 123, 168; David W. Kidner, *Nature and Psyche: Radical Environmentalism and the Politics of Subjectivity* (Albany: State University of New York Press, 2001), pp. 109, 213, 245, 252.

45. Fisher, *Radical Ecopsychology*, p. 114.

46. Val Plumwood illuminates the differences between distinction and dualism: Val Plumwood, *Feminism and the Mastery of Nature* (London; New York: Routledge, 1993), pp. 58–59, 215.

47. Kidner, *Nature and Psyche*, p. 109.

48. Laura Sewall, "Beauty and the Brain," in *Ecopsychology: Science, Totems, and the Technological Species*, ed. Peter H. Hasbach and Patricia H. Kahn (Cambridge, MA: MIT Press, 2012), pp. 274–75.

49. Ibid.

50. Ibid., p. 280.

51. Ibid., pp. 270, 276.

52. Ibid., p. 281.

53. Ibid., p. 270.

54. J. Baird Callicott, *Earth's Insights: A Survey of Ecological Ethics from the Mediterranean Basin to the Australian Outback* (Berkeley: University of California Press, 1994), p. 84.

55. More recent environments play a stronger role in shaping the organism's qualities than older ones do. Richard Dawkins, "The Evolved Imagination: Animals as Models of Their World," *Natural History* 104, no. 9 (1995): 8.

56. Robert L. Beschta and William J. Ripple, "Wolves, Elk, Willows, and Trophic Cascades in the Upper Gallatin Range of Southwestern Montana, USA," *Forest Ecology and Management* 200, no. 1 (2004): 161.

57. Mary Ellen Hannibal, "Why the Beaver Should Thank the Wolf," *New York Times*, September 28, 2012, http://www.nytimes.com/2012/09/29/opinion/the-world-needs-wolves.html?smid=pl-share (accessed December 28, 2012).

58. Callicott, *Earth's Insights*, p. 204.

59. Aldo Leopold, *A Sand County Almanac, and Sketches Here and There* (New York: Oxford University Press, 1970), pp. 138–39, 262.

60. Gina Kolata, "In Good Health? Thank Your 100 Trillion Bacteria," *New York Times*, June 14, 2012, http://www.nytimes.com/2012/06/14/health/human-microbiome-project-decodes-our-100-trillion-good-bacteria.html?smid=pl-share (accessed December 28, 2012); Peter J. Turnbaugh et al., "The Human Microbiome Project," *Nature* 449, no. 7164 (2007): 805.

61. Alfred North Whitehead, *Process and Reality: An Essay in Cosmology*, ed. David Ray Griffin and Donald W. Sherburne, corrected ed. (New York: Free Press, 1978), pp. 18, 22, 35.

62. Jan Christiaan Smuts, *Holism and Evolution* (New York: Macmillan, 1926).

63. Ibid., pp. 86–87.

64. Ibid.

65. David Bohm, "Postmodern Science and a Postmodern World," in *Key Concepts in Critical Theory: Ecology*, ed. Carolyn Merchant (Atlantic Highlands, NJ: Humanities Press, 1994), p. 343.

66. Ibid.

67. Ibid., p. 349.

68. Ibid., p. 350. Bohm's cosmology is relevant to dissociation not just because it offers a non-dissociating alternative but also because it relates dissociation to destruction. Bohm concludes that "meaning and value are as much integral aspects of the world as they are of us. If science is carried out with an amoral attitude, the world will ultimately respond to science in a destructive way." Ibid.

69. Callicott, *Earth's Insights*, p. 198. Logic is an example of a field that's recently

seen more integrative and relational thinking. The philosopher, environmentalist, and logician Richard Sylvan proposed new systems of logic called sociative logics, as mentioned in chapter 5. They insist on more concrete relationships between antecedent and consequent statements.

70. George W. Morgan, *The Human Predicament: Dissolution and Wholeness* (Providence, RI: Brown University Press, 1968). Morgan was on the philosophy faculty of Brown University through the middle decades of the twentieth century. I could not find his birth and death years.

71. Ibid., pp. 221, 222, 224.

72. Ibid., p. 227.

73. Ibid., pp. 231, 235, 251, 253, 255, 298, 299.

74. Ibid. The *Oxford English Dictionary* entries for *prosaic, a.* include the following phrases: "Lacking poetic beauty, feeling, or imagination; plain, matter-of-fact. Unpoetic, unromantic; commonplace, dull, tame. (Of persons and things.)" John Alexander Simpson and Edmund S. C. Weiner, eds., *Oxford English Dictionary*, 2nd ed., OED Online (Oxford: Oxford University Press, 1989).

75. Morgan, *Human Predicament*, pp. 306, 307, 314.

76. Ibid., pp. 306, 307, 314–16.

77. David Abram, *The Spell of the Sensuous: Perception and Language in a More-Than-Human World*, 1st ed. (New York: Pantheon Books, 1996), p. 262.

78. Ibid., p. 178.

79. Ibid., p. 69.

80. Edward S. Reed, *The Necessity of Experience* (New Haven, CT: Yale University Press, 1996), pp. 32–50.

81. Ibid., p. 104.

82. Ibid., pp. 46, 139.

83. Ibid., p. 127.

84. Anthony Weston, *Back to Earth: Tomorrow's Environmentalism* (Philadelphia: Temple University Press, 1994), p. 85.

85. Ibid., pp. 9, 25, 36, 37, 54, 59, 62, 113.

86. Martin Buber and Walter Arnold Kaufmann, *I and Thou* (New York: Scribner, 1970).

87. Ibid., p. 67.

88. Ibid., p. 76.

89. Ibid., p. 14.

90. Plumwood, *Feminism and the Mastery of Nature*, p. 183.

91. Carolyn Merchant, *Reinventing Eden: The Fate of Nature in Western Culture* (New York: Routledge, 2003), p. 223.

92. James Lovelock, *Gaia: A New Look at Life on Earth* (Oxford: Oxford University

Press, 1995); James Lovelock, *Gaia: The Practical Science of Planetary Medicine* (Oxford; New York: Oxford University Press, 2000).

93. The biologist Lynn Margulis (1938–2011) was a leading advocate of Gaia theory. Her account of symbiosis—the "living together in physical contact of organisms of different species"—deeply influenced evolutionary biology. Of humans' essential interconnectedness with other life, Margulis writes, "We are symbionts on a symbiotic planet, and if we care to, we can find symbiosis everywhere. Physical contact is a nonnegotiable requisite for many differing kinds of life." Lynn Margulis, *Symbiotic Planet: A New Look at Evolution*, 1st ed. (New York: Basic Books, 1998), p. 5.

EPIGRAPH SOURCES

Front Matter

*George Santayana, *Scepticism and Animal Faith: Introduction to a System of Philosophy* (New York: Dover Publications, 1955), p. ix.

Chapter 2: Sense and Connection

*David Abram, *The Spell of the Sensuous: Perception and Language in a More-Than-Human World*, 1st ed. (New York: Pantheon Books), p. 268.
†Greg Egan, *Diaspora: A Novel* (New York: HarperPrism, 1999), p. 67.

Chapter 3: From Dissociation to Destruction

*Stanley Milgram, *Obedience to Authority: An Experimental View*, 1st ed. (New York: Harper & Row, 1974), p. 40.
†Philip G. Zimbardo, "On 'Obedience to Authority,'" *American Psychologist* 29, no. 7 (1974): 566.
‡Louise J. Kaplan, *Oneness and Separateness: From Infant to Individual* (New York: Simon & Schuster, 1978), p. 26.

Chapter 4: Dissociation in Western Psyches

*Edward Twitchell Hall, *Beyond Culture* (Garden City, NY: Anchor Books, 1981), p. 7.

Chapter 5: Ancient Traces of Dissociation

*Geoffrey Ernest Richard Lloyd, *Ancient Worlds, Modern Reflections: Philosophical Perspectives on Greek and Chinese Science and Culture* (Oxford; New York: Clarendon Press; Oxford University Press, 2004), p. 186.

†Edith Hamilton, *The Greek Way* (New York: Time, 1963), p. 23.

‡Alvin Ward Gouldner, *Enter Plato; Classical Greece and the Origins of Social Theory* (New York: Basic Books, 1965), p. 172.

§Alfred North Whitehead, *Process and Reality: An Essay in Cosmology*, ed. David Ray Griffin and Donald W. Sherburne, corrected ed. (New York: Free Press, 1978), p. 39.

**Samuel Sambursky, *The Physical World of the Greeks* (London: Routledge, 1960), p. 80.

††J. Baird Callicott, *Earth's Insights: A Survey of Ecological Ethics from the Mediterranean Basin to the Australian Outback* (Berkeley: University of California Press, 1994), p. 29.

Chapter 6: Modern Traces of Dissociation

*David Bohm, "Postmodern Science and a Postmodern World," in *The Reenchantment of Science: Postmodern Proposals*, ed. David Ray Griffin, SUNY Series in Constructive Postmodern Thought (Albany: State University of New York Press, 1988), p. 65.

†Henri Lefebvre, *The Production of Space* (Oxford, UK; Cambridge, MA: Blackwell, 1991), p. 406.

Chapter 8: Reconnecting and Healing a Planet

*Martin Buber and Walter Arnold Kaufmann, *I and Thou* (New York: Scribner, 1970), p. 69.

†John Muir, *My First Summer in the Sierra: And Selected Essays*, ed. Bill McKibben (New York: Library of America, 2011; original pub. date 1911), p. 157.

‡George W. Morgan, *The Human Predicament: Dissolution and Wholeness* (Providence, RI: Brown University Press, 1968), p. 90.

BIBLIOGRAPHY

Abram, David. *Becoming Animal: An Earthly Cosmology*. 1st ed. New York: Pantheon Books, 2010.

———. *The Spell of the Sensuous: Perception and Language in a More-Than-Human World*. 1st ed. New York: Pantheon Books, 1996.

Alpine Ascents International. "Mount Everest Guided Climbing Expedition with Alpine Ascents International." http://www.alpineascents.com/everest.asp. Accessed September 7, 2012.

American Psychiatric Association, Task Force on DSM-IV. *Diagnostic and Statistical Manual of Mental Disorders: DSM-IV*. 4th ed. Washington, DC: American Psychiatric Association, 1994.

Arendt, Hannah. *Eichmann in Jerusalem; a Report on the Banality of Evil*. New York: Viking Press, 1963.

Aristotle. *The Basic Works of Aristotle*. Edited by Richard Peter McKeon. New York: Random House, 1941.

———. *Selected Works*. Translated by Hippocrates George Apostle and Lloyd P. Gerson. 3rd ed. Grinnell, IA: Peripatetic Press, 1991.

Atiyah, Jeremy. "Ocean Dome: The Best Thing About This Most Perfect of Beaches Is That One Visit Is Sufficient for a Lifetime." *Independent* (London), July 16, 2000.

Barnard, Chester Irving. *The Functions of the Executive*. Cambridge, MA: Harvard University Press, 1938.

Barringer, Felicity. "New Electricity Meters Stir Fears." *New York Times*, January 30, 2011. http://www.nytimes.com/2011/01/31/science/earth/31meters.html. Accessed February 5, 2013.

Bateson, Gregory. *Mind and Nature: A Necessary Unity*. New York: Bantam Books, 1979.

Bateson, Gregory, and Margaret Mead. *Balinese Character, a Photographic Analysis*. New York: New York Academy of Sciences, 1942.

Baudrillard, Jean. *Simulacra and Simulation*. Ann Arbor: University of Michigan Press, 1994.

"Bay Area Creek Restoration Updates." Ecology Center of Berkeley, California. http://www.ecologycenter.org/erc/creeks/creekreport.html. Accessed February 4, 2013.

Bekoff, Marc. *Minding Animals: Awareness, Emotions, and Heart*. Oxford; New York: Oxford University Press, 2002.

Berry, Wendell. *Bringing It to the Table: On Farming and Food*. Berkeley, CA: Counterpoint: Distributed by Publishers Group West, 2009.

Birch, Thomas H. "The Incarceration of Wildness: Wilderness Areas as Prisons." In *The Great New Wilderness Debate*, edited by J. Baird Callicott and Michael P. Nelson, 443–70. Athens: University of Georgia Press, 1998.

Blass, Thomas. "Attribution of Responsibility and Trust in the Milgram Obedience Experiment." *Journal of Applied Social Psychology* 26, no. 17 (1996): 1529–35.

———. "A Cross-Cultural Comparison of Studies of Obedience Using the Milgram Paradigm: A Review." *Social and Personality Psychology Compass* 6, no. 2 (2012): 196–205.

———. *The Man Who Shocked the World: The Life and Legacy of Stanley Milgram*. 1st ed. New York: Basic Books, 2004.

———. "The Milgram Paradigm after 35 Years: Some Things We Now Know About Obedience to Authority." *Journal of Applied Social Psychology* 29, no. 5 (1999): 955–78.

———. *Obedience to Authority: Current Perspectives on the Milgram Paradigm*. Mahwah, NJ: Lawrence Erlbaum Associates, 2000.

———. "Understanding Behavior in the Milgram Obedience Experiment: The Role of Personality, Situations, and Their Interactions." *Journal of Personality & Social Psychology* 60, no. 3 (1991): 398–413.

Bohm, David. "Postmodern Science and a Postmodern World." In *Key Concepts in Critical Theory: Ecology*, edited by Carolyn Merchant, 342–50. Atlantic Highlands, NJ: Humanities Press, 1994.

———. "Postmodern Science and a Postmodern World." In *The Reenchantment of Science: Postmodern Proposals*, edited by David Ray Griffin, SUNY Series in Constructive Postmodern Thought. Albany: State University of New York Press, 1988.

Boice, John D., Jr., Donald E. Marano, Heather M. Munro, Bandana K. Chadda, Lisa B. Signorello, Robert E. Tarone, William J. Blot, and Joseph K. McLaughlin. "Cancer Mortality among US Workers Employed in Semiconductor Wafer Fabrication." *Journal of Occupational and Environmental Medicine* 52, no. 11 (2010):1082–97.

Breuer, Josef, and Sigmund Freud. *Studies on Hysteria*. 2nd ed. New York: Basic Books, 1957.

Buber, Martin. *I and Thou*. Translated by Walter Arnold Kaufmann. New York: Scribner, 1970.

Burnet, John, and Philip Whalen. *Early Greek Philosophy*. New York: Meridian Books, 1957.

Buss, Arnold H. "Instrumentality of Aggression Feedback and Frustration as Determinants of Physical Aggression." *Journal of Personality and Social Psychology* 3, no. 2 (1966): 153–62.

Butler, Peter V. "Destructive Obedience in 1924: Landis' 'Studies of Emotional Reactions' as a Prototype of the Milgram Paradigm." *Irish Journal of Psychology* 19, no. 2–3 (1998): 236–47.

Callicott, J. Baird. *Earth's Insights: A Survey of Ecological Ethics from the Mediterranean Basin to the Australian Outback*. Berkeley: University of California Press, 1994.

Calloway, Colin G. *The World Turned Upside Down: Indian Voices from Early America*, Bedford Series in History and Culture. Boston: St. Martin's Press, 1994.

Celebration Company. "Welcome to Celebration, Florida." http://celebrationfl.com/index .html. Accessed February 4, 2013.

Choi, Incheol, and Richard E. Nisbett. "Situational Salience and Cultural Differences in the Correspondence Bias and Actor-Observer Bias." *Personality and Social Psychology Bulletin* 24, no. 9 (1998): 949–60.

Collingwood, Robin George. *The Idea of Nature*. New York: Oxford University Press, 1960. Reprint, 1967.

Cook, Christopher D. *Diet for a Dead Planet: How the Food Industry Is Killing Us*. New York: New Press, 2004.

Cook, Christopher D., and A. Clay Thompson. "Silicon Hell: The Computer Industry Prides Itself on a 'Clean' Image—but It's Actually Doing Horrible Damage to Its Workers and the Environment." *San Francisco Bay Guardian*, April 26, 2000.

Coonan, Clifford. "Workers Threaten Mass Suicide at Company That Supplies Apple; Fresh Labour Dispute at Foxconn Factory Turns Spotlight on Working Conditions in China." *Independent* (London), January 12, 2012. http://www.independent.co .uk/news/world/asia/workers-threaten-mass-suicide-at-company-that-supplies-apple -6288160.html. Accessed December 28, 2012.

Correa, Adolfo, Ronald H. Gray, Rebecca Cohen, Nathaniel Rothman, Faridah Shah, Hui Seacat, and Morton Corn. "Ethylene Glycol Ethers and Risks of Spontaneous Abortion and Subfertility." *American Journal of Epidemiology* 143, no. 7 (1996): 707–17.

Coyle, Colin. "Celebration: The Town That Mickey Built." *Sunday Times* (London), July 18, 2004.

Crapanzano, Vincent. "Preface." In Maurice Leenhardt, *Do Kamo: Person and Myth in the Melanesian World*, vii–xxv. Chicago: University of Chicago Press, 1979.

Cronon, William. *Changes in the Land: Indians, Colonists, and the Ecology of New England*. 1st ed. New York: Hill and Wang, 1983.

———. "The Trouble with Wilderness; or, Getting Back to the Wrong Nature." In *Uncommon Ground: Rethinking the Human Place in Nature*, edited by William Cronon, 561. New York: W. W. Norton, 1996.

DARA and the Climate Vulnerable Forum. "Climate Vulnerability Monitor, 2nd Edition: A Guide to the Cold Calculus of a Hot Planet." Madrid: DARA International, 2012.

Darley, John M., and Bibb Latané. "Bystander Intervention in Emergencies: Diffusion of Responsibility." *Journal of Personality & Social Psychology* 8, no. 4, pt. 1 (1968): 377–83.

Dasmann, Raymond, and Peter Berg. "Reinhabiting California." *Ecologist* 7, no. 10 (1977): 399–410.

Davis, Mike. *City of Quartz: Excavating the Future in Los Angeles*, Haymarket Series. London: Pimlico, 1998.

———. *Late Victorian Holocausts: El Niño Famines and the Making of the Third World*. London & New York: Verso, 2001.

Davis, Sheila, and Ted Smith. "Corporate Strategies for Electronics Recycling: A Tale of Two Systems." Silicon Valley Toxics Coalition; Computer TakeBack Campaign! 2003. http://svtc.org/wp-content/uploads/prison_final.pdf. Accessed December 28, 2012.

Dawkins, Richard. "The Evolved Imagination: Animals as Models of Their World." *Natural History* 104, no. 9 (1995): 8–11, 22–23.

Descartes, René. *Meditations on First Philosophy.* Translated by Laurence Julien Lafleur. 2nd ed. Library of Liberal Arts. New York: Bobbs-Merrill, 1960.

"Detroit Black Community Food Security Network." http://detroitblackfoodsecurity.org. Accessed January 3, 2013.

Diamond, Jared M. *Collapse: How Societies Choose to Fail or Succeed.* New York: Viking, 2005.

Duhigg, Charles, David Barboza, and Gu Huini. "In China, Human Costs Are Built into an iPad." *New York Times,* January 26, 2012. http://www.nytimes.com/2012/01/26/business/ieconomy-apples-ipad-and-the-human-costs-for-workers-in-china.html?smid=pl-share. Accessed November 14, 2012.

Ecocity-Builders. "Ecocity Builders Projects." http://www.ecocitybuilders.org. Accessed February 4, 2013.

Egan, Greg. *Diaspora: A Novel.* New York: HarperPrism, 1999.

Ellenberger, Henri F. *The Discovery of the Unconscious; the History and Evolution of Dynamic Psychiatry.* New York: Basic Books, 1970.

Elms, Alan C. *Social Psychology and Social Relevance.* Boston: Little, Brown, 1972.

Empedocles. "Purifications." In *An Introduction to Early Greek Philosophy; the Chief Fragments and Ancient Testimony, with Connecting Commentary,* edited by John Mansley Robinson. Boston: Houghton Mifflin, 1968.

Estes, James A., John Terborgh, Justin S. Brashares, Mary E. Power, Joel Berger, William J. Bond, Stephen R. Carpenter, Timothy E. Essington, Robert D. Holt, Jeremy B. C. Jackson, Robert J. Marquis, Lauri Oksanen, Tarja Oksanen, Robert T. Paine, Ellen K. Pikitch, William J. Ripple, Stuart A. Sandin, Marten Scheffer, Thomas W. Schoener, Jonathan B. Shurin, Anthony R. E. Sinclair, Michael E. Soulé, Risto Virtanen, and David A. Wardle. "Trophic Downgrading of Planet Earth." *Science* 333, no. 6040 (2011): 301–306.

Evans, Nigel, David J. Campbell, and George Stonehouse. *Strategic Management for Travel and Tourism.* Oxford: Butterworth-Heinemann, 2003.

Farrington, Benjamin. *Greek Science; Its Meaning for Us.* Harmondsworth, Middlesex: Penguin Books, 1949.

Farrington, Benjamin, and Francis Bacon. *The Philosophy of Francis Bacon; an Essay on Its Development from 1603 to 1609, with New Translations of Fundamental Texts.* Chicago: University of Chicago Press, 1964.

Fisher, Andy. *Radical Ecopsychology: Psychology in the Service of Life,* SUNY Series in Radical Social and Political Theory. Albany: State University of New York Press, 2002.

————. "What Is Ecopsychology? A Radical View." In *Ecopsychology: Science, Totems, and the Technological Species*, edited by Peter H. Kahn Jr. and Patricia Hasbach. Cambridge, MA: MIT Press, 2012.

Fisher, Jim. "Poison Valley, Parts 1 and 2." Salon.com, http://archive.salon.com/tech/feature/2001/07/30/almaden1/print.html and http://archive.salon.com/tech/feature/2001/07/31/almaden2/index.html. Accessed June 21, 2011.

Fishman, Robert. *Bourgeois Utopias: The Rise and Fall of Suburbia*. New York: Basic Books, 1987.

"Flying Dutchmen Travel." http://www.flyingdutchmentravel.com/family_vacations/golf/spa/spa-california.html. Accessed February 4, 2013.

Foderaro, Lisa W. "Privately Owned Park, Open to the Public, May Make Its Own Rules." *New York Times*, October 13, 2011. http://www.nytimes.com/2011/10/14/nyregion/zuccotti-park-is-privately-owned-but-open-to-the-public.html. Accessed January 21, 2013.

Freedman, Jonathan L., and Scott C. Fraser. "Compliance without Pressure: The Foot-in-the-Door Technique." *Journal of Personality and Social Psychology* 4, no. 2 (1966): 195–202.

Friends of Five Creeks. "Who Are the Friends of Five Creeks?" http://www.fivecreeks.org. Accessed February 4, 2013.

Fuhrmann, Andre. "Sociative Logics and Their Applications: Essays (Review)." Review of Sociative Logics and Their Applications: Essays. *Philosophical Quarterly* 53, no. 210 (2003): 137–41.

Galilei, Galileo. *Discoveries and Opinions of Galileo: Including the Starry Messenger (1610), Letter to the Grand Duchess Christina (1615), and Excerpts from Letters on Sunspots (1613), the Assayer (1623)*. Edited by Stillman Drake. New York: Anchor Books, 1957.

————. *Le Opere* di *Galileo Galilei*. Edited by Antonio Favaro. 20 vols. Firenze: Le Monnier, 1890.

Galtung, Johan. *Development, Environment, and Technology: Towards a Technology for Self-Reliance: Study*. Edited by United Nations Conference on Trade and Development Secretariat. New York: United Nations, 1979.

————. "Towards a New Economics: On the Theory and Practice of Self-Reliance." In *The Living Economy: A New Economics in the Making*, edited by Paul Ekins, 97–109. London; New York: Routledge, 1986.

Gardner, Gerald T., and Paul C. Stern. *Environmental Problems and Human Behavior*. Boston: Allyn & Bacon, 1996.

Geiser, Ken. *Materials Matter: Toward a Sustainable Materials Policy*. Cambridge, MA: MIT Press, 2001.

"The G-Force I.E.C. 'Beam of Luxor' History." G-Force International Entertainment Corporation. http://gforceiec.com/luxor_beam.php. Accessed January 22, 2013.

Gibson, Janice T., and Mika Haritos-Fatouros. "The Education of a Torturer." In *Readings in Social Psychology: General, Classic, and Contemporary Selections*, edited by Wayne A. Lesko, 246–51. Needham Heights, MA: Allyn & Bacon, 1991.

Giddens, Anthony. *The Consequences of Modernity*. Stanford, CA: Stanford University Press, 1990.

Giedion, Sigfried. *Mechanization Takes Command: A Contribution to Anonymous History*. New York: Norton, 1969.

Gledhill, Lynda. "Point, Click and Shoot; Live Animals Can Now Be Hunted on the Web, and State Lawmaker Wants Practice Stopped." *San Francisco Chronicle*, March 10, 2005. http://www.sfgate.com/news/article/Point-click-and-shoot-Live-animals-can -now-be-2724455.php. Accessed February 3, 2013.

Glendinning, Chellis. "Technology, Trauma, and the Wild." In *Ecopsychology: Restoring the Earth, Healing the Mind*, edited by Theodore Roszak, Mary E. Gomes, and Allen D. Kanner, 21–40. San Francisco: Sierra Club Books, 1995.

GoodGuide. http://www.goodguide.com. Accessed February 6, 2013.

Gouldner, Alvin Ward. *Enter Plato; Classical Greece and the Origins of Social Theory*. New York: Basic Books, 1965.

Grossman, Dave. *On Killing: The Psychological Cost of Learning to Kill in War and Society*. New York: Little, Brown, 2009.

Hall, Edward Twitchell. *Beyond Culture*. Garden City, NY: Anchor Books, 1981.

Hamilton, Edith. *The Greek Way*. New York: Time, 1963.

Haney, Craig, Curtis Banks, and Philip Zimbardo. "Interpersonal Dynamics in a Simulated Prison." *International Journal of Criminology & Penology* 1, no. 1 (1973): 69–97.

Hannibal, Mary Ellen. "Why the Beaver Should Thank the Wolf." *New York Times*, September 28, 2012. http://www.nytimes.com/2012/09/29/opinion/the-world-needs -wolves.html?smid=pl-share. Accessed December 28, 2012.

Harris, Richard. *Discovering Timber-Framed Buildings*. Aylesbury, England: Shire Publications, 1978.

Harrison, Myron. "Semiconductor Manufacturing Hazards." In *Hazardous Materials Toxicology: Clinical Principles of Environmental Health*, edited by John B. Sullivan and Gary R. Krieger. Baltimore: Williams & Wilkins, 1992.

Hauter, Wenonah. *Foodopoly: The Battle over the Future of Food and Farming in America*. New York: New Press, 2012.

Ho, Andy. "Time to Throw the Book at Child Sex Tour Organisers." *Straits Times* (Singapore), May 6, 2005.

Horkheimer, Max, and Theodor W. Adorno. *Dialectic of Enlightenment*. New York: Seabury Press, 1972.

Hughes, J. Donald. *An Environmental History of the World: Humankind's Changing Role in the Community of Life*. 2nd ed. London; New York: Routledge, 2009.

Humes, Edward. *Garbology: Our Dirty Love Affair with Trash.* New York: Avery, 2012.

Husserl, Edmund. *Cartesian Meditations.* The Hague: M. Nijhoff, 1965.

Huxley, Aldous. *Brave New World, a Novel.* London: Chatto & Windus, 1932.

ICF International. "Electronics Waste Management in the United States through 2009." Edited by US Environmental Protection Agency Office of Resource Conservation and Recovery: US Environmental Protection Agency, 2011. http://www.epa.gov/osw/conserve/materials/ecycling/docs/fullbaselinereport2011.pdf. Accessed October 16, 2012.

International Fund for Animal Welfare. "Caught in the Web: Wildlife Trade on the Internet." London: International Fund for Animal Welfare, 2005. http://www.ifaw.org/sites/default/files/Report%202005%20Caught%20in%20the%20web%20UK.pdf. Accessed February 4, 2013.

Islands, July/August 2003, misc.

Islands, December 2012, misc.

Jefferson, Thomas. *Notes on the State of Virginia.* London: J. Stockdale, 1787.

Jesseph, Douglas M. "Galileo, Hobbes, and the Book of Nature." *Perspectives on Science* 12, no. 2 (2004): 191–211.

Ji, Li-Jun, Kaiping Peng, and Richard E. Nisbett. "Culture, Control, and Perception of Relationships in the Environment." *Journal of Personality and Social Psychology* 78, no. 5 (2000): 943–55.

Johnson, Nathanael. "The 4th Annual Year in Ideas: The Augmented Bar Code." *New York Times,* December 12, 2004. http://select.nytimes.com/gst/abstract.html?res=F0091 0F73E550C718DDDAB0994DC404482&smid=pl-share. Accessed December 28, 2012.

Kanner, Allen D., and Mary E. Gomes. "The All-Consuming Self." In *Ecopsychology: Restoring the Earth, Healing the Mind,* edited by Theodore Roszak, Mary E. Gomes, and Allen D. Kanner, 77–91. San Francisco: Sierra Club Books, 1995.

Kaplan, Louise J. *Oneness and Separateness: From Infant to Individual.* New York: Simon & Schuster, 1978.

Kashi, Ed, and Michael Watts. *Curse of the Black Gold: 50 Years of Oil in the Niger Delta.* Brooklyn, NY: PowerHouse Books, 2008.

Kay, Lily E. *Who Wrote the Book of Life?: A History of the Genetic Code,* Writing Science. Stanford, CA: Stanford University Press, 2000.

Kidner, David W. *Nature and Psyche: Radical Environmentalism and the Politics of Subjectivity.* Albany: State University of New York Press, 2001.

Kilham, Wesley, and Leon Mann. "Level of Destructive Obedience as a Function of Transmitter and Executant Roles in Milgram Obedience Paradigm." *Journal of Personality and Social Psychology* 29, no. 5 (1974): 696–702.

Kloppenburg, Jack Ralph. *First the Seed: The Political Economy of Plant Biotechnology, 1492–2000.* Cambridge; New York: Cambridge University Press, 1988.

Kolata, Gina. "In Good Health? Thank Your 100 Trillion Bacteria." *New York Times*, June 14, 2012. http://www.nytimes.com/2012/06/14/health/human-microbiome-project -decodes-our-100-trillion-good-bacteria.html?smid=pl-share. Accessed December 28, 2012.

Kraut, Robert, Sara Kiesler, Bonka Boneva, Jonathon N. Cummings, Vicki Helgeson, and Anne M. Crawford. "Internet Paradox Revisited." *Journal of Social Issues* 58, no. 1 (2002): 49–74.

Kraut, Robert, Michael Patterson, Vicki Lundmark, Sara Kiesler, Tridas Mukopadhyay, and William Scherlis. "Internet Paradox: A Social Technology That Reduces Social Involvement and Psychological Well-Being?" *American Psychologist* 53, no. 9 (1998): 1017–31.

Kunda, Ziva. *Social Cognition: Making Sense of People*. Cambridge, MA: MIT Press, 1999.

Kurzweil, Ray. *The Singularity Is Near: When Humans Transcend Biology*. New York: Viking, 2005.

Laforest, Thomas John, and Gérard Lebel. *Our French-Canadian Ancestors*. Vol. 2. Palm Harbor, FL: LISI Press, 1984.

Latané, Bibb, and John M. Darley. "Group Inhibition of Bystander Intervention in Emergencies." In *Readings in Social Psychology: The Art and Science of Research*, edited by Steven Fein, Steven Spencer, and Sharon S. Brehm, 135–41. Boston: Houghton Mifflin, 1996.

Latour, Bruno. *Science in Action: How to Follow Scientists and Engineers through Society*. Cambridge, MA: Harvard University Press, 1987.

Le Clercq, Chrestien, William Francis Ganong, and Champlain Society. *New Relation of Gaspesia: With the Customs and Religion of the Gaspesian Indians*, Publications of the Champlain Society, Vol. 5. Toronto: Champlain Society, 1910.

Leenhardt, Maurice. *Do Kamo: Person and Myth in the Melanesian World*. Chicago: University of Chicago Press, 1979.

Lefebvre, Henri. *The Production of Space*. Oxford, UK; Cambridge, MA: Blackwell, 1991.

Lefebvre, Henri, and Michel Trebitsch. *Critique of Everyday Life*. Vol. 1. London; New York: Verso, 1991.

Leopold, Aldo. *A Sand County Almanac, and Sketches Here and There*. New York: Oxford University Press, 1970.

Leuchtag, Alice. "Merchants of Flesh: International Prostitution and the War on Women's Rights." *Humanist* 55, no. 2 (1995): 11.

Levin, Steve. "Internet Hunting Comes under Fire; Legislators Here and Elsewhere Aim to Ban It." *Pittsburgh Post-Gazette*, June 10, 2005. http://www.post-gazette.com/ stories/local/uncategorized/internet-hunting-comes-under-fire-586493. Accessed February 5, 2013.

Levine, Robert V. "Cities with Heart." In *Readings in Social Psychology: General, Classic, and*

Contemporary Selections, edited by Wayne A. Lesko, 270–75. Needham Heights, MA: Allyn & Bacon, 1991.

Lewis, Jonathan. "Shifting Nature." In *China from the Inside*. BBC/Granada/PBS, 2007.

Lewis-Fernández, Roberto. "Culture and Dissociation: A Comparison of Ataque De Nervios among Puerto Ricans and Possession Syndrome in India." In *Dissociation: Culture, Mind, and Body*, edited by David Spiegel, 123–67. Washington, DC: American Psychiatric Press, 1994.

Linebaugh, Peter, and Marcus Rediker. "The Many-Headed Hydra: Sailors, Slaves and the Atlantic Working Class in the Eighteenth Century." In *Gone to Croatan: Origins of North American Dropout Culture*, edited by Ronald B. Sakolsky and James Koehnline, 129–60. Brooklyn; Edinburgh: Autonomedia, 1993.

Live-shot.com. "Live-Shot: Real Time, On-Line, Hunting and Shooting Experience." http://www.live-shot.com/demo.shtml. Accessed October 7, 2005.

Lloyd, Geoffrey Ernest Richard. *Adversaries and Authorities: Investigations into Ancient Greek and Chinese Science*. Cambridge; New York: Cambridge University Press, 1996.

———. *Ancient Worlds, Modern Reflections: Philosophical Perspectives on Greek and Chinese Science and Culture*. Oxford; New York: Clarendon Press; Oxford University Press, 2004.

———. *Science and Morality in Greco-Roman Antiquity: An Inaugural Lecture*. Cambridge; New York: Cambridge University Press, 1985.

Lo, Shao-Kang, Chih-Chien Wang, and Wenchang Fang. "Physical Interpersonal Relationships and Social Anxiety among Online Game Players." *CyberPsychology & Behavior* 8, no. 1 (2005): 15–20.

Locke, John, and Peter Laslett. *Two Treatises of Government; a Critical Edition with an Introduction and Apparatus Criticus*. 2nd ed. Cambridge: Cambridge University Press, 1970.

Logan, Robert K. *The Alphabet Effect: The Impact of the Phonetic Alphabet on the Development of Western Civilization*. 1st ed. New York: Morrow, 1986.

Louv, Richard. *Last Child in the Woods: Saving Our Children from Nature-Deficit Disorder*. Updated and expanded ed. Chapel Hill, NC: Algonquin Books, 2008.

Lovelock, James. *Gaia: A New Look at Life on Earth*. Oxford: Oxford University Press, 1995.

———. *Gaia: The Practical Science of Planetary Medicine*. Oxford; New York: Oxford University Press, 2000.

Lubin, Gus. "Satellite Pictures of the Empty Chinese Cities Where Home Prices Are Crashing." *Business Insider*, December 10, 2011. http://www.businessinsider.com/china-ghost-cities-2011-11. Accessed January 29, 2013.

Luhmann, Niklas. *Ecological Communication*. Chicago: University of Chicago Press, 1989.

———. "Familiarity, Confidence, Trust: Problems and Alternatives." In *Trust: Making and Breaking Cooperative Relations*, edited by Diego Gambetta, 94–107. New York: Blackwell, 1990.

———. *Observations on Modernity* (Writing Science). Stanford, CA: Stanford University Press, 1998.

Macy, Joanna. "Working through Environmental Despair." In *Ecopsychology: Restoring the Earth, Healing the Mind*, edited by Theodore Roszak, Mary E. Gomes, and Allen D. Kanner, 21–40. San Francisco: Sierra Club Books, 1995.

Mahler, Margaret S. *Infantile Psychosis and Early Contributions.* Edited by Marjorie Harley and Annemarie Weil. 2 vols. Vol. I, "The Selected Papers of Margaret S. Mahler, M.D." New York: J. Aronson, 1979.

Mahler, Margaret S., Fred Pine, and Anni Bergman. *The Psychological Birth of the Human Infant: Symbiosis and Individuation.* New York: Basic Books, 1975.

Malo, David. *Hawaiian Antiquities (Moolelo Hawaii).* Translated by Dr. N. B. Emerson. Honolulu: Hawaiian Gazette, 1903.

"Management of Electronic Waste in the United States: Approach Two." United States Environmental Protection Agency, 2007. http://www.epa.gov/wastes/conserve/materials/ecycling/docs/app-2.pdf. Accessed December 28, 2012.

Margulis, Lynn. *Symbiotic Planet: A New Look at Evolution.* 1st ed. New York: Basic Books, 1998.

Markus, Hazel Rose, and Shinobu Kitayama. "Culture and the Self: Implications for Cognition, Emotion, and Motivation." In *The Self in Social Psychology*, edited by Roy F. Baumeister, 339–71. Philadelphia: Psychology Press/Taylor & Francis, 1999.

Marx, Karl, and Friedrich Engels. *Marx and Engels on Ecology.* Edited by Howard L. Parsons. Westport, CT: Greenwood Press, 1977.

Mauss, Marcel. "A Category of the Human Mind: The Notion of the Person; the Notion of the Self." In *The Category of the Person: Anthropology, Philosophy, History*, edited by Michael Carrithers, Steven Collins, and Steven Lukes. Cambridge; New York: Cambridge University Press, 1985.

Mazalin, Dennis, and Susan Moore. "Internet Use, Identity Development, and Social Anxiety among Young Adults." *Behaviour Change* 21, no. 2 (2004): 90–102.

McClintock, Anne. *Imperial Leather: Race, Gender, and Sexuality in the Colonial Contest.* New York: Routledge, 1995.

McGinnis, Michael Vincent, ed. *Bioregionalism.* London; New York: Routledge, 1999.

McNally, David. *Against the Market: Political Economy, Market Socialism, and the Marxist Critique.* London: Verso, 1993.

Meeus, Wim H. J., and Quinten A. W. Raaijmakers. "Administrative Obedience: Carrying out Orders to Use Psychological-Administrative Violence." *European Journal of Social Psychology* 16, no. 4 (1986): 311–24.

———. "Obedience in Modern Society: The Utrecht Studies." *Journal of Social Issues* 51, no. 3 (1995): 155–75.

Merchant, Carolyn. *The Death of Nature: Women, Ecology, and the Scientific Revolution.* 1st ed. San Francisco: Harper & Row, 1980.

————. *Major Problems in American Environmental History: Documents and Essays.* 2nd ed. Major Problems in American History Series. Boston: Houghton Mifflin, 2005.

————. *Reinventing Eden: The Fate of Nature in Western Culture.* New York: Routledge, 2003.

Merleau-Ponty, Maurice. *Phenomenology of Perception.* London; New York: Routledge, 1962.

————. *The Primacy of Perception, and Other Essays on Phenomenological Psychology, the Philosophy of Art, History, and Politics,* Northwestern University Studies in Phenomenology & Existential Philosophy. Evanston, IL: Northwestern University Press, 1964.

Milgram, Stanley. "Behavioral-Study of Obedience." *Journal of Abnormal and Social Psychology* 67, no. 4 (1963): 371–78.

————. "Issues in the Study of Obedience: A Reply to Baumrind." *American Psychologist* 19, no. 11 (1964): 848–52.

————. *Obedience to Authority: An Experimental View.* 1st ed. New York: Harper & Row, 1974.

————. "Some Conditions of Obedience and Disobedience to Authority." *Human Relations* 18, no. 1 (1965): 57–76.

Millennium Ecosystem Assessment Program. *Ecosystems and Human Well-Being: Synthesis,* Millennium Ecosystem Assessment Series. Washington, DC: Island Press, 2005.

Miller, Arthur G. *The Obedience Experiments: A Case Study of Controversy in Social Science.* New York: Praeger, 1986.

Miller, Arthur G., Barry E. Collins, and Diana E. Brief. "Perspectives on Obedience to Authority: The Legacy of the Milgram Experiments." *Journal of Social Issues* 51, no. 3 (1995): 1–19.

Monbiot, George. "On the 12th Day of Christmas ... Your Gift Will Just Be Junk." *Guardian* (London), December 10, 2012. http://gu.com/p/3cdnx. Accessed February 4, 2013.

Moody, Eric J. "Internet Use and Its Relationship to Loneliness." *CyberPsychology & Behavior* 4, no. 3 (2001): 393–401.

Morgan, George W. *The Human Predicament: Dissolution and Wholeness.* Providence, RI: Brown University Press, 1968.

Morris, Michael W., and Kaiping Peng. "Culture and Cause: American and Chinese Attributions for Social and Physical Events." *Journal of Personality and Social Psychology* 67, no. 6 (1994): 949.

Mourelatos, Alexander P. D. "Plato's Science—His View and Ours of His." In *Science and Philosophy in Classical Greece,* edited by Alan C. Bowen and Institute for Research in Classical Philosophy and Science. New York: Garland, 1991.

Muir, John. *My First Summer in the Sierra: And Selected Essays.* Edited by Bill McKibben. New York: Library of America, 2011.

Munro, Donald J. *Individualism and Holism: Studies in Confucian and Taoist Values.* Ann Arbor: Center for Chinese Studies, University of Michigan, 1985.

National Institute for Occupational Safety and Health, Education and Information Division. "Methyl Iodide." US Centers for Disease Control and Prevention. http://www.cdc.gov/niosh/npg/npgd0420.html. Accessed December 4, 2012.

Needham, Joseph. "Poverties and Triumphs of the Chinese Scientific Tradition." In *The "Racial" Economy of Science toward a Democratic Future*, edited by Sandra G. Harding, 30–46. Bloomington: Indiana University Press, 1993.

Neeson, Janet M. *Commoners: Common Right, Enclosure and Social Change in England, 1700–1820*. Cambridge; New York: Cambridge University Press, 1993.

Nisbett, Richard E. *The Geography of Thought: How Asians and Westerners Think Differently—and Why*. New York: Free Press, 2003.

Nisbett, Richard E., Incheol Choi, Kaiping Peng, and Ara Norenzayan. "Culture and Systems of Thought: Holistic Versus Analytic Cognition." *Psychological Review*, April 2001: 291.

Nixon, Rob. *Slow Violence and the Environmentalism of the Poor*. Cambridge, MA: Harvard University Press, 2011.

Norgaard, Kari Marie. *Living in Denial: Climate Change, Emotions, and Everyday Life*. Cambridge, MA: MIT Press, 2011.

O'Connor, Terrance. "Therapy for a Dying Planet." In *Ecopsychology: Restoring the Earth, Healing the Mind*, edited by Theodore Roszak, Mary E. Gomes, and Allen D. Kanner, 149–55. San Francisco: Sierra Club Books, 1995.

Orbach, Israel. "The Role of the Body Experience in Self-Destruction." *Clinical Child Psychology & Psychiatry* 1, no. 4 (1996): 607–19.

O'Rourke, Dara. *Shopping for Good* (Boston Review Books). Cambridge, MA: MIT Press, 2012.

Orwell, George. *Nineteen Eighty-Four, a Novel*. London: Secker & Warburg, 1949.

Pachirat, Timothy. *Every Twelve Seconds: Industrialized Slaughter and the Politics of Sight*, Yale Agrarian Studies Series. New Haven, CT: Yale University Press, 2011.

Papler, Roger, and California Regional Water Quality Control Board–San Francisco Bay Region. "Third Five-Year Review: Hewlett-Packard (620–640 Page Mill Road) Superfund Site; Palo Alto, Santa Clara County, California." Edited by Stephen A. Hill and Kathleen Salyer, 2010. http://yosemite.epa.gov/r9/sfund/r9sfdocw.nsf/3dc283e6c5d6056f88257426007417a2/351973b3feacb7a8882577af0073b60e!OpenDocument. Accessed October 10, 2012.

Parr, Martin. "What the Photographer Saw." *Independent* (London), February 18, 2001.

Pastides, Harris, E. J. Calabrese, D. W. Hosmer Jr., and D. R. Harris Jr. "Spontaneous Abortion and General Illness Symptoms among Semiconductor Manufacturers." *Journal of Occupational Medicine* 30, no. 7 (1988): 543–51.

Peet, Richard, and Michael Watts. *Liberation Ecologies: Environment, Development, Social Movements*. London; New York: Routledge, 1996.

Pellow, David N., and Lisa Sun-Hee Park. *The Silicon Valley of Dreams: Environmental Injustice, Immigrant Workers, and the High-Tech Global Economy* (Critical America). New York: New York University Press, 2002.

Peng, Kaiping, Daniel R. Ames, and Eric D. Knowles. "Culture and Human Inference: Perspectives from Three Traditions." In *The Handbook of Culture & Psychology*, edited by David Ricky Matsumoto. New York: Oxford University Press, 2001.

Pimentel, Benjamin. "Big Blue Settles Lawsuits; Sick Employees Blamed Chemicals." *San Francisco Chronicle*, June 24, 2004.

———. "Ex-IBM Workers Lose Toxics Case: Jury Says Company Not Responsible for Cancer They Claimed Was Caused by Chemicals." *San Francisco Chronicle*, February 27, 2004. http://www.sfgate.com/business/article/Ex-IBM-workers-lose-toxics -case-Jury-says-2817137.php. Accessed April 22, 2012.

Pirsig, Robert M. *Lila: An Inquiry into Morals*. New York: Bantam Books, 1992.

———. *Zen and the Art of Motorcycle Maintenance: An Inquiry into Values*. Toronto; New York: Bantam Books, 1981.

"Planting Justice: Grow Food. Grow Jobs. Grow Community." http://plantingjustice.org. Accessed November 30, 2012.

Plato. *The Collected Dialogues of Plato: Including the Letters*. Edited by Edith Hamilton and Huntington Cairns. New York: Pantheon Books, 1961.

Plumwood, Val. *Environmental Culture: The Ecological Crisis of Reason*. New York: Routledge, 2001.

———. *Feminism and the Mastery of Nature*, Opening Out. London; New York: Routledge, 1993.

Pollan, Michael. *Cooked: A Natural History of Transformation*. New York: Penguin Press, 2013.

———. *The Omnivore's Dilemma: A Natural History of Four Meals*. New York: Penguin Press, 2006.

———. "Power Steer." *New York Times Magazine*, March 31, 2002.

Puckett, Jim, Leslie Byster, Sarah Westervelt, Richard Gutierrez, Sheila Davis, Asma Hussain, and Madhummitta Dutta. "Exporting Harm: The High-Tech Trashing of Asia." Seattle, WA: Basel Action Network, Silicon Valley Toxics Coalition, 2002. http:// www.ban.org/E-waste/technotrashfinalcomp.pdf. Accessed September 12, 2012.

Reed, Edward S. *The Necessity of Experience*. New Haven, CT: Yale University Press, 1996.

Revkin, Andrew. Dot Earth. *New York Times*. http://dotearth.blogs.nytimes.com. Accessed February 5, 2013.

Ripple, William J., and Robert L. Beschta. "Wolves, Elk, Willows, and Trophic Cascades in the Upper Gallatin Range of Southwestern Montana, USA." Forest Ecology and Management 200, no. 1 (2004): 161.

Rissler, Jane, and Margaret G. Mellon. *The Ecological Risks of Engineered Crops*. Cambridge, MA: MIT Press, 1996.

Rochat, François, and Andre Modigliani. "Authority: Obedience, Defiance, and Identification in Experimental and Historic Contexts." In *A New Outline of Social Psychology*, edited by Martin Gold and Elizabeth Douvan, 235–46. Washington, DC: American Psychological Association, 1997.

Rogers, Heather. *Gone Tomorrow: The Hidden Life of Garbage*. New York: New Press: Distributed by Norton, 2005.

Roszak, Theodore. *The Voice of the Earth*. New York: Simon & Schuster, 1992.

Roszak, Theodore, Mary E. Gomes, and Allen D. Kanner. *Ecopsychology: Restoring the Earth, Healing the Mind*. San Francisco: Sierra Club Books, 1995.

Royte, Elizabeth. *Garbage Land: On the Secret Trail of Trash*. 1st ed. New York: Little Brown, 2005.

Russell, Nestar J. C., and Robert J. Gregory. "Spinning an Organizational 'Web of Obligation'? Moral Choice in Stanley Milgram's 'Obedience' Experiments." *American Review of Public Administration* 41, no. 5 (2011): 495–518.

Sabini, John, and Maury Silver. *Moralities of Everyday Life*. Oxford; New York: Oxford University Press, 1982.

Sambursky, Samuel. *The Physical World of the Greeks*. London: Routledge & Kegan Paul, 1960.

Santayana, George. *Scepticism and Animal Faith: Introduction to a System of Philosophy*. New York: Dover Publications, 1955.

Schivelbusch, Wolfgang. *Disenchanted Night: The Industrialization of Light in the Nineteenth Century*. Berkeley: University of California Press, 1988.

Schlosser, Eric. *Fast Food Nation: The Dark Side of the All-American Meal*. Boston: Houghton Mifflin, 2001.

Schrödinger, Erwin. *What Is Life?: The Physical Aspect of the Living Cell; with, Mind and Matter: & Autobiographical Sketches*. Canto ed. Cambridge; New York: Cambridge University Press, 1992.

Sewall, Laura. "Beauty and the Brain." In Ecopsychology: Science, Totems, and the Technological Species. Edited by Peter H. Kahn Jr. and Patricia H. Hasbach, 265–84. Cambridge, MA: MIT Press, 2012.

Shapiro, Judith. *Mao's War against Nature: Politics and the Environment in Revolutionary China*, Studies in Environment and History. Cambridge; New York: Cambridge University Press, 2001.

Shapiro, Judith S. "Loneliness: Paradox or Artifact?" *American Psychologist* 54, no. 9 (1999): 782–83.

Shelley, Mary Wollstonecraft. *Frankenstein; or, the Modern Prometheus*. London: Printed for Lackington, Hughes, Harding, Mavor, & Jones, 1818.

Shepard, Paul. "Nature and Madness." In *Ecopsychology: Restoring the Earth, Healing the Mind*, edited by Theodore Roszak, Mary E. Gomes, and Allen D. Kanner, 21–40. San Francisco: Sierra Club Books, 1995.

Sheridan, Charles L., and Richard G. King. "Obedience to Authority with an Authentic Victim." *Proceedings of the Annual Convention of the American Psychological Association* 7, no. 1 (1972): 165–66.

Shlain, Leonard. *The Alphabet Versus the Goddess: The Conflict between Word and Image.* New York: Viking, 1998.

Simpson, John Alexander, and Edmund S. C. Weiner, eds. *Oxford English Dictionary.* 2nd ed., OED Online. Oxford: Oxford University Press, 1989.

Smith, Neil. *Uneven Development: Nature, Capital, and the Production of Space.* New York: Blackwell, 1984.

Smuts, Jan Christiaan. *Holism and Evolution.* New York: Macmillan, 1926.

Snyder, Gary. *The Practice of the Wild: Essays.* San Francisco: North Point Press, 1990.

———. *Turtle Island.* New York: New Directions, 1974.

Solnit, Rebecca. *River of Shadows: Eadweard Muybridge and the Technological Wild West.* New York: Viking, 2003.

———. *Wanderlust: A History of Walking.* New York: Viking, 2000.

Soper, Kate. *What Is Nature?: Culture, Politics and the Non-Human.* Oxford; Cambridge, MA: Blackwell, 1995.

Spence, Mark David. *Dispossessing the Wilderness: Indian Removal and the Making of the National Parks.* New York: Oxford University Press, 1999.

Speth, James Gustave. *The Bridge at the Edge of the World: Capitalism, the Environment, and Crossing from Crisis to Sustainability.* New Haven, CT: Yale University Press, 2008.

Spiegel, David. *Dissociation: Culture, Mind, and Body.* 1st ed. Washington, DC: American Psychiatric Press, 1994.

"Spiral Gardens Community Food Security Project." http://www.spiralgardens.org. Accessed November 30, 2012.

Staub, Ervin. "Individual and Societal (Group) Values in a Motivational Perspective and Their Role in Benevolence and Harmdoing." In *Social and Moral Values: Individual and Societal Perspectives,* edited by Nancy Eisenberg, Janusz Reykowski, and Ervin Staub, 45–61. Hillsdale, NJ: Lawrence Erlbaum Associates, 1989.

———. *The Roots of Evil: The Origins of Genocide and Other Group Violence.* Cambridge; New York: Cambridge University Press, 1989.

Steinberg, Marlene. "Systematizing Dissociation: Symptomatology and Diagnostic Assessment." In *Dissociation: Culture, Mind, and Body,* edited by David Spiegel. Washington, DC: American Psychiatric Press, 1994.

Steinberg, Theodore. *Down to Earth: Nature's Role in American History.* Oxford; New York: Oxford University Press, 2002.

Stirling, Ian, and Andrew E. Derocher. "Effects of Climate Warming on Polar Bears: A Review of the Evidence." *Global Change Biology* 18, no. 9 (2012): 2694–706.

"Stress in America: Our Health at Risk." American Psychological Association, 2012. http://

www.apa.org/news/press/releases/stress/2011/final-2011.pdf. Accessed December 17, 2012.

"SVTC Eco-Map Family." Silicon Valley Toxics Coalition. http://www.mapcruzin.com/svtc_ecomaps. Accessed December 28, 2012.

Sylvan, Richard. *Sociative Logics and Their Applications: Essays.* Edited by Dominic Hyde and Graham Priest. Aldershot; Burlington, VT: Ashgate, 2000.

Szasz, Andrew. *Shopping Our Way to Safety: How We Changed from Protecting the Environment to Protecting Ourselves.* Minneapolis: University of Minnesota Press, 2007.

Takooshian, Harold. "How Stanley Milgram Taught About Obedience." In *Obedience to Authority: Current Perspectives on the Milgram Paradigm,* edited by Thomas Blass, 9–24. Mahwah, NJ: Lawrence Erlbaum Associates, 2000.

Tam, Fiona. "Foxconn Factories Are Labour Camps: Report." *South China Morning Post* (Hong Kong), October 11, 2010. http://www.scmp.com/article/727143/foxconn-factories-are-labour-camps-report. Accessed September 19, 2012.

Taylor, Alex H., Douglas Elliffe, Gavin R. Hunt, and Russell D. Gray. "Complex Cognition and Behavioural Innovation in New Caledonian Crows." *Proceedings of the Royal Society B-Biological Sciences* 277, no. 1694 (2010): 2637–43.

Thompson, Edward P. "Time, Work-Discipline and Industrial Capitalism." In *Customs in Common.* New York: New Press: Distributed by W. W. Norton, 1991.

―――. *Whigs and Hunters: The Origin of the Black Act.* 1st American ed. New York: Pantheon Books, 1975.

Tilker, Harvey A. "Socially Responsible Behavior as a Function of Observer Responsibility and Victim Feedback." *Journal of Personality and Social Psychology* 14, no. 2 (1970): 95–100.

Tillyard, Eustace Mandeville Wetenhall. *The Elizabethan World Picture.* New York: Vintage Books, 1959.

"Tobacco Farmers to Access Data through Cellphones." *Citizen* (Dar es Salaam), June 13, 2011. http://www.thecitizen.co.tz/business/13-local-business/11884-tobacco-farmers-to-access-data-through-cellphones.html. Accessed December 28, 2012.

Tokar, Brian. "Monsanto: A Checkered History." *Ecologist* 28, no. 5 (1998): 254–61.

Treehugger. http://www.treehugger.com. Accessed February 5, 2013.

Turkle, Sherry. *Alone Together: Why We Expect More from Technology and Less from Each Other.* New York: Basic Books, 2012.

Turnbaugh, Peter J., Ruth E. Ley, Micah Hamady, Claire M. Fraser-Liggett, Rob Knight, and Jeffrey I. Gordon. "The Human Microbiome Project." *Nature* 449, no. 7164 (2007): 804–10.

US Global Change Research Program. "Third National Climate Assessment (Draft)." Washington, DC: US Global Change Research Program, 2013. http://ncadac.globalchange.gov/download/NCAJan11-2013-publicreviewdraft-fulldraft.pdf. Accessed February 1, 2013.

US National Research Council. *Hidden Costs of Energy: Unpriced Consequences of Energy Production and Use*. Washington, DC: National Academies Press, 2010.

Vallette, Jim. "Larry Summers' War against the Earth." Global Policy Forum. http://www .globalpolicy.org/socecon/envronmt/summers.htm. Accessed February 4, 2013.

Walker, Richard. "Landscape and City Life: Four Ecologies of Residence in the San Francisco Bay Area." *Ecumene* 2, no. 1 (1995): 33–64.

Wan, William. "To Analysts, Foxconn Brawl Is No Anomaly." *Washington Post*, September 26, 2012.

Ware, Chris. "Thanksgiving.com." *New Yorker*, November 27, 2000, Cover.

Weiser, Matt. "State Commission Outlaws Internet Hunting, Fishing." *Sacramento Bee*, August 20, 2005.

Weston, Anthony. *Back to Earth: Tomorrow's Environmentalism*. Philadelphia: Temple University Press, 1994.

White, Richard. "Are You an Environmentalist, or Do You Work for a Living?" In *Uncommon Ground: Rethinking the Human Place in Nature*, edited by William Cronon. New York: W. W. Norton, 1996.

Whitehead, Alfred North. *Process and Reality: An Essay in Cosmology*. Edited by David Ray Griffin and Donald W. Sherburne. Corrected ed. New York: Free Press, 1978.

Williams, Eric D., Robert U. Ayres, and Miriam Heller. "The 1.7 Kilogram Microchip: Energy and Material Use in the Production of Semiconductor Devices." *Environmental Science & Technology* 36, no. 24 (2002): 5504–10.

Wilson, Elizabeth. *The Sphinx in the City: Urban Life, the Control of Disorder, and Women*. London: Virago, 1991.

Wilson, Sara. "Internet Is Blamed for Increase in Sex-Tours." *Scotsman* (Edinburgh), July 3, 1996.

Winthrop, John. "Conclusions for the Plantation in New England." In *Old South Leaflets, No. 50. (1629)*. Boston: Directors of the Old South Work, 1897.

Wirion, M., N. Zdanowicz, Ch. Pull, and M. Hildgen. "Is the Internet a Pathological Communication Tool or Not? Where the Weakest Are the Most Vulnerable." *Annales Medico-Psychologiques* 162, no. 6 (2004): 477–82.

Wolf, Eric Robert. *Europe and the People without History*. Berkeley: University of California Press, 1982.

Worthy, Kenneth. "Modern Institutions, Phenomenal Dissociations, and Destructiveness toward Humans and the Environment." *Organization & Environment* 21, no. 2 (2008): 148–70.

Worthy, Kenneth A., Richard C. Strohman, Paul R. Billings, Jason A. Delborne, Earth Duarte-Trattner, Nathan Gove, Daniel R. Latham, and Carol M. Manahan. "Agricultural Biotechnology Science Compromised: The Case of Quist and Chapela." In *Controversies in Science and Technology: From Maize to Menopause*, edited by Daniel Lee

Kleinman, Abby J. Kinchy, and Jo Handelsman. Madison: University of Wisconsin Press, 2005.

Xia Huo, Lin Peng, Xijin Xu, Liangkai Zheng, Bo Qiu, Zongli Qi, Bao Zhang, Dai Han, and Zhongxian Piao. "Elevated Blood Lead Levels of Children in Guiyu, an Electronic Waste Recycling Town in China." *Environmental Health Perspectives* 115, no. 7 (2007): 1113–17.

"Yale Environment 360: Opinion, Analysis, Reporting & Debate." Yale University. http://e360.yale.edu. Accessed February 5, 2013.

Zehner, Ozzie. *Green Illusions: The Dirty Secrets of Clean Energy and the Future of Environmentalism* (Our Sustainable Future). Lincoln: University of Nebraska Press, 2012.

Zhang, Kejing. "Rough Times in Guiyu—E-Waste Recycling in China Has Become a Serious Threat to Human Health and the Environment. Now the Authorities Are Stepping In." *Recycling m@gazine*, 2007. http://www.recyclingmagazin.de/epaper/rm0005/default.asp?ID=8. Accessed October 8, 2012.

Zimbardo, Philip G. "On 'Obedience to Authority.'" *American Psychologist* 29, no. 7 (1974): 566–67.

Zoetmulder, Petrus Josephus, Stuart Owen Robson, and Koninklijk Instituut voor Taal-Land- en Volkenkunde (Netherlands). *Old Javanese-English Dictionary*. 's-Gravenhage: Nijhoff, 1982.

Zukin, Sharon. *Landscapes of Power: From Detroit to Disney World*. Berkeley: University of California Press, 1991.

INDEX

San Francisco East Bay Area, 29, 261
daylighting of creeks in, 333n11
(*see also* Friends of Five
Creeks)
San Jose, California, 41
Santa Clara, California, 41
Santayana, George, 13
São Paulo, mechanization of, 218
Saurman, Jeffrey, 41
Scepticism and Animal Faith (Santayana), 13
Schivelbusch, Wolfgang, history of the
lamp by, 236
Schlosser, Eric, 263
Schrödinger, Erwin, on entropy and
negentropy, 303n14
science, ancient Greek theoretical, 169
science, as institution, intermediation
between people and nature, 73
Scientific Revolution, 147, 193
continued influence of, 212
scope of *Invisible Nature*, 17
secondhand information, versus
primary experience, 203. *See also*
primary experience (Reed), loss of,
in Western life
Second Life, virtual living in, 240
self
community and, 150
independent sense of, 150
interdependent sense of, 150
sense of (*see chapter 4*)
self-destruction, and childhood caregiving, 307n9
self-enhancement bias, 316n56
self-perpetuating harm, 118
self-reflexivity, 90

semiconductors, 39
manufacture of, chemicals in, 40
Semitic alphabet, 170
senses
atrophy of, 126
in encounters, importance of, 74
in relation, role of, 286
sensuality
industrialization of, 93
separation-individuation phase of psychological birth, 150
sequential nature, of everyday actions
and harms, 128
Sequoia sempervirens, ecological relationships of, 30
seventeenth-century, philosopher-scientists of the
and Plato, 195
and Pythagorean cosmology, 195
Sewall, Laura, 125–26, 276, 313n90
and deep, sensual knowledge, 275
and looking courageously at ecological distress, 275
sex tours, 96
Shanghai, mechanization of, 218
Shapiro, Judith, 295n4
Shapiro, Judith S., on Internet use and
psychopathology, 313n95
Shelley, Mary, 72
Shenzhen, China, 46
Sherif, Muzafer, 105
Shlain, Leonard, on subjugation of
women and nature alienation, 319n20
shoji (Japanese rice-paper screens), 242
Silicon Valley, 41, 45, 48, 66
Silicon Valley Toxics Coalition
(SVTC), 48, 51, 60